Nature and Design

WITPRESS
WIT Press publishes leading books in Science and Technology.
Visit our website for the current list of titles.
www.witpress.com

WITeLibrary

Home of the Transactions of the Wessex Institute, the WIT electronic-library provides the international scientific community with immediate and permanent access to individual papers presented at WIT conferences. Visit the WIT eLibrary at www.witpress.com

International Series on Design and Nature

Objectives

Our understanding of the modern world is largely based on an ever increasing volume of scientific knowledge. Engineering designers have at their disposal a vast array of relationships for materials, mechanisms and control, and these laws have been painstakingly assembled by observation of nature. As space activity accustoms us to cosmic scales, and as medicine and biology to the molecular scale of genetics, we have also become more aware of the rich diversity of the structural world around us.

The parallels between human design and nature has inspired many geniuses through history, in engineering, mathematics and other subjects. Much more recently there has been significant research related to design and invention. Even so, current developments in design engineering, and the huge increase in biological knowledge, together with the virtual revolution in computer power and numerical modelling, have all made possible more comprehensive studies of nature. It is these developments which have led to the establishment of this international book series.

Its rationale rests upon the universality of scientific laws in both nature and human design, and on their common material basis. Our organic and inorganic worlds have common energy requirements, which are of great theoretical significance in interpreting our environment.

Individual books in the series cover topics in depth such as mathematics in nature, evolution, natural selection, vision and acoustic systems, robotics, shape in nature, biomimetics, creativity and others. While being rigorous in their approach, the books are structured to appeal to specialist and non-specialist alike.

Series Editors

M.W. Collins
Department of Engineering Systems
London South Bank University
London, SE1 0AA
UK

M.A. Atherton
Department of Engineering Systems
London South Bank University
London, SE1 0AA
UK

J.A. Bryant
Dept. of Biological Sciences
University of Exeter
Exeter, EX4 4PS
UK

Associate Editors

I. Aleksander
Imperial College of Science, Technology & Medicine, UK

J. Baish
Bucknell University
USA

G.S. Barozzi
Universita Degli Studi di Modena E Reggio Emilia, Italy

C.D. Bertram
The University of New South Wales
Australia

D.F. Cutler
Royal Botanical Gardens, Kew
UK

S. Finger
Carnegie Mellon University
USA

M.J. Fritzler
University of Calgary
Canada

J.A.C. Humphrey
Bucknell University
USA

D. Margolis
University of California
USA

J. Mikielewicz
Polish Academy of Sciences
Poland

G. Prance
Lyme Regis
UK

D.M. Roberts
The Natural History Museum
UK

X. Shixiong
Fudan University
China

T. Speck
Albert-Ludwigs-Universitaet Freiburg
Germany

J. Stasiek
Technical University of Gdansk
Poland

J. Thoma
Zug
Switzerland

J. Vincent
Bath University
UK

Z.-Y. Yan
Peking University
China

K. Yoshizato
Hiroshima University
Japan

G. Zharkova
Institute of Theoretical and Applied Mechanics, Russia

Nature and Design

M.W. Collins, M.A. Atherton and J.A. Bryant

WITPRESS Southampton, Boston

Nature and Design
The International Series on Design in Nature

M.W. Collins
Department of Engineering Systems, London South Bank University, UK.

M.A. Atherton
Department of Engineering Systems, London South Bank University, UK.

J.A. Bryant
Department of Biological Sciences, University of Exeter, UK.

Published by

WIT Press
Ashurst Lodge, Ashurst, Southampton, SO40 7AA, UK
Tel: 44 (0) 238 029 3223; Fax: 44 (0) 238 029 2853
E-Mail: witpress@witpress.com
http://www.witpress.com

For USA, Canada and Mexico

WIT Press
25 Bridge Street, Billerica, MA 01821, USA
Tel: 978 667 5841; Fax: 978 667 7582
E-Mail: infousa@witpress.com
http://www.witpress.com

British Library Cataloguing-in-Publication Data

A Catalogue record for this book is available
from the British Library

ISBN: 1-85312-852-X
ISSN: 1478-0585

Library of Congress Catalog Card Number: 2003116915

No responsibility is assumed by the Publisher, the Editors and Authors for any injury and/or damage to persons or property as a matter of products liability, negligence or otherwise, or from any use or operation of any methods, products, instructions or ideas contained in the material herein.

© WIT Press 2005.

Printed in Great Britain by Athenaeum Press Ltd, Gateshead.

All rights reserved. No part of this publication may be reproduced, stored in a retrieval system, or transmitted in any form or by any means, electronic, mechanical, photocopying, recording, or otherwise, without the prior written permission of the Publisher.

Contents

Introduction to the series xv

Preface xxxi

Chapter 1 What is design?   1
C. Dowlen & M. Atherton

1 Design is… 1
2 Design is about changing the world 2
3 Design is making dreams a reality 4
4 Design is meeting a need 6
5 Design is about interaction with the senses 7
6 Design is about function 9
7 Design is about the process of arriving at the solution 10
8 Summary: a design overview 12
9 An example 13

Chapter 2 Mathematics in the natural world   17
J. Gomatam & F. Amdjadi

1 Introduction 17
2 Greek philosophy and mathematical foundations 18
    2.1 Pythagoras 18
    2.2 Plato and Neo-Platonism 19
    2.3 Aristotle 20
3 Irrational numbers 20
    3.1 Mathematics representation 20
    3.2 An engineer's description 20
4 Mathematics of biological form 22
    4.1 Spirals and phyllotaxis 22
    4.2 Reaction-Diffusion waves 23
    4.3 Unduloids 23
5 Golden section, architecture and music 24
6 Conclusion 26

Chapter 3 The Laws of Thermodynamics: cell energy transfer
J. Mikielewicz, J.A. Stasiek & M.W. Collins

1 Introduction 29
2 The Zero'th Law of Thermodynamics 30
    2.1 Definitions 30
    2.2 The system 30
    2.3 Properties 31
    2.4 State 31
    2.5 Thermal equilibrium 31
    2.6 The Zero'th Law 32
    2.7 Temperature 32
    2.8 Scales of temperature 33
    2.9 Work 33
    2.10 Heat 33
    2.11 Work output for a p-V process 33
    2.12 Reversibility 34
3 The First Law of Thermodynamics 35
    3.1 Introduction 35
    3.2 Joule's experiments 35
    3.3 The cycle 35
    3.4 Statement of Law 35
4 Flow process – the control volume 36
5 The Second Law of Thermodynamics 37
    5.1 Historical: Carnot's brilliant insight 37
    5.2 General 38
    5.3 Second Law statement (Planck) 39
    5.4 First Corollary (or Clausius statement) 40
    5.5 More Corollaries 40
    5.6 The Thermodynamic Temperature scale 40
    5.7 The definition of entropy 41
6 The Third Law of Thermodynamics 42
7 The central equation of thermodynamics 42
8 The Carnot Cycle and entropy 43
9 The Rankine Cycle and global warming 43
10 Combustion processes 44
    10.1 Conservation of mass 45
    10.2 Conservation of energy – the First Law of Thermodynamics 46
11 The concept of maximum work 48
12 Analogy between combustion and energy release in cells 49
13 Fuels and food 50
    13.1 Combustion of hexane 50
    13.2 Combustion of glucose 50
    13.3 Aerobic metabolism with glucose 50
14 Cell energy release – general 52
    14.1 Cell energy release – glycolysis 52
    14.2 Cell energy release – citric acid cycle 53
    14.3 Cell energy release – respiratory chain 54
    14.4 Aerobic metabolism 55

   14.5 Anaerobic metabolism and fermentation 55
   14.6 Photosynthesis – general..56
   14.7 Photosynthesis – photophosphorylation 57
   14.8 Photosynthesis – the Calvin-Benson cycle 57
   14.9 The food chain – conclusion 58
15 Carbon dioxide and global warming...58
16 Conclusions 59

Chapter 4 Robustness and complexity   63
M. Atherton & R. Bates

1 Introduction: Robustness of function..63
2 Robustness through noise interactions...66
   2.1 Noise in pattern formation 66
   2.2 Interactions and systematic organisation 73
3 Robustness out of transformation 76
   3.1 Ideal function and noise...77
   3.2 Phase space in low-dimensional non-linear dynamics............................78
4 Summary..82

Chapter 5 D'Arcy Thompson: nature and design through growth and form   85
M.A.J. Chaplain

1 Introduction 85
2 A brief biography..87
3 On growth and form, style and substance 88
4 D'Arcy's legacy and influence today 89
5 Conclusions 92

Chapter 6 Design in plants   95
D.F. Cutler

1 Introduction 95
   1.1 Move to land from water 96
   1.2 Competition for light 97
   1.3 Economy...98
2 What holds a plant up? 98
   2.1 Water 98
   2.2 Strong skin..99
   2.3 Xylem 99
   2.4 Fibres 103
   2.5 Lignified parenchyma..103
   2.6 Fibres in composite structures 103
   2.7 Sclereids..104
   2.8 Collenchyma 105
3 Micro-engineering 105
   3.1 Fan vaulting and leaf architecture..105
   3.2 Guttering or thin tubes: the petiole 106

  3.3 Girders, I beams 107
  3.4 Reinforced concrete, rod reinforcement and the peripheral placement
     of stem vascular bundles 108
  3.5 Tubes 109
  3.6 Ropes and roots; central location of strengthening material 109
  3.7 Gelatinous fibre, tracheids and inclined trunks......................................110
  3.8 Self-shading – keeping cool..110
4 Surface features 111
  4.1 Introduction 111
  4.2 Boundary layer, wind speed and turbulence: morphological considerations...112
  4.3 Heat and mass transfer, turbulence, and the boundary layer..................113
  4.4 Overall assessment..118
  4.5 Leaf margins – dentate and air flow 119
  4.6 Chimneys and prairie dogs 119
  4.7 Reflection: shiny cuticle, hairs, wax, self-cleaning 120
5 Windows and light pipes 120
6 Light acceptors and internal mirrors..120
7 UV absorption 121
8 Fire retardants 121
9 Preservatives 122
10 Floating devices and air chambers...122
11 Velcro 123
12 Conclusion 123

Chapter 7 The tree as an engineering structure 125
D. Hunt

1 Introduction 125
2 Structure and growth of a tree 126
  2.1 Growth..126
  2.2 Chemistry, transport, storage..127
  2.3 Wood macro- and microstructure 127
  2.4 Cell walls 129
  2.5 Ultra-structure...130
  2.6 Grain orientations at knots..131
3 The mechanical properties of the wood structure described above................131
  3.1 Strength...131
  3.2 Crack avoidance in tension...132
  3.3 The effects of growth stresses on bending failures 133
  3.4 Stiffness: juvenile vs. mature wood 134
  3.5 Reduction of stress concentration by change of grain direction 134
4 Tree geometry...135
  4.1 Types of growth adjustments..135
5 Engineering structures 141
  5.1 Traditional methods 141
  5.2 Modern methods 142
6 Conclusions 144

Chapter 8 The homeostatic model as a tool for the design and analysis of shell structures 145
O.A. Andrés, N.F. Ortega & J.C. Paloto

1 Introduction 146
2 Physical models 146
3 The homeostatic model 147
    3.1 Principles 147
    3.2 The homeostatic technique 147
4 Design model..148
5 Analysis model 149
    5.1 Experimental method 149
    5.2 Hybrid method 152
6 Conclusions 155

Chapter 9 Adaptive growth 157
J. Platts

1 Introduction 157
2 Graphic statics 158
3 Medical insights..160
4 Achieving structural efficiency...161
5 Field theory...162
6 Modelling growth 163
7 Honouring nature 165

Chapter 10 Optical reflectors and antireflectors in animals 169
A.R. Parker & N. Martini

1 Introduction 169
2 Mechanisms causing reflection/antireflection in animals 171
    2.1 Multilayer reflectors (interference)...171
    2.2 Diffraction gratings...176
    2.3 Scattering..181
3 Functions of animal reflectors in behavioural recognition and camouflage 184
    3.1 Terrestrial invertebrates 185
    3.2 Aquatic invertebrates 185
4 Mirror and antireflection functions of animal structures 187
    4.1 Mirrors in photophores 187
    4.2 Mirrors in eyes...188
    4.3 Mirrors in other body parts 190
5 Evolution of animal reflectors 191

Chapter 11 A medical engineering project in the field of cardiac assistance: a lumped-parameter model of the Guldner muscle-powered pump trainer and its use with a ventricular assist device   **199**
C.D. Bertram & J.P. Armitstead

1 Introduction 200
   1.1 Heart transplantation and the need for 'assist' devices.................................200
   1.2 The skeletal muscle graft: historical background 201
   1.3 Skeletal muscle: application alternatives and the SMV................................202
   1.4 The muscle-powered pump..203
   1.5 The 'Frogger' device 203
   1.6 Modelling using computational fluid dynamics (CFD) and lumped-parameter approaches...204
2 Methods 206
   2.1 The Frogger barrel 206
   2.2 The VAD diaphragm 207
   2.3 The blood in the VAD and beyond 208
   2.4 Muscle performance 209
   2.5 Overall model 210
3 Results 211
4 Discussion..217
5 Conclusion 218

Chapter 12 Leonardo da Vinci   **223**
W. Grassi & M. Collins

1 Introduction 223
2 Life and works...225
3 Leonardo the polymath 233
   3.1 Background...233
   3.2 Leonardo's breadth of achievement...235
4 Leonardo's holistic approach...238
5 The scientific revolution: the academic context of Leonardo..................................239
6 Leonardo the designer 243
7 Was Leonardo a genius?...247
8 Conclusion 248

Chapter 13 The evolution of land-based locomotion: the relationship between form and aerodynamics for animals and vehicles with particular reference to solar powered cars   **253**
D. Andrews, A. Shacklock & P.D. Ewing

1 Introduction 253
2 Aero and hydrodynamics in the animal kingdom for high-speed species...............256
   2.1 The Cheetah..256
   2.2 The Peregrine Falcon...257
   2.3 The Sailfish...257
   2.4 Summary...258

- 3 Human locomotion ... 258
  - 3.1 Human evolution ... 258
  - 3.2 Gait patterns ... 258
  - 3.3 Walking ... 259
  - 3.4 Running ... 259
  - 3.5 Summary ... 259
- 4 Assisted locomotion – the bicycle ... 260
  - 4.1 Development of the bicycle ... 260
  - 4.2 The aerodynamics of the upright bicycle ... 262
  - 4.3 Recumbent bicycles ... 263
  - 4.4 Maximum achievable speeds ... 263
  - 4.5 Summary ... 265
- 5 Assisted locomotion – the automobile ... 265
  - 5.1 The popularity of the car ... 265
  - 5.2 The future of fossil fuels for transportation ... 265
  - 5.3 Fossil fuels and the environment ... 266
  - 5.4 'Alternative' and emerging technologies ... 266
- 6 Vehicle aerodynamics ... 267
- 7 Automotive styling ... 268
  - 7.1 Design constraints ... 268
  - 7.2 Aerodynamics and early vehicle design ... 269
  - 7.3 Consumer expectation and contemporary vehicle design ... 271
  - 7.4 Contemporary vehicle nomenclature ... 272
  - 7.5 The influence of aerodynamics on contemporary car styling ... 274
  - 7.6 Summary ... 276
- 8 The future of automotive aerodynamics ... 277
  - 8.1 Concept cars ... 277
  - 8.2 ICE – electric hybrid vehicles ... 277
  - 8.3 Electric vehicles ... 278
  - 8.4 The Hypercar concept ... 279
  - 8.5 Summary ... 280
- 9 Solar powered cars ... 280
  - 9.1 History ... 280
  - 9.2 Solar car design principles ... 281
  - 9.3 Solar cars speeds ... 286
  - 9.4 Rolling resistance ... 287
  - 9.5 Solar cars aerodynamics ... 287
  - 9.6 Streamlining in the animal world ... 290
  - 9.7 The evolution of solar car form ... 291
  - 9.8 Summary ... 293
- 10 Conclusion ... 294

Chapter 14 Creativity and nature   301
C. Dowlen

- 1 Creativity ... 301
- 2 The scope of creativity ... 301
- 3 The nature of creativity ... 308
- 4 Creativity history ... 310

- 5 Definitions of creativity ... 311
  - 5.1 Research definition ... 312
  - 5.2 Artistic definition ... 312
  - 5.3 Survival definition ... 312
  - 5.4 Definitions from a list collated by Taylor ... 313
  - 5.5 A range of definitions ... 313
  - 5.6 My own preference ... 313
- 6 Measures of creativity ... 314
- 7 Creativity in practice ... 317
  - 7.1 A creativity development exercise ... 317
  - 7.2 The assignment ... 318
  - 7.3 The methods ... 319
  - 7.4 The results ... 321
  - 7.5 Natural analogies ... 322
  - 7.6 Analogical thinking for creative products ... 322
- 8 Creativity and nature ... 323

Design in nature – introduction to the series

Michael W. Collins
London South Bank University, UK.

Prologue

> *almost a miracle* [Cecil Lewis, 1, p. 126]

> *almost miraculously* [Stuart Kauffman, 2, p. 25]

'It was a beautiful evening' wrote Cecil Lewis [1] of the day in 1917 when he took a new SE5 on a test flight. 'At ten thousand feet the view was immense, England quartered on its northern perimeter at twenty two thousand feet, Kent was below me …… for a second the amazing adventure of flight overwhelmed me. Nothing between me and oblivion but a pair of light linen-covered wings and the roar of a 200-hp engine! …… It was a triumph of human intelligence and skill – almost a miracle' (See Plate 1).

Cecil Lewis was only 19 years old at the time, having left the English public school Oundle, in order to join the Royal Flying Corps in the First World War.

Almost 40 years later, in happier times than those of Cecil Lewis, as another ex-schoolboy I was 'filling in time' with a Student Apprenticeship before going to University. My very first job was as 'D.O. Librarian' in an aeronautical engineering drawing office. The circumstances may have been prosaic, but one feature always intrigued me. At the apex of the very large pyramid, at the top of every document distribution list, was the Chief Designer. Of course, I never met him or even saw him, but to me his title expressed the fount of authority, intelligence and creativity, the *producer* of 'almost miracles' for the 1950's.

'Almost miracles' mean different things to different people. Another 40 years brings us to a new millennium, to Stuart Kauffman [2] writing in 2000. Kauffman, a highly regarded American biologist 'is a founding member of the Santa Fe Institute, the leading centre for the emerging sciences of complexity' [2, cover blurb]. In discussing DNA symmetry and replication, he says [2, page 25]: "It seems to most biologists that this beautiful double helix aperiodic structure is almost miraculously pre-fitted by chemistry and God for the task of being the master molecule of life. If so, then the origin of life must be based on some form of a double-stranded aperiodic solid" (see Plate II). Yes, Kauffman is in the heady business of studying life starting 'from non-life here, or on Mars'.

We have reflected Cecil Lewis's and Kauffman's near miracles in Plates I and II. In the case of Cecil Lewis he was still in the first flush of man's ability to fly. Not for him the necessity of filing a flight plan. Like the *natural* fliers, the birds, he could move at will in three-dimensional space.

However, even then, he could far out-fly them, whether in speed or in height. Stuart Kauffman, however, moves about in multi-dimensional space. *His* is a 'fitness landscape in ... thirteen-dimensional parameter space' [2, p. 70].

We do need, at the same time, our sense of wonder to be well informed. Of course, Cecil Lewis's near miracle has been totally replaced, by Jet Propulsion, by travel to the moon, and now by planetary exploration. In the same way, while the thirteen-dimensional space of Kauffman may well impress many of his readers, and the eleven-dimensional space of Stephen Hawking [3] was obviously expected to impress the average UK Daily Telegraph readers, in engineering terms this is a standard practice. Two of the Series Editors [MAA and MWC] started to consider [4] the problem of *visualisation* of complex data. This included reference to the optimisation of nuclear power station design [the UK Magnox system], which used a contour-tracking procedure focusing on 30 major parameters out of about 100 parameters in total [Russ Lewis, 5].

We conclude this prologue with the realisation that nature, nature's laws and the use of nature's laws in human design all have the capacity to enlighten, inform and inspire us. This series will have achieved its end if it demonstrates only a small part of that capacity.

Plate I: 'It was a triumph of human intelligence and skill - almost a miracle'. 'View from an aeroplane' [1, p182-183] (Reproduced by permission of the Victoria and Albert Museum Picture Library).

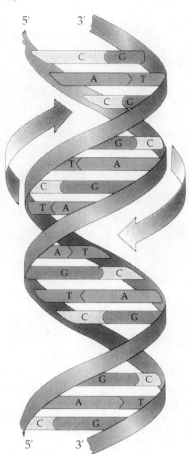

Plate II: 'This beautiful double helix.....structure is almost miraculously pre-fitted....'

DNA is a double helix
Plate II: Watson and Crick proposed that DNA is a double-helical molecule. The helical bands represent the two sugar-phosphate chains, with pairs of bases forming horizontal connections between the chains. The two chains run in opposite directions; biochemists can now pinpoint the position of every atom in a DNA macromolecule.
(Reproduced by permission from *Life. Volume 1 the Cell and Heredity*. 4th Edition by W.K. Purves, G.H. Orians and H.C. Heller, p246, Sinauer Associates, W.H. Freeman & Co.)

Research Field of John Bryant, Series Co-Editor
John Bryant's research is mainly concerned with the initiation of DNA replication - the very start of the process of copying the genome. It is far more complicated than we envisaged even ten years ago.....indeed it has a beautiful and almost awe-inspiring complexity. Each stage is tightly regulated so as to ensure that the cell only duplicates its genome at appropriate times. As we understand more about these control mechanisms we can only wonder at, and about, the evolutionary processes through which they developed.

Nature and engineering

The beavers have practised civil engineering since they became a species
[Eric Laithwaite, 6, p. 231]

Intellectually, the engineer and the artist are not far apart [Michael French, 7, p. 179]

The subject area of our series has great public interest and popularity, if we take the increasing number of publications as evidence. But this needs clarifying. Like Eric Laithwaite having to make a choice at Grammar School [6, p. xi] we might be forgiven for supposing that *our* subject is either biology or physics. On thinking more carefully, we could define our subject as the commonality of the laws of physics, in the natural [biological] and man-made [engineered] worlds. This is nearer the truth.

In the event, Eric Laithwaite chose physics. He went on to become a noted engineer and inventor, being awarded, in 1966, the Royal Society S.G. Brown Medal for Invention. So, for *him*, the beavers were engineers, not scientists.

In the same way, Michael French compares biologists, not with physicists, but with engineers and architects [7, pp. 1–2]. His book, like Laithwaites', is engineering-oriented – 'about design for function, and invention' [7, p. xvii].

So, despite so many of the recent publications being by biologists and physicists, we have chosen two engineers to start our Introduction. In fact, their approach represents a relatively new exploitation of the laws of physics, and materials science, as used in the biological design of living organisms. This points us in the direction of 'biomimetics' which is a recent concept involving the application of biological materials in engineered systems [p. xvii, Vol. 4 of this Series].

Laithwaite and French raise other issues. The first is noticeable by its absence. Those readers whose discipline is chemistry or chemical engineering might wonder if the subject has been 'airbrushed out'. Of course not – if no chemistry, then there is no DNA, no design in nature. We have already quoted Kauffman in this regard.

The next, lightly touched on by French [7, p. 235], as also by Kauffman [2, p. 24], is the question of what is meant by 'beauty'. While French strictly connects it to function in design, we will connect it to art in general, and find it is an integral part of our overall study. As French implies, the engineer and the artist are good friends.

The final issue is the question of mathematics. Whereas French [7, p. 291] rather pejoratively quotes from Bondi that 'mathematicians are not particularly good at thinking … good rather at avoiding thought', Laithwaite is obviously fascinated by the whole thing. For instance, he, like me, is highly intrigued [who isn't] by the identity.

$$e^{i\pi} = -1$$

In the same vein, he deals in some detail with the topics of 'ideal shape' in the form of the golden section [6, pp. 199–202] of Fibonacci numbers, and of helices in plants. He points out that the logarithmic spiral [6, pp. 201–202] retains its shape with growth, coinciding with French's reference to gnomonic shell growth [7, p. 273].

So then, two engineers have introduced our discussion. We now have to explain what *we* mean by the word '*design*'.

Design in the mainstream

> *The buttercups and the locomotive show evidence of design*
> [Michael French, 7, p. 1]
>
> *I am, in fact, not so much concerned with origins or reasons as with relations or resemblances*
> [Theodore Cook, 8, p. 4]

We can best describe our use of the word 'design' by the acronym *wysiwyg* – what you see is what you get. Our ambition is to explore fully the richness of the 'design of the buttercup' and the comparison of the designs of nature and engineering, all in the same spirit of Michael French. We shall avoid all issues like 'despite there being 'evidence of design' we do not believe …' on the one hand, or 'because there is evidence of design, we therefore believe …' on the other. The point has been put more elegantly by Theodore Cook, as long ago as 1914 [8, above].

So, we do not, as does Richard Dawkins, use the expression *designoid*. In 'Climbing Mount Improbable' [9, p. 4], he addresses this very point. 'Designoid objects look designed, so much so that some people – probably, alas, most people – think that they are designed'. So he uses designoid because of his antipathy to theism – 'no sane creator … would have conceived on his drawing board' he says on p. 121 [9]. In our use of the word design, however, we retain Richard Dawkins' friendship, with his pitcher plant giving 'every appearance of being excellently well designed' [9, p. 9], and his approbation of engineers – 'often the people best qualified to analyse how animal and plant bodies work' [9, p. 16].

However, in using the word design, neither do we mean *conscious design, intelligent design, [intelligent] design or [   ] design* … merely design.

Typical use of these explanations is given as follows, with the understanding that 'conscious design' is rather an archaic description:

i.	Conscious design	[Cook, 8, p. 4]	[Ruse, 10, p. 44]
ii.	Intelligent design	[Miller, 11, p. 93]	
iii.	Intelligent – design	[Ruse, 10, p. 120]	
iv.	design	[Miller, 11, pp. 92, 126]	[Ruse, 10, p. 121]
		[Behe, 12, p. 209]	[Dembski, 13, Title]

The last-mentioned author, William Dembski has, sadly, suffered for his beliefs, as explained in 'The Lynching of Bill Dembski' [14]. Nevertheless, in fairness, Dembski separates out the ideas of 'design' and 'designer', as this extended quote makes clear:

> 'Thus, in practice, to infer design is typically to end up with a 'designer' in the classical sense. *Nevertheless, it is useful to separate* [MWC's italics] design from the theories of intelligence and intelligent agency' [13, p. 36].

While the use of the word 'design' here may not be coincidental with that of Dembski, yet the act of separation is crucial, and consistent with the rationale for this Series. By using *wysiwyg* we are trying to retain the friendship of both Dawkins and Dembski and, further, to retain and parallel their common enthusiasm and commitment. In the Series, then, we seek to stay in the mainstream of all aspects of design in the natural and man-made worlds, stressing commonality rather than controversy and reconciliation of differences rather than their sharpening. In that spirit, where necessary, current controversies will be openly discussed and separate issues carefully identified.

Even this brief discussion has shown that the concept of 'design' is both subtle and wide-ranging in its connotations. We now address three specific aspects which are sometimes ignored or even avoided, namely, *mathematics*, *thermodynamics* and *history*.

Mathematics

We like to think mathematics was discovered, not invented
   Prof. Tim Pedley, verbal, Salford, 1998

The universe appears to have been designed by a pure mathematician
   [James Jeans, 15, p. 137]
   quoted in   [Paul Davies, 16, p. 202]

Now while the commonality of scientific laws in the natural world is generally accepted, the fact that the world is also mathematically *oriented* is less well understood. Of course, the concept of mathematics being somehow 'built in' to nature's structure is highly significant in terms of our rationale – nature's designs being parallel to man-made designs. Paul Davies expressed this concept in various telling phrases. In 'The Mind of God' we read '… all known fundamental laws are found to be mathematical in form' [16, p. 84]. 'To the scientist, mathematics … is also, astonishingly, the language of nature itself' [p. 16, 93], and as the heading for Figure 10 [p. 109] 'The laws of physics and computable mathematics may form a unique closed cycle of existence'.

In fairness, it should be added, as does Davies, that this approach is not universally accepted, and mathematicians have 'two broadly opposed schools of thought' [16, p. 141]. In the chapter on mathematics in nature' in *this* Volume the issue is dealt with more fully. However, the point we make here is that the overall detailed study of mathematics is essential for our rationale, which cannot be done in more general single-authored books. Paul Davies himself [16, p. 16] starts the reader with 'no previous knowledge of mathematics or physics is necessary'. Philip Ball, in his beautiful exposition of pattern formation in nature, likewise, restricts the mathematical treatment – 'I will not need to use in this book' (he says [17, p. 14]) 'any more mathematics than can be expressed in words rather than in abstruse equations'. Despite this restriction, however, Ball eulogizes mathematics – 'the natural language of pattern and form is mathematics … mathematics has its own very profound beauty … mathematics is perfectly able to produce and describe structures of immense complexity and subtlety ' [17, pp. 10–11].

The conclusion is straightforward – mathematics is an essential part of the design 'spectrum'.

Thermodynamics

The second law, like the first, is an expression of the observed behaviour of finite systems
   [Gordon Rogers and Yon Mayhew, 18, p. 809]

Thus the second law is a statistical law in statistical mechanics
   [Stuart Kauffman, 2, p. 86]

In seeking to understand thermodynamics there is not so much an obstacle to be surmounted, as ditches to be avoided. This is because thermodynamics uses concepts in common English use like 'energy', 'work', 'heat' and 'temperature', and because the First Law is an expression of the well-accepted 'conservation of energy' principle. However, these concepts are very closely defined in thermodynamics, and it is essential to understand their definitions. When we reach the Second Law, the problem is all too clear. What does entropy *really* mean? Why do different statements of the Law look completely different? So an 'amateur' understanding of thermodynamics can lead to an absence of appreciation of the Zeroth Law [to do with equilibrium and temperature] an erroneous confidence in First Law issues, and greater or lesser confusion regarding the Second Law! These are ditches indeed.

The other key aspect of thermodynamics is that it is part of the warp and weft of our industrial society. It was through the French engineer Carnot's brilliant perceptions, leading to the Second Law, the procedures for optimising work-producing heat engine design became clear. The same Law, with its stated necessity for heat rejection and reversibility, was the explanation of what otherwise looked like rather low heat engine efficiencies. In fact, essentially as a consequence of the Second Law, best practice power station efficiencies were of the order of 30% over a long period of time. As a major consequence of the enormous consumption of fossil fuels [coal and oil for example] in those power stations, and including internal combustion engines, carbon dioxide concentration in the atmosphere has increased dramatically. Over the two centuries 1800–2000 the increase has been some 28%, with approximately half that figure occurring since 1960. This is shown by Fig. 3.3 of John Houghton [19, p. 31]. Such is a major part of the background to the Greenhouse effect.

Carnot perceived that a crucial factor in achieving higher efficiencies was for the heating source to be at the *highest possible temperature*, which led in its turn to the definition of the Absolute Temperature Scale by the British engineer, Lord Kelvin.

It was then the German physicist Clausius who defined entropy – 'a new physical quantity as fundamental and universal as energy' [Kondepudi and Prigogine, 20, p. 78]. It was not just the heat that was important, but the *heat modified by the absolute temperature*, the entropy, that was needed. As a consequence, quantitatively low values of entropy are 'good', and perhaps this has led to conceptual difficulties. Similarly, entropy increases are caused by the individual processes in the heat engine operation [irreversibilities]. Finally, the Austrian physicist Boltzmann developed a theory of molecular statistics and entropy, leading to the association of entropy with *disorder* [20, p. xii]. Altogether then non-scientific [and even scientific and engineering] readers might be forgiven for viewing entropy as a sort of 'spanner in the thermodynamic works' – to be kept as low as possible.

Now it is not fully appreciated that the Laws of Thermodynamics are *empirical* – so [write Rogers and Mayhew] 'the Second Law, like the First, is an expression of … observed behaviour'. That empirical prevalence extends to the statistical mechanics interpretation – 'the macro state is a collection of microstates … the Second Law can be reformulated in its famous statistical mechanics incarnation' [Kauffman, 2, p. 86]. Post World War II, Shannon's information theory, has caused entropy to be associated formally with information. 'The conclusion we are led to' [Paul Davies, 21, p. 39] 'is that the universe came stocked with information, or negative entropy, from the word go'. Incidentally, our 'forgiven' readers might feel well justified by the expression negative entropy!

So much for the classical past of thermodynamics. Davies's quote points us to a new look at the subject. *What we are now seeing is an almost overwhelming desire to systematise the application of thermodynamics to biology.*

… vast amounts of entropy can be gained through the gravitational contraction of diffuse gas into stars … we are still living off this store of low entropy   [Roger Penrose, 22, p. 417].

… far from equilibrium states can lose their stability and evolve to one of the many states available to the system … we see a probabilistic Nature that generates new organised structure, a Nature that can create life itself   [Dilip Kondepudi and Ilya Prigogine, 20, p. 409].

The sequence of the application of thermodynamics to biology can be traced back to Erwin Schrödinger's lectures given at Trinity College, Dublin, Ireland, at the height of the Second World War, currently published as 'What is Life?' [23a, 23b]. In the chapter 'Order, Disorder and Entropy' Schrödinger postulates the following sequence: that living matter avoids the decay to equilibrium [or maximum entropy] by feeding on negative entropy from the environment, that is by 'continually sucking orderliness from its environment', and that the plants which form the ultimate source of this orderliness, themselves 'have the most powerful supply of negative entropy in the sunlight' [23a, pp. 67–75].

To take things further, we turn from the more readily available Reference 23a, to 23b, where Roger Penrose's original Foreword has evolved into a substantial Introduction. This latter Introduction is an important source in itself as it takes up Schrodinger's postulation of the sun's negentropic effect. Using Penrose's own words, [23b, p. xx]: the Sun is not just an energy source, but … a very hot spot in an otherwise dark sky … the energy comes to us in a low-entropy form … and we return it all in a high entropy form to the cold background sky. Where does this entropy imbalance come from? … the Sun has condensed from a previous uniform distribution of materials by gravitational contraction. We trace this uniformity … to the Big Bang … the extraordinary uniformity of the Big Bang … is ultimately responsible for the entropy imbalance that gives us our Second Law of Thermodynamics and upon which all life depends. So, too, we repeat Davies [21, p. 39]]…'the universe came stocked with information, or negative entropy, from the word go'.

Briefly, we mention that the detailed evolutionary procedures involved can be based on Prigogine's far-from-equilibrium thermodynamics [see above quote] on dissipative structures [20, chapter 19] and on the resultant order and the growth of complexity, or information. This latter, exemplified by Kauffman [2] is described by Ian Stewart [24, p366] as 'a kind of converse to chaos theory'.

I regard the concept of 'gnergy' as one of the most important results of my theoretical investigations in biology over the past two decades
[Sungchal Ji, 25, p. 152]

Such a law could be my hoped-for fourth law of thermodynamics for open self-constructing systems
[Stuart Kauffman, 2, p. 84]

We pass rapidly on to Sungchal Ji, with the proposed concept of 'gnergy' encompassing both energy and information, and to Kauffman with his hoped-for Fourth Law of Thermodynamics. At least they cannot be accused of lack of ambition! Ji's rather beautiful graphical interpretation of the evolutions of density and information since the Big Bang [25, p. 156] is reproduced here, as Plate III. [In doing so, however, it may be noticed that Ji's zero initial information density is hardly consistent with Davies' initial stock of information. This point will be addressed in the chapter on thermodynamics in Volume 2 of the Series]. Eric Chaisson's more concise research paper approach [26] should be noted, as it elegantly combines and quantifies some of the key issues raised by both Ji and Kauffman. It forms a nice introduction to the subject area.

We are about to bury our thermodynamics 'bone'. However, it must be appreciated that other 'dogs' still prefer non-thermodynamics 'bones', for example Stephen Boyden [27] and Ken Wilber [28]. The latter's ambitious 'A Theory of Everything' is sub-titled 'An Integral Vision for Business, Politics and Spirituality'. In his Note to the Reader he makes what to him is a conclusive remark about the 'second law of thermodynamics telling us that in the real world disorder always increases. Yet simple observation tells us that in the real world, life creates order everywhere: the universe is winding up, not down' [28, p. x]. For readers who, like me, cannot put their 'bones' down, this statement cannot be allowed to rest, and represents another issue for Thermodynamics in Volume 2. However, my comment is not meant to be pejorative. Ken Wilber seeks, as do so many writing in this subject area, a mastery almost painful to appreciate!

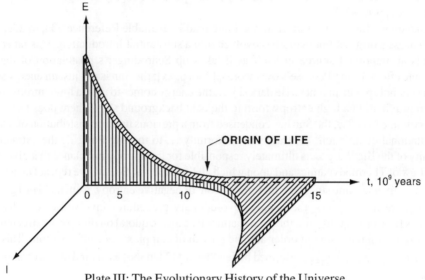

Plate III: The Evolutionary History of the Universe.

In this graph, E and I represent energy and information densities. t is time, on an approximately logarithmic scale, with the origin an estimated 15 billion years ago at the Big Bang. The substantial increase in I occurs following biological 'emergence of the first self-replicating systems ... about 3 billion years ago' [after Sungchal Ji].

The final point here is the most interesting of all, namely that the origin of life remains a question. 'How this happened we don't know' said Stephen Hawking recently [29, p. 161]. Somewhat differently, Ilya Prigogine some ten years ago [30, p. 24] – 'we are still far from a detailed explanation of the origins of life, notwithstanding we begin to see the type of science which is necessary ... mechanisms which lead from the laws of chemistry to "information". However, where there's a bone there's a dog, [if the reader will forgive this final use of the metaphor] and in this case our dog is Michael Conrad. Conrad's essential thesis contrasts with that of Schrödinger [31, p. 178], and is rather that – to quote the Abstract [31, p. 177] – 'the non-linear self-organising dynamics of biological systems are *inherent* [my italics] in any ... theory ... requirements of both quantum mechanics and general relativity'. Conrad's line [ten of the twenty four references in 31 are by himself as sole author] is termed the fluctuon model, and is particularly interesting in relating to 'nanobiological phenomena and that might be detected through nanobiological techniques'. Stuart Kauffman [2, Chapter 10] similarly surveys quantum mechanics and general relativity, but more in the nature of questioning than Conrad's tighter theorising ....

History

'This concept of an ideal, perfect form behind the messy particulars of reality is one that is generally attributed to Plato'
    [Philip Ball, 17, p. 11]

'Leonardo da Vinci was my childhood hero, and he remains one of the few great geniuses of history'
    [Michael White, 32, p. xi]

'The only scientific book I read that summer was Charles Darwin's 'The Origin of Species' ... but we do still read Darwin'
    [Kenneth Miller, 11, p. 6]

Having sought to show that a fuller understanding of 'Design in Nature' needs to be both mathematically and thermodynamically oriented, we will now point out its historical aspects. We will focus on thee principal characters – Plato, Leonardo da Vinci and Darwin. It is not so easy to give reasons for the choice of these three, but I believe that they represent timeless flashes of genius. Somewhat unexpectedly, they can be viewed in the context of *engineering design*.

So, Plato is associated with one of the key aspects of design, namely form [cf quote by Ball above]. Leonardo epitomizes the ideal of the engineering designer, namely a 'universal man' at home in any branch of knowledge and able to conceptualise almost limitlessly. So we read [33, p. 488] ... 'Italian painter, sculptor, architect, engineer and scientist ... of immensely inventive and enquiring mind, studying aspects of the natural world from anatomy to aerodynamics'. Finally, Darwin can be associated with the idea of progress and adaptation with time [namely, evolution]. It is difficult to overemphasise this, the point being made explicitly in the *titles* of two recent books. Michael French's [7] title is 'Invention and Evolution'. Design in Nature and Engineering'. Similarly, Norman Crowe on architecture [34] 'Nature and the idea of a man-made world. An Investigation into the Evolutionary Roots of Form and Order in the Built Environment'. These are but two examples.

We mentioned flashes of genius. These flashes also possess mathematical connotations.

    ... Plato esteemed the science of numbers highly ... [David Smith, 35, p. 89].

In that Plato postulated transcendent [non-earthly] form, he must have been close in approach to the multi-dimensional character of the studies we have already discussed in our Prologue. Platonism *per se* is dealt with at some length by Roger Penrose [22, pp. 146–151] whose 'sympathies lie strongly with Platonistic view that mathematics truth is absolute, external and eternal ...'. Paul Davies [16, p. 145] carries Penrose's sympathies forward as ... 'Many physicists share his Platonic vision of mathematics'.

'Norway builds Da Vinci's 500 year-old bridge ... it conformed with the laws of mathematics'
    [Roger Boyes, 36]

Turning to Leonardo, Michael White freely admits his hero's deficiency in this area. And yet, despite Leonardo's being 'barely competent' [32, p. 152] in mathematics and reliant on Pacioli ['he gained a good deal from Pacioli [32, p. 153]], he designed better than he knew. So we have the Norwegian artist Veljorn Sand, who was the persistent catalyst [it took him 5 years] to secure funding for Leonardo's design, paying Leonardo two compliments, firstly to do with his genius ['when you work with geniuses, you work with eternal forms that never go out of fashion'] and secondly his *implicit* mathematical ability [... 'the design was of lasting beauty because it conformed with the laws of mathematics and geometry ... the Mona Lisa of bridges']. To round off Leonardo's relationship with mathematics, he was nothing if not ambitious, and is on record himself as having a very deep commitment. Is it a case of an initial shortcoming being more than subsequently compensated for? So Sherwin Nuland [a surgeon] gives this different picture of Leonardo ... 'for Leonardo, mathematics was the ultimate key to the understanding of the nature he scrutinised so carefully ... to all of science, including the biology of man' [37, p. 53]. Nuland quotes Leonardo as 'no human investigation can be termed true knowledge if it does not proceed to mathematical demonstration'.

Darwin and Mathematics

Inside the sanctum sanctorum they got things done ... to Stokes this was 'flimsy to the last degree' ... But Huxley pulled off the coup ... It was published intact'
[Adrian Desmond, 38, p. 42]

'... Kelvin got very few calculations wrong ... here he understandably failed to include the contribution of the heat of radioactivity'
[Dennis Weaire, 39, p. 61]

'Darwin's view of persistent co-evolution remains by and large unconnected with our fundamental physics, even though the evolution of the biosphere is manifestly a physical process. Physicists cannot escape the problem ... We will search for constructive laws true of any biosphere. We will found a general biology. And we will be spellbound'
[Stuart Kauffman, 2, pp. 245, 269]

Finally, Darwin and mathematics. 'The Origin of Species' [41] is essentially, in engineering terms, and experimental report writ large, unaccompanied by mathematical theory. So we have an amusing account as to why Eric Laithwaite chose physics rather than biology. 'Physics seems to be mostly sums, biology mostly essays ... my best friend is going to do biology, so I can keep asking him about it and keep in touch that way. That does it ... I'll do physics' [6, pp. xi–xii]. Eric Laithwaite's schoolboy choice was a personal reflection of an extremely sharp division in the Royal Society regarding the application of Darwin's work. In fact, Desmond's quote above relates not to a publication of Darwin himself, but an ms submitted on Huxley's suggestion by Kovaleski. The real point here is that the Royal Society's conservative Physical Secretary, George Gabriel Stokes' [38, p. 41] opposed the Kovaleski acceptance because it would make 'speculative Darwinism as axiomatic as Newton's laws' and compromise the rock-like status of knowledge' [38, p. 42]. Now GGS lost, and if Desmond's comment is fair, GGS was spectacularly wrong since Darwin *is* roughly on a par of acceptance with Newton's laws. Not only so, but GGS's close friend Kelvin managed to miscalculate the age of the Earth [second quote above], a scientific *cause celebre* of the time.

GGS is given 'a bad book' by Desmond. In fact, he was an extraordinarily talented and productive physical mathematician and Stokes Summer Schools are run in Ireland, organised by Alastair Wood of Dublin City University [who wrote the parallel section on GGS [39] to that of Kelvin]. I declare a personal interest here. I have an immense regard and affection for Stokes, having worked for decades on numerical studies of convective heat transfer using the Navier-Stokes equations. In fact, Stokes spoke better than he knew, in making an outright comparison [having renamed the word 'speculative'] of Darwinistic [biology] with Newtonian [Physics]. That 1873 assessment was repeated in out anecdotal comment of Laithwaite around 1940, and repeated more tellingly by Kauffman in 2000. Here we remind ourselves that Kauffman is a biologist himself.

Digressing, Darwin was not the only experimentalist to have problems with the Royal Society. Joule [James Prescott Joule 1818–1889] the near-genius who worked assiduously on the equivalence of various forms of energy – notably heat and work – suffered the indignity of having only abstracts of submitted papers published by the Royal Society, on two occasions [J.G. Crowther, 41, pp. 189, 204]. He was young, very young, so despite the setbacks he was still only 32 years old when finally elected to the RS [41, p. 214].

Our final Darwin-related character is Kelvin who, despite the age-of-the-earth *faux pas*, has almost ethereal status of having proposed the Absolute Temperature Scale. In a subsequent volume in this series it is intended to focus on the contributions of [the two Scotsmen] James Clerk Maxwell and Kelvin to thermodynamics, and how this now relates to present day biology – information, complexity and the genome for example. The latter is epitomised by the recent work of Jeffrey Wicken, the full title of a major publication speaking for itself – 'Evolution, Thermodynamics and Information. Extending the Darwinian programme' [43]. So do the titles of some 17 Journal publications that he references [43, p. 233] for example 'A thermodynamic theory of evolution' in 1980 [44].

In all this, out quiet participant is Darwin himself. Part of his genius, I believe, was his caution, and he let his data collection speak for itself. No mathematics *there*, but an immense sub-surface, iceberg-like, volume of mathematics *underneath*, shown for its worth, as the genome unfolds, and interpreted in terms of information, complexity and Shannon entropy by those such as Kauffman and Wicken.

History summarised

So our three examples of Plato, Leonardo da Vinci and Darwin, have been given a brief introduction. Rather improbably, their genius has been introduced in terms of *engineering design* and *mathematical significance*. Above all, their genius was, and is, timeless. How else could Plato's views on form and mathematics be regarded as relevant two and a half *millennia* later? How else could Leonardo's bridge design be accepted half a millennium later? How else could Darwin's conclusions stand the test of exhaustive and sometimes hostile assessment, lasting for almost a century and a half?

A further aspect of this timelessness, which will be merely stated rather than discussed, is that the Renaissance [epitomised by Leonardo] had as one of its sources the rediscovery of the Greek texts ... 'the finding of ancient manuscripts that gave the intellectuals of the Renaissance direct access to classical thought ...' [32, p. 39]. So Michael White gives as Appendix 11: 'Leonardo and his place in the History of Science' [32, pp. 339–342], a chronological sequence running from Pythagoras through to Newton ....

Epilogue

Miraculous harmony at Epidaurus   [Henri Stierlin, 45, p. 168]

At the commencement of the Prologue to this Introduction, two 'almost miracles' were described. We conclude with a final example going back to 330 BC – to the absolute end of Greek classicism [45, p. 227]. 'Miraculous harmony at Epidaurus' is how Henri Stierlin describes the wonderfully preserved Greek 'theatre set into the hill of Epidaurus' [44, pp. 168–169] – see Plate IV.

Plate IV: 'Miraculous harmony at Epidaurus'.

(See page xiv of Optimisation Mechanics in Nature): 'Around the orchestra, the shell-like theatre set into the hill fans out like a radial structure, whose concentric rows of seating are all focused on the stage where the dramatic action would unfold. With its diameter of 120m., the theatre of Epidaurus is one of the finest semi-circular buildings of Antiquity. Its design, the work of Polyclitus the Younger, according to Pausanias, dates from the end of the fourth century B.C. It is based on a series of mathematical principles and proportions, such as the Golden Section and the so-called Fibonacci Sequence. Its harmony is thus the result of a symmetria in the real sense of the term' [45, p168].
(Reproduced by permission of the Greek National Tourism Organisation).

There are three distinct aspects to this piece of architecture by Polyclitus the Younger. The design has a mathematical basis – including what is now termed the Golden Section and the Fibonacci sequence. Secondly, the harmony spoken of by Stierlin is a consequence of the theatre's 'symmetry' – a subtle technical quality originating in Greek ideas of form. Lastly, the combination of what we now call 'the built environment' with its natural environment has a timeless aesthetic attractiveness. In fact, Plate IV is reproduced not from the reference we have discussed, but a Greek Tourist Organisation advertisement.

In concluding, our introduction has covered an almost impossible range of disciplines, but it is only such a range that can possibly do justice to the theme of design in nature. If 'we' is broadened to comprise editors, contributors and publishers, we want to share our sense of inspiration of design in the natural world and man-made worlds that our three authors of near miracles, Cecil Lewis, Stuart Kauffman and Henri Stierlin have epitomised.

References

[1] Lewis, C., *Sagittarius Rising*, 3rd Edition, The Folio Society: London, 1998.
[2] Kauffman, S.A., *Investigations*, Oxford, 2000.
[3] Hawking, S., *Why we need 11 dimensions*. Highlighted paragraph in 'I believe in a 'brane' new world', extract from Ref. 29. Daily Telegraph, p. 20, 31st October 2000.
[4] Atherton, M.A., Piva, S., Barrozi, G.S. & Collins, M.W., Enhanced visualization of complex thermo fluid data: horizontal combined convection cases. *Proc. 18th National Conference on Heat Transfer*, Eds. A. Nero, G. Dubini & F. Ingoli, UIT [Italian Union of Thermo fluid dynamics], pp. 243–257, 2000.
[5] Lewis, R.T.V., *Reactor Performance and Optimization*. English Electric Company [now Marconi] Internal Document, 1960.
[6] Laithwaite, E., *An Inventor in the Garden of Eden*. Cambridge, 1994.
[7] French, M., *Invention and Evolution. Design in Nature and Engineering*. 2nd Edition, Cambridge, 1994.
[8] Cook, T.A., *The Curves of Life*. Reproduced from original Constable edition, 1914, Dover, 1979.
[9] Dawkins, R., *Climbing Mount Improbable*. Penguin, 1996.
[10] Ruse, M., *Can a Darwinian be a Christian*. Cambridge, 2001.
[11] Miller, K.R., *Finding Darwin's God*. Cliff Street Books [Harper Collins], 1999.
[12] Behe, M.J., *Darwin's Black Box*. Touchstone [Simon & Schuster], 1998.
[13] Dembski, W.A., *The Design Inference*. Cambridge, 1998.
[14] Heeren, F., *The Lynching of Bill Dembski*, The American Spectator, November 2000.
[15] Jeans, J., *The Mysterious Universe*, Cambridge, 1931.
[16] Davies, P., *The Mind of God*, Penguin, 1993.
[17] Ball, P., *The Self-Made Tapestry*, Oxford, 1999.
[18] Rogers, G. & Mayhew, Y., *Engineering Thermodynamics, Work and Heat Transfer*, 4th Edition, Prentice Hall, 1992.
[19] Houghton, J., *Global Warming*, Lion, 1994.
[20] Kondepudi, D. & Prigogine, I., *Modern Thermodynamics*, Wiley, 1998.
[21] Davies, P., *The Fifth Miracle*, Penguin, 1999.
[22] Penrose, R., *The Emperor's New Mind*, Oxford, 1989/1999.
[23a] Schrödinger, E., *What is Life?* with *Mind and Matter and Autobiographic Sketches*, and a Foreword by R. Penrose, Canto Edition, Cambridge, 1992.

[23b] Schrödinger, E., *What is Life?* and an Introduction by R. Penrose, The Folio Society: London, 2000.
[Note: these are quite distinct publications. The key section *What is Life?* is type-set differently and the page numbers do not correspond.]
[24] Stewart, I., *Does God Play Dice?* 2nd Edition, Penguin, 1997.
[25] Ji, S., *Biocybernetics: A Machine Theory of Biology*, Chapter 1 in: *Molecular Theories of Cell Life and Death*, Ed. S. Ji, Rutgers, 1991.
[26] Chaisson, E., The cosmic environment for the growth of complexity, *Biosystems*, **46**, pp. 13–19, 1998.
[27] Boyden, S., *Western civilization in biological perspective*, Oxford, 1987.
[28] Wilber, K., *A Theory of Everything*, Gateway: Dublin, 2001.
[29] Hawking, S., *The Universe in a Nutshell*, Bantam Press, 2001.
[30] Prigogine, I., *Schrödinger and the Riddle of Life*, Chapter 2 in: *Molecular Theories of Cell Life and Death*, Ed. S. Ji, Rutgers, 1991.
[31] Conrad, M., Origin of life and the underlying physics of the universe, *Biosystems*, **42**, pp. 117–190, 1997.
[32] White, M., *Leonardo*, Little Brown & Co.: London, 2000.
[33] *The Complete Family Encyclopaedia*, Fraser Stewart Book Wholesale Ltd., Helicon Publishing: London, 1992.
[34] Crowe, N., *Nature and the Idea of a Man-Made World*, MIT Press: Cambridge MA, USA & London, UK, 1995.
[35] Smith, D., *History of Mathematics*, Volume 1, First published 1923, Dover Edition, New York, 1958.
[36] Boyes, R., *Norway builds Da Vinci's 500-year-old bridge*, The Times [UK Newspaper], London, 1November 2001.
[37] Nuland, S., *Leonardo da Vinci*, Weidenfield & Nicolson, London, 2000.
[38] Desmond, A., *Huxley Evolution's High Priest*, Michael Joseph: London, 1997.
[39] Weaire, D., *William Thomson [Lord Kelvin] 1824–1907*, Chapter 8 in: *Creators of Mathematics: the Irish Connection*, Ed. K. Houston, University College, Dublin Press: Ireland, 2000.
[40] Darwin, C., *The Origin of Species*, Wordsworth Classics Edition, Ware, Hertfordshire, UK, 2000.
[41] Crowther, J.G., *The British Scientists of the Nineteenth Century*, Volume 1, Allen Lane/Penguin, Pelican Books, 1940.
[42] Wood, A., *George Gabriel Stokes 1819–1903*, Chapter 5 in: *Creators of Mathematics: the Irish Connection*, Ed. K. Houston, University College, Dublin Press: Ireland, 2000.
[43] Wicken, J.S., *Evolution, Thermodynamics and Information*, Oxford University Press 1987.
[44] Wicken, J.S., A thermodynamic theory of evolution, *J. Theor. Biol.*, **87**, pp. 9–23, 1980.
[45] Stierlin, H., *Greece from Mycenae to the Parthenon*, Series on Architecture and Design by TASCHEN, Editor-in-Chief A. Taschen, Taschen: Cologne, Germany, 2001.

Preface

Michael W. Collins
London South Bank University, United Kingdom

………which have taken many months to complete.
                                                    H.A. Vallance [1]

It may be of interest to readers to learn how this Series came about. It was originally due to a suggestion made by one of the Series Editors (MAA) in the context of the old School of Engineering Systems and Design at London South Bank University. This was for an in-house publication covering all aspects of design. MAA and MWC discussed this with Lance Sucharov, Publishing Editor at Wessex Institute of Technology, who made the far more ambitious proposal of the current series. To make this credible, MWC embarked on discussions with a number of senior UK biologists, one of whom (JB) became the third Series Editor. Additionally, MWC began to study representative books of the entire subject range.

Volumes 1 and 2 together are meant to form a holistic introduction to the subject, focusing respectively on the common scientific laws in the natural and engineered worlds, and then on the growth of information/complexity. DNA studies form a convenient 'peg' for the latter. The way in which Volumes 1 and 2 are assembled may be explained as follows. It is analogous to the original concept of tomography, where to reconstruct three-dimensional objects from two-dimensional optical slices required 360-degree field-of-view data. In that original concept, the reconstruction accuracy degraded rather quickly with loss of any data. To appreciate the meaning of design itself is challenging, let alone the subtleties of combining that meaning in natural and man-made worlds. So Volumes I and II comprise not only straightforward engineering design and biology, but also include mathematics, physics, chemistry, thermodynamics, biomimetics, medical engineering and history of science. The individual chapters are intended to be personal 'flashes' of illumination, taken from different angles and in different ways.

In the case of Volume 3, by contrast, we have a Proceedings-type archive of the first Design in Nature Conference. Subsequent Volumes are more conventional in their approach, each usually covering a kind of sub-set of subject area.

Returning to Volumes 1 and 2, here we, the Editors wish to express our heartfelt thanks to each author. Without their many months of hard work this venture would have been impossible. Finally, we hope that you, the reader, will find each chapter informative and inspiring, and that the whole will be greater than the sum of the parts.

Reference

[1] "The Great North of Scotland Railway", H.A. Vallance (dust jacket blurb). Revised Edition, David St. John Thomas/ TGNSR Association. 1989.

CHAPTER 1

What is design?

C. Dowlen[1] & M. Atherton[2]
[1]*Department of Architecture and Design, London South Bank University, London, UK.*
[2]*Department of Engineering Systems, London South Bank University, London, UK.*

Abstract

This chapter addresses the question 'What is Design?' by placing accepted engineering design concepts within a framework that incorporates the natural world. The natural world is expressed here as a resource with which to make human artefacts and also as something inevitably changed by the artefact produced. Although shedding light on how things function can be fascinating, our intent is to clarify the practice of design as a preparation for understanding its relevance to nature. In gathering together existing knowledge, we avoid formal objective definitions and concentrate on discussion of the activities involved and their implications for design process. Design is thus presented as fundamentally a change activity that involves issues of ordering thoughts in relation to reality, capturing emotions and sparking reactions, meeting needs, and creating function. Getting these things done requires tasks to be put together but not in purely logical convergent steps, as room needs to be made for creativity in the engineering design process. Thus our framework places design on the cusp between the natural and man-made real worlds along one axis and between a related pair of thought worlds on the other. Design is central to the way humans live and therefore the lessons to be learned from nature about function in particular should be of the utmost interest.

1 Design is ...

'Design' is used as a verb and a noun. This chapter explores what is meant by the term *design* and also considers the relationships between *design* and *nature*. Rather than seek strict definitions, we aim for a more general discursive treatment and build a two-dimensional pictorial map suggesting how design and nature interrelate. However, there is a view that design is purely a human phenomenon and that an attempt to enlarge its domain to include the natural world is moving away from the realms of design and engineering expertise and more towards philosophy and theology. It is not the intention of this chapter to delve significantly beyond the philosophies that underpin design as it is conceived, used and understood within the discipline of engineering.

Linguistics research has shown that sometimes formal, objective definitions are not the best way of developing understanding [1]. Some subjects are conceptualised through a series of snapshots, including visual memories and cameos that are categorised for no particular reason except perhaps that they touch in space, time or some other dimension. For example, appreciation of fashion or style is ordered in this way [2]. Even mathematics is conceptualised differently depending upon whether it is delivered to students of engineering, science, or mathematics [3].

Thus, how we view design will affect how we view a multitude of other topics, and vice versa. In particular our perception of nature will have a bearing on the place we give to design. Therefore we recognise that there are many ways of construing and organising the material. Rather than present any radical new wisdom here we amalgamate existing knowledge within the context of an overall understanding and perception of how the world may be organised.

As mentioned already, the term design can refer both to a process of design that takes place and to an object that has been designed. Moreover, it is clear that a design process, perhaps better termed *designing*, has been going on when a product has been designed. The product that has resulted from that process may also be termed the design. It can be difficult and pedantic to determine which terminology is being used at any one time. However, the model that is built up in this chapter generally refers to the process of design, or rather, that of designing.

Design in the context of nature is addressed in two ways. Firstly, design as a means of understanding how the earth and the life on it came to be, in particular the process by which animals and plants have reached their current state of function, complexity and beauty. In science there is an overwhelming naturalistic view that this process has been very long and unintelligent. Observations of nature only show gradual improvement (microevolution or adaptation). Much remains unproven. Thus, in this sense design is seen as a refining, iterative process. Attempts to explain away the differences in life as a product of mutation are unsatisfactory. Secondly, the design of biological configurations, functions and principles which life forms depend on throughout their lives for existence and survival: here the reference is to design as being the object that has been designed.

We are living in times when the detrimental effects of 'humankind's' (or 'man's' for short) technological innovations and developments on the viability of natural systems are reality. Species depletion and extinction occur because of man's influence on both local and global environmental conditions due to use of fossil fuels and the production of pollutants. This prompts us to question the relationship between man and nature.

The approach that will be taken is to develop an overall picture of how design might be viewed in relation to the world – or perhaps rather a way in which our understanding of the world may be categorised around the topic of design.

2 Design is about changing the world

Design, the process, can be described as an agent of change [4]. Design is either the initiation of change in man-made things or the thinking that drives that process of change. Understanding this we can use a very simple diagram to see what might go on in our picture of reality (Figure 1).

On one side of a line, say the left, we ultimately have the natural world. On the other, the right, we have the man-made world. Standing in between them we have some sort of change process: man changing the natural world into the man-made world. Here is man the maker. *Homo Faber*, as has sometimes been stated [5].

What is design? 3

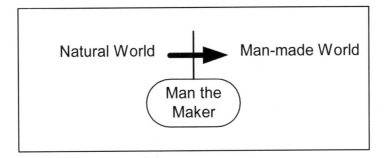

Figure 1: Change through man the maker.

Homo Faber changes the natural world into his own, man-made world. But *Homo Faber* makes decisions about how to change the natural world. The arrow is added to show the direction of change that has been man's primary concern. (Of course, man the investigator and scientist is now also interested in the opposite direction - the consequent impact of his designs on the environment – but we shall continue with the current line of enquiry).

Conceptually, then, we could consider Figure 1 to represent the physical process between the natural and man-made worlds in terms of making or manufacturing – the agent that *Homo Faber* uses to change the world, as shown in Figure 2.

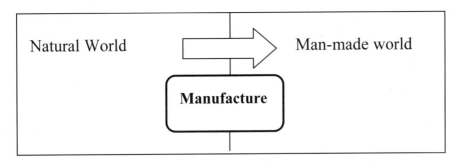

Figure 2: Change through manufacture.

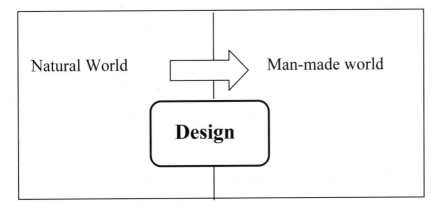

Figure 3: Change through design.

Or we could consider that Figure 1 represents the conceptual intent of what man is doing in this process, showing design as the agent of the change (Figure 3).

Thus in this sense man's design process lies between the natural world and the man-made world and designing is the key to the way in which humans attempt to make the world their own. This is especially true in engineering.

3 Design is making dreams a reality

At another level, design is the process that links the ideas that we have in our heads to the real world where they come to pass in reality (Figure 4). Ideas, thoughts and dreams about how the world could be improved become encapsulated as plans for changing the world through the process of design.

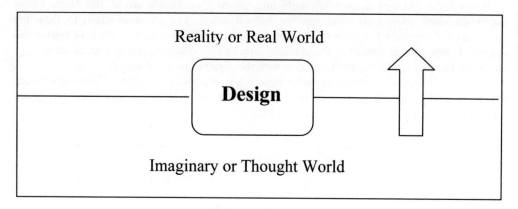

Figure 4: Design transforms the virtual world into the real world.

This design process includes thinking, drawing, modelling, and testing out the ideas in various ways before the final arrangements are in place. Figure 4 above is necessarily a significant simplification of the whole process, which includes elements of testing (in the real world) and thought sequences that are necessarily circuitous and messy. The general intention, however, is clearly that the thoughts and dreams should become an ordered reality, as reflected in the engineering design process literature (e.g. [4], [6] and [7]).

This process contrasts significantly with the way that scientists work. Design and engineering ('science of manipulation') are about changing the world, whilst pure science is about understanding nature. Schumacher [8] spoke of both the science of understanding and the science of manipulation, the latter in a somewhat derogatory manner, highlighting that the essential feature of science is that it is aiming to understand the real world. Thus science is about comprehending the reality, i.e. how the world (both the natural world and the man-made world) works. Science is also about acquiring knowledge through observation and experiment that can ideally be tested, systemised and incorporated into general principles.

Figure 5 summarises this approach.

It may also be convenient to split the thought world into two parts. On the one hand we have thoughts that can be validated in the real world, such as scientific theories and mathematical models. On the other we have the thought world that is not validated so coherently in the real world – that of ideas, ways of thinking, recommendations, political theories and so on. This area deals with human-generated perceptions and emotions such as beauty, artistic merit, marketing concepts, and tacit processes. This division is fraught with problems: it is on this debate that objectivist philosophies and relativists divide – not as the split line – with one party holding to the line's existence and the other disputing it. Such a description considerably simplifies the notions of both objectivism and relativism, which can each take on a multitude of different forms.

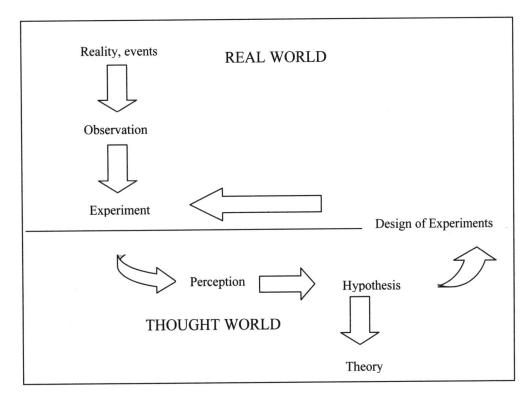

Figure 5: Scientific process of thought.

The direction of scientific thinking draws away from reality and towards theory: with the design process the direction is towards reality. Science is aiming at a more coherent theoretical definition of what is; design is aiming at a real embodiment of a thought idea. Design is a process aimed at making thinking concrete and realising this thinking process within the existing world. The manufacture or incorporation of a design idea into an artefact confirms that the idea is valid: it confirms that the idea works and functions in reality. If this function is not included as an outcome of the design process then we have no real product of any sort. Let us be quite clear, however, that this so-called validation is considerably different in character to the scientific concept of validation.

However, scientific understanding of the real world ought to be incorporated within the perception of the product function in order to minimise the timescale of the design process. The product can then be defined more precisely without having to go through the complete trial and error process at every step due to scientifically validated theories. Thus the designer is able to validate a significant proportion of the product within the thought or virtual world. Things such as mathematical calculations and modelling are what are meant here.

Ultimately, design is of no use if no manufactured reality or change of materials into something else takes place. It is about as good as the composer who could never get round to writing down his grandiose schemes for symphonies, but kept them in his head until the grave, unrealised.

4 Design is meeting a need

A significant amount of design activity is not simply to see ideas becoming realised but also to meet a need of some sort.

What exactly is meant by a need, how such a need is determined and how it is defined is the subject of a considerable amount of energy on the part of market researchers and psychologists.

In general, the term *need* is used by marketers in a different manner to the way in which it is perhaps used in daily life, where needs are perceived as the fundamental physiological drives such as the needs for food, air and water without which most organisms would die. Psychologists have proposed other fundamental drives that must be met by humans such as achievement, activity, affection, affiliation, curiosity, elimination, exploration, manipulation, maternity, pain avoidance, sex, and sleep [9].

Whilst design is clearly able to meet needs that are fundamental ones – such as in the development of food products – the marketing premise is that a need is something that is more akin to a desire, wish or dream that is elicited from the potential customer in some way.

The basic premise is that some product will be able to be designed if some sort of customer requirement can be discovered. This provides a ready market for whatever is produced. So determination of needs is important for designers – particularly those who are interested in being able to make a living from their design work.

There is considerable overlap between what marketers construe to be a need and what customers construe to be a desire or a wish on their part. If a product can be designed so that someone can say, 'That's just what I wanted!' does that mean that they absolutely need that product? Or is it something they enjoy because it meets their dreams and fits their general lifestyle and ambitions? This is difficult to determine, and in general, within market research, it is perhaps slightly irrelevant. If you are able to determine what it is that someone desires and fill that gap with a product, then you should be able to sell your product whether it is satisfying a desire or something more fundamental.

A considerable amount of literature within the marketing and design area (e.g. [10] and [11]) is devoted to determining exactly what these marketing needs are and how to elicit them. These generally seek to find ways of establishing the 'voice of the customer', exploring ways that people think when they want to buy products, exploring the shortcomings in existing products and generally trying to understand the way people behave when they want to buy something. From this exploration the needs can be determined, as can the less strongly expressed customer desires.

A need is determined as that purpose that a product is required to perform, such that if this particular thing is not met, there is no point in developing the product. But in the marketing arena, customers are also encouraged to voice and develop what are known as latent needs. These develop effective product gaps and can determine spaces where designers can develop products that they didn't know could be developed – producing the products that you always wanted but never knew it.

These needs are developed into a 'design problem definition' and the design process then becomes one of satisfying the needs through developing a suitable product. In this process the design input, through the manufacturing process, turns some sort of raw material into a manufactured reality, and then that reality is then offered to sales as the answer to the need. Whilst design is the agent of change, the final product user is usually uninterested in the design process, or in the problem formulation. The only concern is that the product meets their felt and latent needs and that it works well for them. For them the proof of the pudding is in the eating – does the product work? This, then, is the real-world validation of the design.

This need drives the design process in addition to the way the designer's imagination does. Thus we can develop an overall diagram, as in Figure 6, showing the way that this consumption and product use drives the way that products are designed.

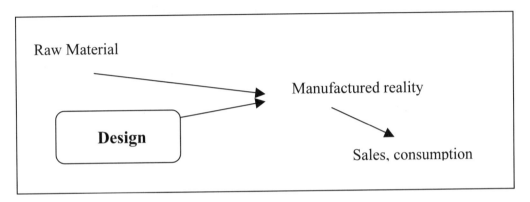

Figure 6: the pull of need on design.

5 Design is about interaction with the senses

The user is also interested in how the product fulfils not just their physical needs, but their emotional and psychological needs. Design is also about meeting these needs, which are even harder for the customer to be able to articulate than their functional ones.

This appeal to and interaction with the senses generally comes under the heading of 'aesthetics', a term that comes from the Greek '*aisthetica*' meaning 'that which is perceivable through the senses'. The term has developed from the simple sensory perception to include the complexity and variability of emotional responses resulting from that perception, modulated further by human experiences, memories, and developed behavioural patterns.

This means that although visual sense is paramount, recognising design includes sensory experiences derived from hearing, touch, movement, balance, muscle tone feedback and to a lesser extent smell and taste; the last term being used in a literal sense rather than a metaphorical one. This, then, is how a product form interacts with the person, both psychologically and also physically. Design is also concerned with how those sensations are developed and modified by the human psyche to evoke sensuality and emotions in their broadest context.

For instance, beauty has been said to be in the eye of the beholder. How does this relate to our perception of a product, a designed object or even a natural object? Just what does this statement mean? We don't actually sense beauty directly. We only perceive the object's form. The beauty of the form is a modulated perception created by our various experiences and human make-up. We cannot test for beauty in the same way that we can test whether an object exists, although it is quite possible to measure our reactions to products and determine our response to aesthetics [12].

Our full picture comes from innumerable different sources. Into the melting pot of human psyche come previous experiences to produce product recognition, sensual desires and aspirations to define whether what and how something is perceived; say, in a sophisticated or whimsical manner, or whether it has a mystique, beauty, simplicity, honesty, truth to materials and so on.

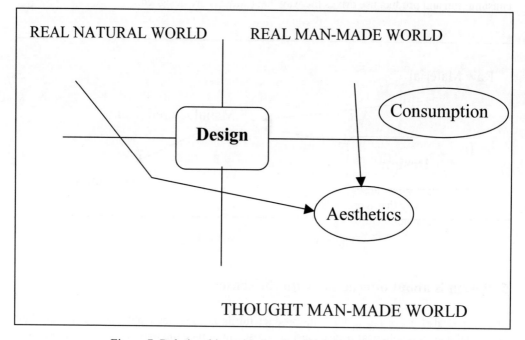

Figure 7: Relationships with the thought man-made world.

Of course the experienced designer will immediately understand that we have these emotional reactions to natural things and not just to things in the man-made world. Indeed, the natural world demanded a reaction to it before humans ever got around to making any product. Design in this sense is about capturing particular emotions within the reality of a human creation, within something that is man-made. Designers have the opportunity to select the emotions they wish to capture.

Design is also about mimicking perceived reactions to nature, even to create man-made reactions. A product will always produce some sort of emotional reaction, and some of these will be reactions that are deliberately created to mimic our reaction to some part of nature: in this way they can be called man-made. So the product or environment that is the result of the design process includes not only the real product but also the human thoughts derived from the product. The picture becomes even more complex, as in Figure 7, in that there needs to be an extra arrow down from the 'consumption' area back into the aesthetics.

Design is about developing psychological reactions to a human creation within the beholder – who may be user, critic, owner, manager or a mixture of such 'stakeholders'.

6 Design is about function

As well as design being about how things are perceived by our senses and then translated into an emotional response, design is about getting a job done. It is only as we get to art-based products where the function can be totally ignored (or alternatively where function is limited to that producing an appropriate artistic or emotional effect). Hubka and Eder [13] in their text *Design Science* have a useful list of product categories that mention the interplay of appearance and operability in a range of product categories. In all but the artistic product (and possibly the consumer products where the product itself is consumed and the design emphasis is on the product packaging rather than the product) they conclude that some form of operability is essential for the product's existence.

Design is about function in that designed products have to interact with the real world. They have to do things. In this case it doesn't really matter whether the product relates to the natural world or to the man-made world. Products need to be able to relate to both of these, and it is seldom that the two can be separated coherently. In whatever way scientific theories are incorporated and utilised by the designer in developing the product, it stands or falls on the way it relates to the real world.

In many instances, the product function will be about the way the product works to incorporate a system. That is, to do with the way that the product changes materials of some sort into something else, about the way that it uses or changes energy, and about the way that it is controlled as it does these. Function may also be about the way that the product supports loads, about the way that the material is distributed and about the choice and type of material that is used in its manufacture.

Design is also about how well a product functions, about the effectiveness of the product. For instance, does the washing machine clean the clothes effectively? Are they clean when they come out of it? And, to some extent at least, how much work do we have to do to ensure that the machine works well? And what is the sort of effort we'd like to put in to control it? Continuing with our product theme of the washing machine, apparently it was this particular product in the 1940s that was most noticeably ineffective. It was little developed from the 1930s machine and needed an incredible amount of work on the part of the operator to get it to produce good results. This involved moving the machine around, filling and emptying it using hoses, adding soap powder, being sure that fingers were nowhere near the electric mangle and so on [14]. With some products, skill in their use is appreciated: this is unlikely in the case of a washing machine. Washing machines now function better. There is no doubt that being automatic makes them more effective – not necessarily in terms of simply the cleanliness of the clothes at the end, but in terms of consistently doing what purchasers want.

Products do not actually have to function well. Some function appallingly badly, but nonetheless they still function and are still designed. It seems incredible that someone would actually design things that don't function well, but for various reasons this seems to take place from time to time. Rather than concentrate on such products it would be more profitable to ask what processes have to be in place in order to design something so it functions well. We might surmise that it is in the way that scientific theories that act as the designer's input data correctly model the real world behaviour. But that doesn't seem to be the whole story. There is something also about the way in which the heuristics and human recommendations of the design methods and processes of design that seems to be able to make products better. Baxter [15] investigates success and failure in products and, whilst a product functioning well and a successful product are not the same thing, there is a similarity. A product that doesn't function well is unlikely to be a successful one. In particular, Baxter highlights that sharply defined products and those where they are defined precisely early on in their gestation periods are more likely to be more successful. We could extrapolate and suggest that defining the product well could be related to its functioning well.

Functional design is also about efficiency. This is not the same as effectiveness, although the two are very often confused. A product that functions well may not in fact be efficient. For a systems-type product the definition can be easily determined: it is the ratio between the useful outputs of the product and the inputs required to produce the outputs. In this simple case both input and output are normally measured in energy units and the efficiency is thus normally a numerical figure, a ratio of comparison of one with the other.

In the case where the product includes some form of heat engine, the efficiency will normally include a measure of how much fuel is used to produce the desired effect, and may not be defined in such precise terms, people rather loosely speaking of efficiency in a comparative manner, such as when comparing the fuel economy in family cars. This use of fuel extends the term, again in loose parlance, to mean the avoidance of using any material wastefully, also particularly relevant in systems-type products, those where material transformation of some type is carried out.

That is not the only extension of the term. It is also used in structural products to denote how well the structure carries the load and how the load is distributed around the structure. This extension may have roots in how well the strain energy is distributed, but it probably also has roots from the comparisons of energy use in moving structures such as vehicles and aircraft, where the lighter a given structure is, the less fuel will be used to obtain a given performance. Hence the car or aircraft is actually more efficient – but the efficiency is loosely carried over to describe the sense that the lighter the structure for a given load, the more efficient it must be. A structure that removes unnecessary redundancy, avoids padding and where every component does something useful is perceived as efficient, even though there may be no way to measure that efficiency as a numerical ratio in conventional terms.

Functional design is also about practicality. This is defined as being concerned with actually doing rather than with the theory. Sometimes this can be perceived as being practical in addition to theory, in order to test out the theory, in the same sense as one might in, say, a physics experiment, and sometimes it is taken to be actually doing, in place of developing the theory. If a product is not practical then it isn't a useful product in the real world – but here again the principle is extended, and when we say that a product isn't practical we don't necessarily mean that it doesn't work at all, merely using the term to mean a qualitative measure of its effectiveness or a measure of whether it is likely to succeed or to be a realistic approach to the problem posed.

If design is not about practicality then it must be about non-functional products that don't work, that can't be put together and that just look as if they might do something. If it isn't practical then it effectively isn't about meeting the needs or wishes that were posed in the first place.

7 Design is about the process of arriving at the solution

Not only is design about the aesthetic side of life, our interaction with our senses leading to an emotional response, together with function, but it is also about this process of achieving some sort of man-made product or object, or some alteration in the surroundings. For some, the term *designing* is more appropriate than the shorter term *design*, which is nevertheless commonly used. We have thoughts about what we wish to see in the world, and we use our practical abilities or maybe our management skills to enable this vision to be realised in the concrete, real world. This process is not the simple one of draw it and it happens [16], but is much more complex, more iterative, convoluted and fraught with disasters. We find that the real world doesn't behave the way we thought it would. We find that the people we thought would love our product are ambivalent to it. We find that this real, physical issue of time doesn't allow us to get the job done properly and that the human interactions that have to take place in order to make it happen aren't managed as well as they might be. After all, the people involved in the design team have their own visions of what the product might be – witness the well-known contrasting pictures of swings or aircraft 'designed' by different members of the company, with different ideas of how the product should be put together [17].

Some designing is about getting things done in the most obvious way. The process does not always have to be mystical or mysterious. It is simply about getting the product on the road in a state that is acceptable for the purpose to which it has to be put. So a lot of the process is common sense, even humdrum. Drawings may have to be done, tests carried out and so on. Edison is quoted as saying that genius is 99% perspiration and only 1% inspiration [18]. This could sometimes be said about the design process. If the perspiration isn't done, the product doesn't fulfil its requirements in terms of consumer needs, aesthetics or functions – not because these couldn't be achieved physically, but because the effort had not been put in to make the process work. So design can also be about common sense.

And a lot of that sort of common sense has gone into the development of texts looking at the design process and about how to get the whole of the product design and development process happening so that the timescales are managed better, the various tasks are put together well and so that the product is both what is required by the end clientele and is a step forward in the market. Some texts (for example [15] and [11]) concentrate on the parts of the process that enhance the aesthetic side of the product, while others (such as [6] and [7]) dwell on those aspects of the process that develop the functional aspects.

There is considerable debate about which processes can be labelled design processes and which are labelled as something else. Some, (e.g. [6]), take the view that the design process starts with the stages after the initial task has been determined, only including a brief task clarification phase. Others, (e.g. [7] and [10]), on the other hand, include a significant determination of what the design task is as part of the design process. In the later stages of the process, the Total Design core [7] continues onwards to include product development stages until the product is being sold in the market, whereas [6] concludes the design process with the completion of the documentation for production.

The purpose of investigating the design process is to put forward recommendations for the use of particular methods to be applied to the design problem on hand, at whatever level. These recommendations are not normally derived from scientific observation, but are developed as empirical methods that appear to produce some sort of result when applied to a significant number of individual cases [19]. They can be described as emanating from our man-made thoughts rather than from any scientific reasoning, and therefore they can be placed on the man-made thought quarter of our diagram as shown in Figure 8.

Generally, design methodologies not only recommend ways to manage the process of design, but they all also highlight the necessity to ensure that there is sufficient space within the process for the illogical and iterative aspects of design associated with creative thinking. Whilst there are certain stages that will need to be reached in order for the product to be produced, there is inevitably a large amount of backtracking, management decision, re-thinking and unsuccessful validation that has to go on. Not to do this and to pretend that it doesn't happen or need to happen will limit the innovation and creativity the product demonstrates and almost inevitably results in something that could be significantly improved. The overall time and process schedule, however, within its method framework, will ensure that at the end of the process, something tangible is developed that is in a position to be sold, demonstrated or whatever the final outcome of the process is intended to be in the particular case.

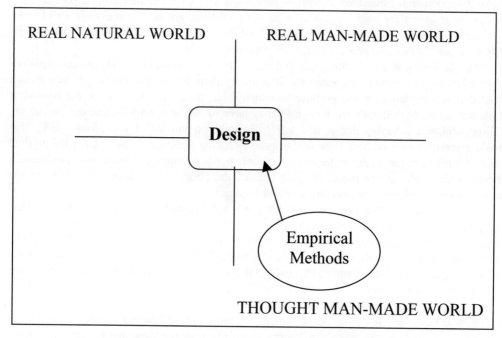

Figure 8: The position of empirical design methods.

8 Summary: a design overview

We have attempted to build up an introductory picture explaining the place that the topic of design has within our world. Perhaps due to our particular and individual backgrounds we see design as being a central plank of the way in which humans work, changing and shaping the world to be in the form that man wants, adapting and modifying it for his own ends. We also see design as being that crucial link between the way humans explore their individual thoughts and ideas and make these concrete in the real world.

We see that design is something that relates to the human expression in terms of art, in terms of emotion, and which has links with the continual scientific human attempt to explain what the world around us is like. We see a process that encapsulates not only these ideas, but also our understanding of the real world, our emotions, our aesthetic sensibilities and our propensity to use up resources in both essential and trivial ways.

Figure 9 shows the overall picture of what we have been building up, and demonstrates exactly why we see design as being central to the way humans live.

This rather bland and colourless description is of course not adequate to contain the fullness of the relationships that design has both to the human world of creativity, art, aesthetics and emotions or to the scientific and cultural understanding of what is going on in the world and how the world is put together.

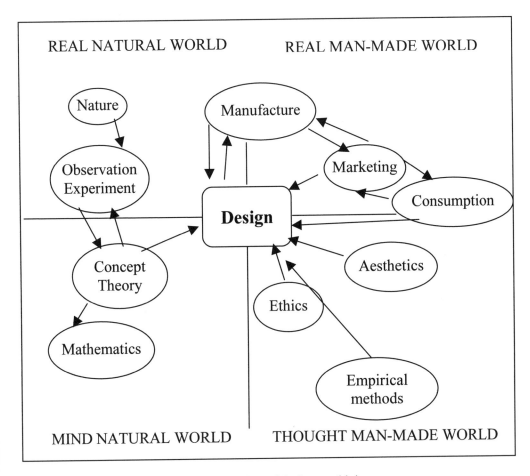

Figure 9: Overview of design worldview.

9 An example

This theoretical treatment is somewhat divorced from reality. It would be helpful to see an example in order to understand the overview better.

Typically, one might look at some form of structure – a bridge.

Structures exist in the real natural world. Things such as trees, bones, tortoiseshells and so on. They perform functions of carrying, supporting, protecting and so on. Structures can be designed by humans: things such as aircraft, buildings, armour plating, and so on perform these same functions of carrying, supporting and protecting.

A fallen log becomes a bridge over a stream: if it falls naturally it is in the real natural world; if someone put it there it becomes part of the real man-made world. If someone looked at the fallen tree over the stream and decided to make something that looked similar over another stream, they designed that bridge, as it has now become. There was an idea that formed in the human mind world as an analogy with the fallen tree. There was some form of what might be described as a rudimentary design management process, taking the idea from just that, realising and embodying it as 'bridge'. There was even the human concept of 'bridge' that implies such things as crossing a gap, making it so that someone or something can go from one side of the gap to the other, and so on.

Specifically, Figure 9 now becomes Figure 10.

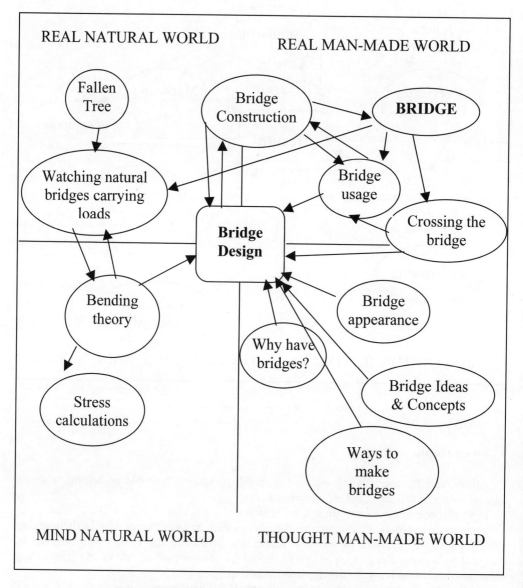

Figure 10: Bridge: an example of the design worldview.

References

[1] Lakoff, G., *Women, Fire and Dangerous Things: What Categories Reveal about the Mind*, Chicago Press, 1987.
[2] Douglas, M., *Thought Styles: Critical Essays on Good Taste*, Sage, 1996.
[3] Hall, A.G., *Models & Mathematics: The Effect of Design on Theory in Engineering*, Engineering Education Conference: Increasing Student Participation, PAVIC Publications, 1994.
[4] Jones, J.C., *Design Methods: Seeds of Human Futures*, Van Nostrand Reinhold, 1992.
[5] Tiles, M.E., *The Concept of Human Nature*. In: *Philosophical Anthropology*, Encyclopaedia Britannica CD, 2000.
[6] Pahl, G. & Beitz, W., *Engineering Design: a Systematic Approach*, Tr. K. Wallace, L. Blessing & F. Bauert, Springer, 1996.
[7] Pugh, S., *Total Design*, Addison Wesley, 1991.
[8] Schumacher, F., quoted by HRH Prince Charles during the *2000 Reith Lecture*, BBC Radio 4, 17 May 2000.
[9] Petri, H.L., *Drive* in *Motivation*. Encyclopaedia Britannica CD, 2000.
[10] Hollins, G. & Hollins, W.J., *Total Design: Managing the Design Process in the Service Sector*, Pitman, 1991.
[11] Otto, K. & Wood, K., *Product Design: Techniques in Reverse Engineering and New Product Development*, Prentice Hall, 2001.
[12] Eysenck, H.J., *Sense and Nonsense in Psychology*, Penguin, 1958.
[13] Hubka, V. & Eder, W.E., *Design Science: Introduction to the Needs, Scope and Organization of Design Knowledge*, Springer, 1996.
[14] Sturgis, M., Number 17; The Unsung Hero of the War, *Daily Telegraph* property section, 30 Dec. 2000.
[15] Baxter, M., *Product Design: Practical Methods for the Systematic Development of New Products*, Chapman & Hall, 1995.
[16] Johnson, C., [pseudonym for Leisk, D.J.] *Harold and the Purple Crayon*, Harper & Bros: New York, 1955.
[17] Hollins & Hollins, *op cit.*, page 138.
[18] Petty, G., *How to be Better at...Creativity*. Kogan Page, page 136, 1997.
[19] Dowlen, C.M.C., Development of a Cognitive Framework for Design Science. *International Conference on Engineering Design*, ICED'97, Tampere: Finland, 1997.

CHAPTER 2

Mathematics in the natural world

J. Gomatam & F. Amdjadi
School of Computing and Mathematical Sciences
Glasgow Caledonian University, UK.

Abstract

Mathematics relies on logic and creativity, and it is pursued both for a variety of practical purposes and for its intrinsic interest. For some people, and not only professional mathematicians, the essence of mathematics lies in its beauty and its intellectual challenge. For others, including many scientists and engineers, the chief value of mathematics is how it applies to their own work. Because mathematics plays such a central role in modern culture, some basic understanding of the nature of mathematics is requisite for scientific literacy. Mathematics reveals hidden patterns that help us understand the world around us. Now much more than arithmetic and geometry, mathematics today is a diverse discipline that deals with data, measurements, and observations from science; with inference, deduction, and proof; and with mathematical models of natural phenomena, of human behaviour, and of social systems. From rainbows, river meandering, and shadows to spiders' webs, honeycombs, and the markings on animal coats, the visible world is full of patterns that can be described mathematically. Examining such readily observable phenomena, this article introduces readers to the beauty of nature as revealed by mathematics and the beauty of mathematics as revealed in nature.

1 Introduction

> "Time present and time past
> Are both perhaps present in time future,
> And time future contained in time past"
>
> T.S. Eliot [1]

The discipline of mathematics can be defined as "a group of related subjects, including ALGEBRA, GEOMETRY, TRIGONOMETRY and CALCULUS, concerned with the study of number, quantity, shape, and their inter-relationships, applications, generalisations and abstractions" (Borowski [2], page 365). According to historians "the beginning" of mathematics has its inception in the prehistoric period in the sense that the geometry of spiral nebulae or snow crystals existed as well as, later, Syriac numerals, the ornament on the Egyptian predynastic pottery or the Early Bronze Age Cyprus jug of the period 3000–2000 B.C. or the extensive astronomical observations of the Fuh-hi period 2852–2738 B.C. or alluded to in the numerical skills displayed by central characters in the Ramayana and the Mahabharata, the two Hindu epics (Smith [3], pages 1–18). Without any offence to claims of diverse citizens of the world to early mathematical knowledge we will confine ourselves to Pythagoras (c. 580–500 B.C.), Plato (c. 428–347 B.C.) and Aristotle (c. 384–322 B.C.) as providing the beginnings of mathematics. This facilitates a working definition of the natural world: the worlds we perceive all around us, including deeper space, past and future, by our innate senses aided by our technology. In a sense the above quotation from T.S. Eliot has a ring of truth about it.

2 Greek philosophy and mathematical foundations

2.1 Pythagoras

Pythagoras apotheosised arithmetic as opposed to logistics. He developed a prototype curriculum, later known as the *quadrivium* culminating in arithmetic and music, geometry and astronomy as shown in Fig. 1.

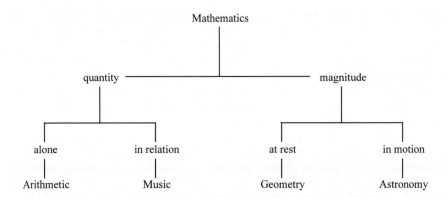

Figure 1: The quadrivium of Pythagoras

"The quadrivium eventually evolved into the seven *artes liberales* with the addition of the trivium (grammar, rhetoric, and logic), but music's position as a subset of mathematics remained constant throughout the Middle Ages" (Garland [4], pages 63 and 64). "The Pythagoreans believed that all regularities in nature were musical, and that the study of numbers and their relationship to musical harmony was the way to reach divine spiritual understanding and the purity of the soul. This was the beginning of an entirely mathematically based music theory". The Pythagoreans neither understood the physics of acoustics nor the effect of aural anatomy on the pitch perception. They were overawed by the discovery that the ratios of integers produced consonant intervals (Garland [4]). Now we turn from music to architecture, but it is still within the closely integrated conception of the quadrivium. Hence for the ancient Greeks, architecture was "Frozen Music" (Schelling [5]) but at the same time, living music. This reflected the mathematical insight that notions of periodicity and proportion, and their interplay, can be used for succession in time as well as for spatial associations (Ghyka [6], Chapter 1). There are a number of examples of the so-called golden section defined by

$$\frac{\text{Length of a rectangle}}{\text{Its breadth}} = \frac{1+\sqrt{5}}{2} = 1.6180339.$$

Within architecture there are examples among Georgian buildings in England, including the west elevation of Saint Paul's cathedral by Wren (Martin [7]). More on the ubiquity of the golden section in the natural world later.

2.2 Plato and Neo-Platonism

Plato was a student of Socrates and a philosopher with the name Eucledes of Megara. During his extensive travels he encountered a vast amount of knowledge of mathematics and philosophy. "Plato's work has bequeathed many famous tenets of western philosophy to us, including the immortality of the soul, the theory of knowledge as recollection and most notably, the theory of 'forms' which suggests that there is a timeless, universal world beneath the every day, transient material world". (De Botton [8]). On the mathematical side Plato was in awe of the science and meta-science of numbers, but records very little of what was taught in his school. More than any who preceded him he appreciated the scientific possibilities of geometry (Smith [3], page 89). Plato was averse to experimentation and hence this aspect of scientific inquiry was never a natural concomitant to Greek natural philosophy. The "timeless, universal world" of Plato had a considerable impact on the Cambridge Platonists in the mid seventeenth century, including the philosopher Henry More. He believed "that the world was permeated by the Spirit of Nature. This esoteric 'force' mediated between God – who controlled all actions, all purpose, and all outcomes – and the purely mechanical universe – the mundane physical world in which we live and conduct our lives". This was also the key influence guiding Isaac Newton (White [9], page 56). To the two worlds that Platonists are familiar with, the physical world of objects, fauna and flora and the world of mathematical forms, Penrose adds a third world, the world of our conscious perceptions (Penrose [11], page 412). Penrose points out that each world appears to "emerge" mysteriously from a small part of its predecessor: the Platonic world of mathematics describing laws that govern the physical world which in turn "'create' mathematical concepts out of some kind of mental model". "These mental tools ... with which our mental world seems to come equipped, appear nevertheless mysteriously able ... to conjure up abstract mathematical forms". While on the subject of abstraction, the place of Godel's incompleteness theorem in the world of Platonic mathematical truths is aptly explained by Penrose (ref. [10], pages 146 and 147, ref. [11], pages 418 and 419). Godel's theorem is not about the existence of inaccessible mathematical truths, but a plea for power of human insight over formal logic and computable procedures. It is, in fact, "an argument for the very existence of the Platonic mathematical world. Mathematical truth is not determined arbitrarily by the rules of some 'man-made' formal system, but has an absolute nature, and lies beyond any such system of specific rules. Support for the Platonic viewpoint (as opposed to the formal one) was an important part of Godel's initial motivations. On the other hand, the argument from Godel's theorem serves to illustrate the deeply mysterious nature of our mathematical propositions. We do not just 'calculate', in order to form these propositions, but something else is profoundly involved - something that would be impossible without the very conscious awareness that is, after all, what the world of perception is all about" (Penrose [11], page 419).

It is extremely intriguing, at this juncture, to review introductory aspects of Paul Davies' work, *The Mind of God* ([12], pages 34 and 35). Our primary perceptions are about ourselves, but we also have the ability to be aware of an outside world, however flawed. We are imprisoned in Plato's cave with our backs to the light. Facing the wall, we merely interpret the distorted shadows of objects passing the entrance to the cave. According to Plato there are two gods who have dominion over our worlds. An eternal and immutable being existing beyond space and time and another god, a so-called Demiurge, with a remit for fashioning order in the material world. No attempt to reconcile the two was made by Plato.

2.3 Aristotle

Plato's disciple Aristotle, while rejecting the existence of timeless Forms, construed the world as a living organism, "developing like an embryo towards a definite goal" (Davies [12], page 36). Aristotle's writing encompassed logic, philosophy, biology, astronomy and physics. His fortes were logic and biology (White, page 31). "The morphological approach to biology is through structure viewed as form" (Arber [13], page 125). The approach employs "bodily eye and the mind's eye" with all its interconnections. Thus the morphologist follows Aristotle's conception combining "pure thought and the empirical study of individuals" (Arber [13], page 125). Aristotle, the Father of Biology, integrates the Democretean notion of primary position of sensations and Plato's emphasis on conceptual thought (Arber [13], page 125).

3 Irrational numbers

3.1 Mathematics representation

As is generally known, "Any real or complex number that cannot be expressed as a ratio of two integers" is called an irrational number (Borowsky [2]). There are a number of examples of irrational numbers in the real world around us: the number $e = \lim_{x \to \infty} (1+1/x)^x$ in the radio-active decay phenomenon or simple Malthusian population growth models, the quantity π as the area of the unit circle, $i = \sqrt{-1}$ in the solution of the quadratic equation $x^2 + 1 = 0$. The quantity i is also known as a Gaussian integer (Borowsky [2]). An elementary consequence of Euler's formula is the relationship $e^{i\pi} = -1$. "It is said that were civilization, as we know it, to be wiped out, any new civilisation would 'rapidly' (say in a thousand years!) discover e, π and the square root of -1" (Laithwaite [14], page 184). To this list we add the golden ratio, which we will discuss later.

3.2 An engineer's description

3.2.1 Golden ratio
Given a rectangle of length L and breadth B, $L > B$ it is easily shown that

$$\Phi = \frac{L}{B} = \frac{1+\sqrt{5}}{2} = 1.6180339.$$

Statistical tests show that the rectangular shape closely approximating the golden rectangle enjoyed popular preference as being innately attractive. An obvious extension of this experiment is to the golden ellipse, the ratio of the major axis to the minor being 1.62. Three out of four observers preferred the golden ellipse (Huntley [15], pages 64 and 65). There are two separate issues connecting the occurrence of the golden ratio in nature and in man-made design. The ratio may be viewed as so innate to the whole world in which we live that the artist or the designer unconsciously uses it. This is consistent with the earlier cited preference for the golden ratio shape by arbitrary observers. On the other hand, the architect or the composer (namely Bartok, as we will see later) may be so convinced of the fundamental character of the golden ratio, that he/she deliberately uses it. Presumably, in the classic ancient Egyptian New Kingdom tomb of Nakht, it was the former, with a retro-interpretation in its description [31]. Turning back to mathematics, a number of properties of Φ are easily established (Ghyka [6], pages 7 and 8):

$$\Phi^n = \Phi^{n-1} + \Phi^{n-2}$$

$$\frac{1}{\Phi^m} = \frac{1}{\Phi^{m+1}} + \frac{1}{\Phi^{m+2}}$$

$$\Phi = \sum_{k=1}^{\infty} \frac{1}{\Phi^k}$$

$$\Phi = 1 + \cfrac{1}{1 + \cfrac{1}{1 + \cfrac{1}{1 + \cfrac{1}{1+\ldots}}}}$$

$$\Phi = \sqrt{1+\sqrt{1+\sqrt{1+\sqrt{+}}}}$$

3.2.2 Binet's number (Huntley [15], page 56)

The linear second-order difference equation for generating the Fibonacci sequence $u_{n+2} = u_{n+1} + u_n$ with $u_0 = 0$, $u_1 = 1$

$$u_n = \frac{1}{\sqrt{5}}\left(\frac{1+\sqrt{5}}{2}\right)^n - \frac{1}{\sqrt{5}}\left(\frac{1-\sqrt{5}}{2}\right)^n,$$

known as the Binet formula. The golden ratio

$$\Phi = \lim_{n\to\infty} \frac{\frac{1}{\sqrt{5}}\left(\frac{1+\sqrt{5}}{2}\right)^{n+1}}{\frac{1}{\sqrt{5}}\left(\frac{1+\sqrt{5}}{2}\right)^n} = \frac{1+\sqrt{5}}{2}.$$

It is not difficult to see that the exponential law is recovered for large n from

$$u_n = \frac{1}{\sqrt{5}}\left(\frac{1+\sqrt{5}}{2}\right)^2 = 0.4472 \times 1.61180^n.$$

4 Mathematics of biological form

Propagation of helical waves (see Fig. 2), other forms of toroidal wave forms and surface waves in excitable media such as the Belousov-Zhabotinsky {B-Z} reagent (Welsh [16], Murray [18], *Mathematical Biology*, Springer-Verlag) or the cardiac muscle or the nerve fibre (Winfree [17]) have been well documented. Some of these latter wave forms and their diffraction properties are modelled by Reaction-Diffusion (R-D) equations and their eikonal version (Grindrod [19], Gomatam [20], Carter [22, 26] – (note that diffraction refers to the effect of obstacles on the propagation of waves. The eikonal approximation of a partial differential equation is a geometrical approach to deal with travelling waves on the three-dimensional surfaces). The three-dimensional geometries of the astounding variety of shapes found in nature are documented by D'Arcy Thompson [23] (see also Chaplain [24]).

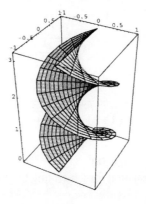

Figure 2: A typical helical surface.

4.1 Spirals and phyllotaxis

The subject of the relationship of the logarithmic spiral to shells has been extensively reviewed (Ghyka [6]). Helical geometry plays a significant role in three-dimensional chemical wave propagation (Gomatam [21]) as well as in relatively more static world flora.

The study of the arrangement of leaves around the stem of a plant or the location of the petals in a flower is referred to as phyllotaxis. The leaves of a plant are generally arranged around the stem in a helical fashion and this strikes one as a cunning adaptation to gain more exposure to sunlight. A closer examination reveals that there are two systems of spirals (Ball [26], pages 105–106). The leaf pattern appears to be determined by the tip of the stem where buds are initiated in a growing plant. An angle of 137.5 degrees between successive leaves facilitates optimal packing. The number of leaves per pitch along the two spiral systems constitutes the adjacent pairs of the Fibonacci sequence starting with 0 and 1. The same laws hold for the arrangement of petals in flowers.

4.2 Reaction-Diffusion waves

The evolution of an approximate involute spiral wave in a Petri dish containing the B-Z reagent is an elegant phenomenon (see Fig. 3). Rotation of the planar spiral about an axis parallel to the y-axis will generate a toroidal scroll surface. These waves were observed *in situ* in the B-Z reagent (Welsh [16]). The significance of the B-Z medium is that it provides a crude analogue medium to a variety of excitable biological media in so far as the geometry of wave forms is concerned. In fact, this geometrical similarity of observations is one of the many reasons that the so-called eikonal approach to three-dimensional R-D systems was developed (Grindrod [20], Gomatam [21]). The eikonal equation is given by: normal velocity of the wave surface = velocity of the plane wave in the medium - (diffusion coefficient) (mean curvature of the wave surface).

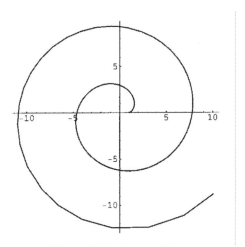

Figure 3: Involute spiral.

4.3 Unduloids

Diffraction by obstacles in the context of wave propagation is observed in biological systems. Winfree [22] shows that a damaged heart tissue acts as an obstacle to reaction-diffusion waves. It is known that if a region of the heart is damaged, such as may happen during a heart attack, then this region acquires altered electrical conductivity and becomes an obstacle to wave propagation. Some types of obstacles can even block waves. The eikonal equation offers a natural framework for the investigation of diffraction phenomena. According to the eikonal equation increased mean curvature implies decreased wave speed. The disks or tori present obstacles which increase the curvature of a wave passing between them. Depending on the specific configuration, a wave might actually be blocked and unable to propagate. The usefulness of predictions based on the eikonal approach is extensively documented (Carter [27]). Stable stationary patterns in three dimensions in the presence of obstacles have been investigated (Mullholand [28], Carter [27], Carter [23]). According to the eikonal equation, the stationary wave surfaces are given by surfaces of constant mean curvature, viz unduloids. We present here one such stable stationary unduloid in the presence of obstacles as tori in Fig. 4.

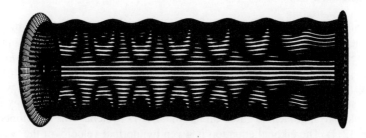

Figure 4: Unduloid generated by the eikonal equation.

A variety of unduloids are presented in the book by D'Arcy Thompson ([24], pages 52–54 and pages 82–85) and compared to metamorphosis in vorticella. The general point made here is the prevalence of mathematically defined geometrical forms in excitable systems, some of which show up in organisms.

5 Golden section, architecture and music

According to the architectural historian Henri Stierlin ([29], page 64), "Pythagoras played a role that greatly resembled that of the founder of a religion. With his motto 'all things are numbers' as a point of departure, he would develop a sort of rational mysticism which was to have a considerable influence in the artistic arena". The very structure of architecture is based not on attraction (eros) but resemblance (mimesis) to nature. Vitruvius's first century B.C. treatise on architecture mentions the works of Greek writings which are not extant. There is frequent reference to the "Symmetry" of Sacred Temples, meaning a set of relations based on reciprocity and balance governed by numerical values. The Pythagoreans believed that the "number is the principle behind all things" (Stierlin [29], page 67), Greek builders employed the Pythagorean right-angled triangle, arithmetical and geometric sequences of integers, the sequence later known as the Fibonacci sequence and the golden section or the divine proportion. The buildings created in Magna Graecia in Southern Italy find expression of the Pythagorean thinking on "Symmetry". "The interplay of symmetria, based on whole numbers, is strictly observed in the 'Basilica' of Posidonia (Paestum) (Stierlin [29], page 73). During its construction, Pythagoras was teaching at Croton, 200 miles away. The second Temple of Hera, replaced the old Basilica, and is contemporaneous with the Parthenon". "This is a peripteral hexastyle, with fourteen columns lengthwise; the dimensions of the stylobate are 24.31 by 59.93 m, that is 72 by 180 feet. The result is a proportion of 2:5, again, whole numbers which relate back to the teachings of Pythagoras" (Stierlin [29], page 79). In his unpublished 1939 prize-winning essay, Bruce Martin [7] states, "It is recognized that by arranging the parts in a certain relationship to each other and to the whole, they appear beautiful". This is of great importance for building since it implies that if due attention is given to the positioning of the parts, a building can appear beautiful. There are plenty of examples of Greek "Symmetry" reproduced in this article. The one branch of science where the ancient Greeks did relevant experiment was in (musical) acoustics. In his experiments with the monochord, Pythagoras noticed that plucking a string of half the length produced a similar sound but higher in pitch.

We would refer to this as sounding an octave higher. An enormous number of pairs of pitches, or frequency ratios, called intervals, are possible. A subset of these intervals sound pleasant; these are called consonants, and unpleasant sounding intervals are called dissonants. Pythagoras found three intervals to be consonants: they are known as the diapason, the diapente, and the diatesaron (Garland [4]). In modern parlance these are referred to as the octave, the fifth and the fourth, respectively. They correspond to the eighth, the fifth, and the fourth notes of the Pythagorean diatonic scale. Consider the scale

```
                                                        do
                                                ti
                                        la
                                sol
                        fa  (5th note)
                mi  (4th note)
        re
do  (1st note)
```

Figure 5: An octave.

If the full string produced the first note of a scale do 2/3 of the length will generate the note of sol, the fifth note. As was pointed out in the beginning, the Pythagoreans understood neither acoustics nor the aural perception of sound. The numbers in this ratio sum to 10, the holy tetractys of Pythagoras (Garland [4]):

```
1-point                         0
2-line                    0          0
3-plane              0         0          0
4-solid         0         0         0         0
```

Figure 6: The holy tetractys.

These experiments with the monochord led to the Pythagorean diatonic scale employed in the Western world. The Greeks termed the ratio between the frequencies of fifth and the fourth note $(3/2)/(4/3) = 9/8 = 1.125$, the whole tone. The ratio between the fourth and the third $(4/3)/(81/64) = 256/243 = 1.0535$. Similarly the ratio between the eighth note and the seventh note $= 2/(243/128) = 256/243 = 1.0535$. (For the intricate difference between Pythagorean tuning and the even-tempered scale see Garland [4]). The review article by Howat [30] examines the claim that Bartok organized many pieces around the golden section in time. Consider the Fibonacci sequence 0, 1, 1, 2, 3, 5, 8, 13, 21, 34, 55, 89, 144, ... According to Howat, Ledvai supplies a logical architectural model for the fugal first movement of the Fugue Music for Strings, Percussion and Celeste: The principal point of the golden section occurs after 55 bars, the climax, the first Fugue episode at bar 21. He concludes the article by saying, "Remarkable proportions can be found in a surprisingly large range of music, in some cases suggesting only subconscious application by the composer". Further critical appraisal of this golden section(ism) in art and music can be found in the article by Nagy [31].

6 Conclusion

In this chapter we have addressed several aspects of the mathematical character of the natural world. Perhaps at its purest level, our own research has shown that the eikonal approach can unify such diverse phenomena as the BZ reaction and the unduloid nature of the fluid flow in human arterial blockages. Studying this close identity of mathematically-defined shapes with nature follows in the footsteps of D'Arcy Thompson. Also, harking back to Pythagoras, we see that the human creativeness involved in architecture and music again have mathematical roots; these connect what appear to be innate human perceptions of visual and aural attractiveness. So the golden section diversely appears in the highest classic ancient Egyptian tomb art, in Greek temple designs and in Bartok's musical composition. The first preceded the Greek ideas of harmony and symmetry, the second was inspired by it, and the third involved conscious choice on the composer's part. We return to mathematics proper. It is most intriguing that Plato's idea of a timeless universal world so stood the test of civilisation's development that we find it having a formative influence on Newton himself. Moreover, it can still constitute a description of the mathematical enterprise, notably as propounded by Penrose. This brings us to Davies's discussion regarding mathematics as a discovery rather than an invention. He does admit that mathematicians are divided on this issue. As practising mathematicians ourselves, we tend to the view that mathematics is a discovery at the fundamental level because the basic truth was already there; that mathematics is an abstracted invention at the engineering and applications level as these involve several formal discovered techniques, such as solution of differential equations, Fourier transforms, stability analysis, bifurcation theory and the like working in unison, just like clockwork, to produce a final result. As a kind of postscript, the multi disciplinary character of this whole subject area should be noted. The BZ reaction itself, via chemistry, leads us to the new far-from-equilibrium thermodynamics, and with it biology. It could well be that fuller understanding of reaction-diffusion processes, will lead to new man-made methods of formation of materials – in other words, an extension of what is known as biomimetics.

Acknowledgement

We wish to thank Professor Michael Collins, the Series Editor, for great help with the source material and constructive suggestions during the preparation of this article.

References

[1] Eliot, T.S., Burnt Norton, *Four Quartet*, 10th Impression, Faber & Faber: London, 1978.
[2] Borowski, E.K. & Borwein, J.M., *Dictionary of Mathematics*, Collins: London & Glasgow, 1989.
[3] Smith, D.E., *History of Mathematics*, Vol. I, Dover Publications, Inc.: New York, 1951.
[4] Garland, T.H. & Khan, C.V., *Maths and Music, Harmonious Connections*, Dale Seymour Publications, 1995.
[5] Schelling. As quoted by Bruce Martin [7].
[6] Ghyka, M., *The Geometry of Aet and Life*, Dover Publications, Inc.: New York, 1977.
[7] Martin, B., *Proportions in Architecture*, Essay Prize, Architectural Association, 1939.
[8] Botton, A. (ed.), *The Essential Plato*, Details incomplete.
[9] White, M., *Isaac Newton, the Last Sorcerer*, Fourth Estate Limited: London, 1998.

[10] Penrose, R., *Shadows of the Mind, A Search for the Missing Science of Consciousness*, Vintage: London, 1995.
[11] Penrose, R., *The Emperor's New Mind*, Oxford University Press: Oxford, 1999.
[12] Davies, P., *The Mind of God, Science and the Search for Ultimate Meaning*, Penguin Group: London, 1993.
[13] Arber, A., *The Mind and the Eye, A Study of the Biologist's Standpoint*, Cambridge University Press, 1954.
[14] Laithwaite, E., *An Inventor in the Garden of Eden*, Cambridge University Press, 1994.
[15] Huntley, H.E., *The Divine Proportion, A Study in Mathematical Beauty*, Dover Publications, Inc.: New York, 1970.
[16] Sheclid, A.G. & Seidel, M., *The Tomb of Nakeht*, Verlag Philipp vonZabe, Mainz, Germany, 1996.
[17] Welsh, B.J., Gomatam, J. & Burgess, A.E., Three-Dimensional Chemical Waves in the Belousov-Zhabotinsky Reaction, *Nature*, **304**, pp. 611–614, 1983.
[18] Murray, J.D., *Mathematical Biology*, Springer-Verlag, 1989.
[19] Winfree, A.T., *When Time Breaks Down*, Princeton University Press, 1987.
[20] Grindrod, P. & Gomatam, J., The Geometry and Motion of Reaction-Diffusion Waves on Closed Two-Dimensional Manifolds, *J. Math. Biol.*, **25**, pp. 579–610.
[21] Gomatam, J. & Grindrod, P., Three-Dimensional Waves in Excitable Reaction-Diffusion Systems, *J. Math. Biol.*, **25**, pp. 611–622.
[22] Winfree, A.T., Understanding the onset of fibrillation in the heart muscle: two-dimensional vortices in healthy myocardium. *Science at the John van Neumann National Computer Centre*. Consortium for Scientific Computing, Priceton, NJ, USA, pp. 125–130, 1987.
[23] Carter, M., Amdjadi, F. & Gomatam, J., Diffraction of Reaction-Diffusion Waves: The Conformal-Mapped Eikonal Equation. Pre-Print Glasgow Caledonian University, 2003.
[24] Thompson, D.W., *On Growth and Form*, Cambridge University Press, 1961.
[25] Chaplain, M., The Main Contributions of D'Arcy Thompson, *Nature and Design*, **1**, International Series on Design and Nature, WIT Press: Southampton, 2004.
[26] Ball, P., *The Self-Made Tapestry, Pattern Formation in Nature*, Oxford University Press: Oxford, 1999.
[27] Carter, M., Diffraction, Interaction and Core Dynamics of Reaction-Diffusion Waves: Eikonal Solutions, PhD Thesis, Glasgow Caledonian University, 1999.
[28] Mullholand, A.J., Gomatam, J. & McQuillan, P., Diffraction of Spherical Waves by a Toroidal Obstacle: Eikonal Approach to Reaction-Diffusion Systems, *Proc. R. Soc. Lond.*, **452**, pp. 2785–2799.
[29] Stierlin, H., *From Mycenae to the Parthenon*, Taschen, GmBH, Koln, 2001.
[30] Howat, R., Review Article: Bartok, Levandy and the Priciples of Proportional Analysis, *Music Analysis* 2:1, pp. 69–95, 1983.
[31] Nagy, D., Symmetrospective, Golden Section(ism): From Mathematics to the Theory of Art and Musicology, Part I, *Symmetry, Culture and Science*, **7**, pp. 413–442, 1996.

CHAPTER 3

The Laws of Thermodynamics: cell energy transfer

J. Mikielewicz[1], J.A. Stasiek[2] & M.W. Collins[3]
[1]*Institute of Fluid Flow Machinery of the PASci, J. Fiszera, 80-952, Gdańsk, Poland.*
[2]*Faculty of Mechanical Engineering, Technical University of Gdańsk, Narutowicza 11/12, 80-952, Gdańsk, Poland.*
[3]*Faculty of Engineering, London South Bank University, 103 Borough Rd., London SE1 0AA, UK.*

Abstract

The Laws of Thermodynamics are among the most universal of all scientific relationships, and in this chapter they are presented in a typical engineering thermodynamics manner. The First Law, relating to the energy conservation principle, is applied, firstly to combustion of conventional hydrocarbon fuels, and then to the energy cycles in generic animal and plant cells. This enables a unified comparison to be made on the bases of natural sources of energy, and on the use or generation of oxygen, carbon dioxide and (liquid or vapour) H_2O, our three great naturally occurring chemicals.

A companion chapter in the Second Introductory Volume will treat the Second Law of Thermodynamics more fully, together with the implications for transport of information.

1 Introduction

Thermodynamics addresses many of the key issues of our day, issues such as energy sources and their use, the efficient generation of usable energy in the form of say electricity, and the operation of prime-moving transport such as cars. It is the explosion in the worldwide use of energy that led to the global warming effect. Energy considerations are at the heart of the natural world in another way, because generic animal and plant cells operate energy-related cycles, based on food consumption and photosynthesis respectively.

In this chapter, we will define the Laws (Zero, First, Second and Third) of Thermodynamics in classical terms. This treatment will be sufficient for us to give an overview of the energy processes in animal and plant cells.

While the present treatment concentrates on energy it is crucial, too, to address the key issues of <u>entropy</u> and <u>information transfer</u>. These topics will complete the relevance of thermodynamics to the biological world and to engineering design.

Hence the Second Law of Thermodynamics will be treated more fully in a companion chapter in the Second Introductory Volume of this series. The chapter will include a full consideration firstly of entropy and information theory, and secondly of non-equilibrium thermodynamics and its biological consequences.

Taking the subject still further, however, the implications, of <u>far from equilibrium</u> thermodynamics, are very wide, for both the natural world and engineering design. Concepts such as order out of chaos and robustness in design are encompassed. In order to address all these issues adequately, a subsequent volume in this series will be devoted to the overall theme of Thermodynamics, Biology and Design.

We conclude this Introduction with a historical note. Returning to the present chapter, the First Law of Thermodynamics stems from the research of James Prescott Joule (1818–1889), who studied the relationship between electrical, mechanical and chemical forms of energy. He was a brewery owner and a very careful experimentalist, who determined the Mechanical Equivalent of Heat, or Joule's equivalent. The SI unit of energy is named after him, and the Mechanical Equivalent is, in SI, a <u>defined</u> value.

2 The Zero'th Law of Thermodynamics

2.1 Definitions

Thermodynamics as a subject rests on a careful progression of precise definitions. While a 'nodding acquaintance' with the First Law is possible as an expression of Conservation of Energy, the subtleties of the Second Law, and the crucial applications to the natural world, will escape the reader unless this progression is followed. The precision of definition applies just as much to terms in conventional English usage (such as 'work' and 'heat') as to new terms like 'enthalpy' and 'entropy'.

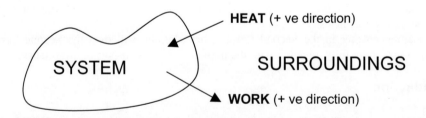

Figure 1: The system and the surroundings.

2.2 The system

A system is a fixed quantity of matter, surrounded by an arbitrary, invisible, three-dimensional envelope called the System Boundary. Energy transfers, namely heat and work, may cross the boundary. Anything outside, of potential influence, is called the surroundings. No mass transfer can take place.

2.3 Properties

Thermodynamic parameters relating to the system, such as pressure and volume, are termed properties, intensive (when mass dependent), extensive when independent of mass.

For example, pressure p is extensive

volume V is intensive,

where

$$V = m \cdot v$$

m, being mass
v, being specific volume.

2.4 State

This is the thermodynamic condition of a system, the totality of its properties. In practice, a useful empirical rule (the Two Property Rule. Ref. [1], pp. 118–119) means that two properties are normally sufficient to define the state. It is a Rule, rather than a Law, because under certain conditions, any two arbitrarily chosen properties will not be enough.

One advantage of the Rule, now perhaps dated, is that states may be represented by points on a two-dimensional graph.

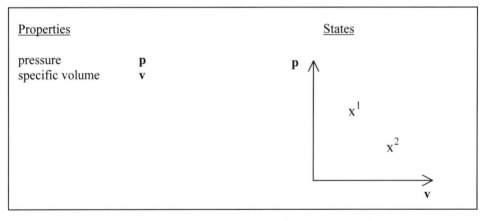

Figure 2: Some common thermodynamic properties.

2.5 Thermal equilibrium

Consider two systems in contact with each other, whose boundaries are insulating except where they are in contact.

Initially, systems A and B may be at different states, but ultimately their thermodynamic properties cease to change. The systems A and B are then said to be in <u>thermal equilibrium.</u>

Although the systems have been shown as having common parts of boundaries, in fact they may be

        i) physically separate

but   ii) in thermal contact. Radiation heat transfer is one practical example of this.

Finally, it is worth appreciating that these definitions are wide in their applications.

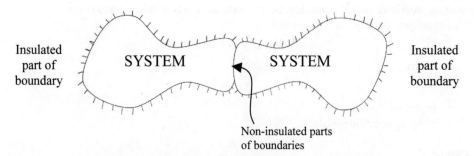

Figure 3: Illustration of thermal contact between systems.

2.6 The Zero'th Law

This may be stated as follows: two systems which are in thermal equilibrium with a third system are in thermal equilibrium with each other.

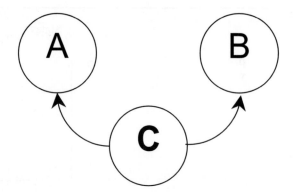

Figure 4: Illustration of Zero'th Law – system C is thermometer.

2.7 Temperature

The Zero'th Law is the basis for the concept of the property 'temperature'. Let us construct an instrument with some property which varies widely with temperature (length of mercury column in a glass stem). This we call System C. Let 'A' and 'C' reach thermal equilibrium. Then let 'B' and 'C' be brought into thermal contact. If the state of C remains unchanged (the property 'mercury column length' is the same), it means by the Zero'th Law, that A and B have the same temperature.

2.8 Scales of temperature

All temperature scales are, in principle, arbitrary but to minimise this generally we will use the gas scale of temperature.

The size of the degree is fixed by defining the normal melting point of ice as having the value 273.15°.

2.9 Work

The customary definition of mechanical work is that it occurs when the point of application of a force moves a certain distance. Quantitatively, for a force F moving through a small distance dx, the work is given by:

$$dW = F \cdot dx$$

In system terms, work is done when a pressure p results in a volume change dV,

$$\text{or} \qquad dW = p \cdot dV \qquad (1)$$

Further, rotational (or stirring) work arises from a torque T operating over an angle dθ,

$$\text{or} \qquad dW = T \cdot d\theta$$

and the concept of electrical work can be expressed as:

$$dW = V_E \cdot dq$$

where dq Coulombs flow due to a voltage difference V_E [from watts = (volts) × (amps)].

The thermodynamic definition of work includes all energy transfers which may, in principle, be totally converted to potential energy.

2.10 Heat

Heat is defined as an energy transfer due solely to temperature gradients. Heat is

> LIKE work in being an energy transfer, not a property.

> UNLIKE work, in that while a quantity of work may be totally converted into heat, the same quantity of heat cannot be totally converted into work.

2.11 Work output for a p-V process

We have noted that usually states may be represented by points on a 2-dimensional graph. Taking the p-V diagram overleaf, we represented (on the left hand graph) a change of state by a line from 1 to 2. A process is defined as a specified path for a change of state, together with the means whereby the change of state is achieved.

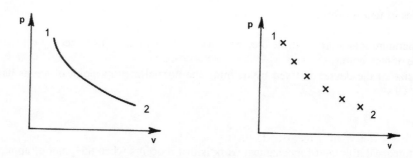

Figure 5: Illustration of a process.

From eqn (1), the work transfer W_{12} for the change of state between 1 and 2 is given by:

$$W_{12} = \int_1^2 p \cdot dV$$

which can be integrated if we know $p = f(V)$.

1) However, the integration can only be made if the path of the change of state is continuous, being composed of innumerable intermediate states, as in the left hand graph. If the system 'jumps' as in the right hand graph, with, say, 4 states between 1 and 2, no integration can be carried out.

2) Hence we say that the system is indefinitely in internal <u>equilibrium</u> for the family of states between 1 and 2.

3) For condition 2) above, we use the expression <u>quasi-static</u> process.

4) Not only the above, but if we return from 2 → 1, then

$$W_{21} = \int_2^1 p \cdot dV = -W_{12}$$

This idea of reversing direction, and returning to the original state, leads to the key thermodynamic concept of <u>reversibility</u>.

2.12 Reversibility

The thermodynamic definition of this term is as follows: A <u>process</u> is <u>reversible</u> if, after it has taken place, means can be found to restore the <u>system</u> and its <u>surroundings</u> to their respective initial conditions.

While reversibility includes the concept of 'reversing' it is much more than that. While it is possible to return the <u>system</u> to its original <u>state</u> for either a perfect (<u>reversible</u>) process, or imperfect (<u>irreversible</u>) process, it is only possible to restore <u>surroundings</u> for the <u>reversible</u> case (in the above, all thermodynamic terms have been underlined).

Now it should fully be appreciated that this is not just a theoretical foible or even an esoteric issue. Because the Laws of Thermodynamics are, as far as we know, applicable throughout the entire organic and inorganic world, then the bases of the subject become of virtual universal interest.

The final comment is related to this relevance. As assiduous reader can possibly guess that the science of thermodynamics will assume processes which are 'reversible', 'quasi-static' and have internal (system) 'equilibrium'. Now, while such are unattainable, they represent an ideal towards which *real processes should approximate as much as possible* – a sort of 'engineering perfect mission statement'!

3 The First Law of Thermodynamics

3.1 Introduction

There are two concepts, embedded in this:

- the well-known one of 'conservation of energy'
- and the similarity between heat and work as forms of energy, namely the 'mechanical equivalent of heat'.

3.2 Joule's experiments

These were carried out in the 1840's as a comprehensive study, leading to an increasingly accurate value for the equivalence between the 'unit for heat' and the 'unit for work' in a system of units. Hence in the old Imperial System this equivalence was expressed as J = 778 [(foot) – (pounds force)]/[British Thermal Unit]. In S.I. the problem is avoided in thermodynamics totally by either

a) the heat unit being the same as the work unit, namely the Joule, or

b) if the heat unit is the calorie, the equivalent is a <u>defined</u> constant.

We describe Joule's experiments in thermodynamic terms, as below:

<u>SYSTEM</u> Fixed mass of water, or other liquids

<u>PROCESS 'A'</u> With the system completely insulated ($Q_A=0$), it was taken by Joule from State 1 to State 2 by a work transfer ($W_A<0$). This W was performed in a variety of ways: i) electrically, ii) stirring, iii) friction. Process 'A', defined by a given temperature change required the same amount of work.

<u>PROCESS 'B'</u> He cooled the system back to its original temperature $Q_B<0$, $W_B=0$.

3.3 The cycle

What Joule did was to take the system through a <u>cycle</u>, that is, a set of <u>processes</u> with identical initial and final <u>states</u>.

3.4 Statement of Law

When any SYSTEM is taken through a CYCLE, the net WORK on the SURROUNDINGS equals the net HEAT taken from the SURROUNDINGS.

$$\Sigma dQ = \Sigma dW \tag{2}$$

Going from a state 1 to a state 2

$$Q - W = U_2 - U_1 \qquad (3)$$

where we call U the internal energy, made up of (internal) kinetic and potential energy of the atoms. U is a <u>property</u>. Eqn (3) is called the Non-Flow Energy Equation.

For the sake of completeness, Planck's First Law Statement is also given as:

It is impossible to construct an engine which will work in a cycle and produce continuous work or kinetic energy from nothing.

This Statement may be shown to be a consequence (a 'Corollary') of the previous Statement. Also it introduces the idea of impossibility of 'perpetual motion', referred to by Spalding and Cole (Ref. [1], p. 213) as a Perpetual Motion Machine of the first kind or PMM1.

FIRST LAW OF THERMODYNAMICS

- Conservation of energy

- Relates TEMPERATURE BASED energy transfer Q

  to   MECHANICAL BASED energy transfer W
  ELECTRICAL BASED energy transfer W
  CHEMICAL etc BASED energy transfer W

- State 1 to state 2: $Q - W = U_2 - U_1$

- For a cycle $\Sigma Q - \Sigma W = 0$

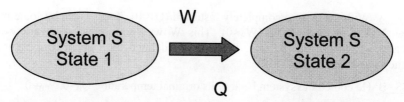

Figure 6: Illustration of First Law of Thermodynamics for a system undergoing a change of state.

4 Flow processes – the control volume

To enable us to accommodate mass transfers an equivalent definition to the system is made.

A <u>control volume</u> is a rigid arbitrary invisible three-dimensional envelope across whose boundaries can flow mass, heat and work.

Alternatively, the Control Volume is termed 'open system' in thermodynamic texts.

In engineering terms, the Control Volume concept can apply to a plethora of structures such as turbines, compressors, condensers and, for our purposes, furnaces and combustion chambers.

Without presenting the analysis (which involves making a system firstly non-coincident, then coincident, then non-coincident with a Control Volume: that is, it 'flows through' the Control Volume. See Ref. [2]) the First Law becomes

$$Q - W = (U + pV + \text{kinetic energy} + \text{potential energy})_2 -$$
$$(U + pV + \text{kinetic energy} + \text{potential energy})_1$$

or, ignoring kinetic and potential energy effects a restricted form of the Steady Flow Energy Equation is

$$Q - W = H_2 - H_1 \tag{4a}$$

where

$$H \text{ (the enthalpy)} \equiv U + pV, \text{ and is a property.}$$

For a combustion process in the absence of work

$$Q = H_2 - H_1 \tag{4b}$$

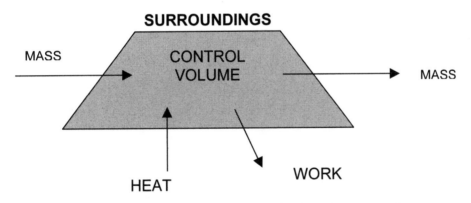

Figure 7: Thermodynamics of a control volume (open system).

5 The Second Law of Thermodynamics

5.1 Historical: Carnot's brilliant insight

The father of Sadi (actually Nicolas-Léonard-Sadi) Carnot (1796–1832) was Lazare Carnot (1753–1823), an engineer who became a very successful general in Napoleon's army, and a politician. As a general he wrote a treatise on fortifications in 1810 which became a military textbook. As an engineer Lazare was a good mathematician, and studied and wrote about the efficiency of mechanical systems, such as pulleys. He correctly understood the need for what we could describe as internal reversibility in mechanics, namely the securing of minimal friction in processes. Sudden jolts should be avoided, with movements by 'imperceptible degrees'. Sadi Carnot, following in his father's footsteps as a military engineer, studied heat engines. By then steam engines were being widely used, but not of very high efficiency. Sadi formulated an ideal working cycle for a heat engine, which he proved would have an efficiency dependent only on the external temperatures t_1 and t_2 and not on the working fluid used.

Here, the efficiency η is given by

$$\eta = \frac{W}{Q_1} = \frac{Q_1 - Q_2}{Q_1} = 1 - \frac{Q_2}{Q_1} \tag{5a}$$

where Sadi Carnot showed that:

$$\eta = f(t_1, t_2)$$

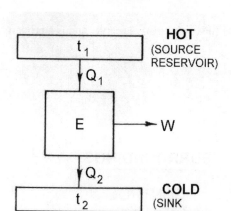

Figure 8: Heat Engine diagram.

5.2 General

Carnot's Theorem led to the formulation of the Second Law of Thermodynamics, which like the First Law, is one of the most universal laws of Nature. However, whereas the First Law has a ready equivalent in popular perception as the Conservation of Energy, there is no such mental 'prop' for the Second Law. 'Disorder always increases' is the best that can be managed – and this will be referred to later. The problem is that the Second Law manifests itself in a number of forms, not immediately obviously related. Major statements of the Law were made by

        Clausius   (1822–1888),

        Kelvin   (1824–1907) and

        Planck   (1858–1947).

It is not surprising that different authors have different statements – Dugdale (Ref. [2], p. 33) uses the Clausius statement as a natural formulation of the observation (p. 27) that 'heat always flows spontaneously from hot to cold'. Kondepudi and Prigogine (Ref. [3], p. 83) state, 'It is impossible to construct an engine which will work in a complete cycle, and convert all the heat it absorbs from a reservoir into mechanical work', hence leading to the impossibility of a particular (Second) kind of Perpetual Motion Machine (also called by Spalding and Cole (Ref. [1], p. 213) a 'PMM2'). This is essentially Planck's statement, which is also used by Rogers and Mayhew (Ref. [4]). In fact, since Rogers and Mayhew give a series of Corollaries (which usefully relate the 2nd Law forms), and since their approach is based on heat engine systems we will follow their sequence, but without proofs.

We mean by a Heat Engine a system operating continuously over a cycle, exchanging heat with the environment and producing work. If we include the concept of a <u>reversed</u> Heat Engine, which would absorb work and supply heat to the environment, then the refrigerator and heat pump could be included as well, as shown in Figure 9.

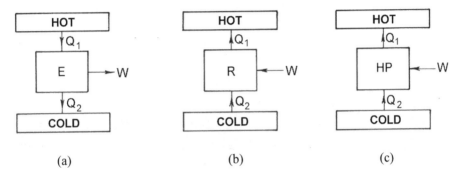

(a)   (b)   (c)

Figure 9: Heat Engine performance diagram; (a) (Direct) Heat Engine, (b) (Reversed Engine) Refrigerator, (c) (Reversed Engine) Heat Pump.

Efficiency   Coefficient of Performance

$$\eta = 1 - \frac{Q_2}{Q_1} \qquad CP = \frac{Q_2}{Q_1 - Q_2} \qquad CP = \frac{Q_1}{Q_1 - Q_2} \qquad (5a, b, c)$$

* Designed for W   * Designed for Q_2 * Designed for Q_1
* 'Cost' is Q_1 * 'Cost' is W   * 'Cost' is W

5.3 Second Law statement (Planck)

The Second Law may be given as: It is impossible to construct a system which will operate in a cycle, extract heat from a reservoir, and do an equivalent amount of work on the surroundings.

If, unlike the Heat Engine arrangement above, only a single (hot) reservoir is involved, the First Law of Thermodynamics gives W=Q_1, or W <u>cannot be greater than</u> Q_1. The Second Law is a further restriction, requiring finite Q_2, or W <u>must be less than</u> Q_1.

5.4 First Corollary (or Clausius statement)

This is given by Rogers and Mayhew as:

> COROLLARY 1. *It is impossible to construct a system which will operate in a cycle and transfer heat from a cooler to a hotter body without work being done on the system by the surroundings.*

5.5 More Corollaries

> COROLLARY 2. *It is impossible to construct an engine operating between only two heat reservoirs, which will have a higher efficiency than a reversible engine operating between the same two reservoirs.*

> COROLLARY 3. *All reversible engines operating between the same two reservoirs have the same efficiency.*

5.6 The Thermodynamic Temperature Scale

> COROLLARY 4. *A scale of temperature can be defined which is independent of any particular thermometric substance, and which provides an absolute zero of temperature.*

We have already noted that Carnot showed that for an ideal (perfect, reversible using only reversible process) heat engine $\eta = f(t_1, t_2)$.

However, since $\eta = 1 - \dfrac{Q_2}{Q_1}$, we may put $\eta = f'(Q_1, Q_2)$

or
$$f'(Q_1, Q_2) = f(t_1, t_2)$$

or
$$\frac{Q_1}{Q_2} = f''(t_1, t_2)$$

Kelvin proposed an absolute scale of Temperature, based on this (see Ref. [3], pp. 76–77) where:

$$\frac{Q_1}{Q_2} = \frac{T_1}{T_2} \tag{6}$$

This means that for a reversible heat engine

$$\eta_R = 1 - \frac{T_1}{T_2}$$

5.7 The definition of entropy

In principle, a Heat Engine can operate between more than two reservoirs.

> COROLLARY 5. *The efficiency of any reversible engine operating between more than two reservoirs must be less than that of a reversible engine operating between two reservoirs which have temperatures equal to the highest and lowest temperatures of the fluid in the original engine.*

What is known as the Clausius Inequality is given as Corollary 6, namely

> COROLLARY 6. *Whenever a system undergoes a cycle,* $\oint (dQ/T)$ *is zero if the cycle is reversible and negative if irreversible, i.e. in general* $\oint (dQ/T) \leq 0$.

Following this, for a non-flow process, 'entropy' is defined by

> COROLLARY 7. *There exists a property of a closed system such that a change in its value is equal to* $\int_1^2 \frac{dQ}{T}$ *for any reversible process undergone by the system between state 1 and state 2.*

That is

$$S_2 - S_1 = \int_2^1 \left(\frac{dQ}{T}\right)_{rev}$$

or

$$dS = \left(\frac{dQ}{T}\right)_{rev}$$

The term 'entropy' was proposed by Clausius in 1856 (Ref. [3], p. 80). A further historical fact is of much more significance. We have already used the absolute temperature scale definition

$$\frac{Q_1}{Q_2} = \frac{T_1}{T_2}$$

However, while this is both logical and simple, it is <u>arbitrary</u>, and in fact Kelvin originally proposed (in 1848) an exponential function, the scale above being his second (1854) definition (see Ref. [2], pp. 35, 36). Clausius' definition of entropy rests upon this scale, so there is a certain circularity of definition between heat transfer (Q), temperature (T) and entropy (S). Further, when the dimensionless group treatment is applied to convective heat transfer (see Ref. [4], pp. 574–575), it may be noted that 'Q' and 'T' do <u>not</u> occur independently.

Finally, Rogers and Mayhew give their eighth corollary

> COROLLARY 8. *The entropy of any closed system which is thermally isolated from the surroundings either increases or, if the process undergone by the system is reversible, remains constant.*

This has relevance to 'design in nature' following Clausius (quoted in Ref. [3], p. 84), for the 1st and 2nd Laws respectively:

> 'The energy of the universe is a constant.

> The entropy of the universe approaches a maximum'.

6 The Third Law of Thermodynamics

This has had a chequered history in terms of its acceptance. It was originally formulated by Nernst (1864–1941), and Dugdale (Ref. [2] p. 145) gives an amusing description of Nernst's personal attitude to it. It relates zero entropy to zero temperature, (ibid, p. 146) as:
The entropy of all systems and of all states of a system is zero at absolute zero.

7 The central equation of thermodynamics

Further, Dugdale (ibid, pp. 48, 49) develops what he calls 'the central equation of thermodynamics' namely from the First Law for non-flow processes:

$$dU = dQ - dW = dQ - p \cdot dV$$

Using the Second Law $dQ = T \cdot dS$, where the concept of temperature comes from the Zeroth Law:

$$dU = T \cdot dS - p \cdot dV \qquad (7)$$

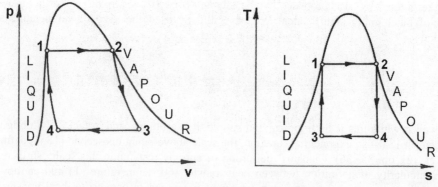

Figure 10: Carnot Cycle for a mixed-phase fluid.

8 The Carnot Cycle and entropy

We now relate the above equation to Carnot's concept of an ideal, or reversible engine. This is composed of reversible non-flow processes, with all heat transfers taking place at constant temperatures. His cycle (described in detail in Ref. [1]) has two isothermal and two isentropic (constant entropy) constituent processes. For a mixed-phase fluid the cycle may be shown as (Figure 10); and for a perfect gas (Figure 11):

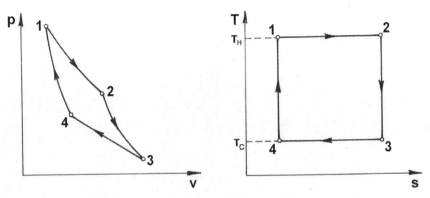

Figure 11: Carnot Cycle for a perfect gas.

From eqn (7), for a cycle:

$$\oint dU = \oint T \cdot dS - \oint p \cdot dV$$

However, the cyclic integral for dU is zero, hence:

$$\oint T \cdot dS = \oint p \cdot dV \qquad (8)$$

This has two implications in terms of entropy. Firstly (see eqn (1)) 'entropy' and 'temperature' together play the same role for <u>heat</u> production as do 'volume' and 'pressure' for work production. Secondly, the two sides of eqn (8) are respectively the areas of the T–s and p–V diagrams of Figs 10 and 11. This means <u>for a cycle</u> the areas of the two diagrams are equal.

9 The Rankine Cycle and global warming

While the Carnot Cycle efficiency given by eqn (5a) is the highest efficiency possible for a heat engine, in its mixed phase form it is extremely sensitive to even small inefficiencies, or friction effects. The UK engineer Rankine proposed a modified ideal cycle using H_2O, which has a lower reversible efficiency because the heat added is over a range of temperatures. However, it is much less susceptible to inefficiencies, and is defined in terms of <u>flow</u> processes, approximating to a 'boiler-turbine-condenser-feed pump' arrangement. This is shown in Fig. 12. Historically, developments and improvements related to the Rankine cycle form the basis for the major central generation of electricity from conventional fossil fuels such as coal or oil.

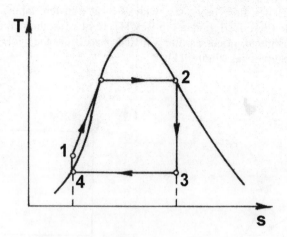

Figure 12: Ideal (reversible) Rankine Cycle.
1 – 2. Heating in boiler
2 – 3. Work production in turbine
3 – 4. Heat rejection in condenser
4 – 1. Feed pump work input to raise H_2O to boiler pressure.

Consideration of these factors and efficiency calculations form part of the second year of a typical 3-year UK course in Thermodynamics (see Ref. 4 for full details).

Now because of the Second Law necessity for heat <u>rejection</u>, overall thermal efficiencies for modern large steam plants generating electricity are just above 30%. Since the Industrial Revolution, then, the burning of fossil fuels for (a) electricity generation, and (b) transport – steam locomotives and then internal combustion engines – has resulted in the production of an enormous quantity of carbon dioxide. This is a major cause of the increased greenhouse effect over the 250 year period (see Houghton, Ref. [5], pp. 29–37).

As we shall see later in this chapter all energy processes for plants <u>absorb</u> CO_2. This forms part of the global balancing for the carbon cycle, but unfortunately the fossil fuel generation of CO_2 has been accompanied by a substantial <u>reduction</u> in the balancing effect. This latter has been caused by deforestation, and other changes, for example in agriculture.

10 Combustion processes

The term 'combustion' refers to the fairly rapid reaction, usually accompanied by a flame, which occurs between a fuel and an oxygen carrier such as air. Combustion is an exothermal chemical reaction. Some important aspects of combustion are handled by the Laws of Thermodynamics. Combustion processes are very complex. They involve knowledge of kinetics of chemical reactions, aspects of fluid dynamics and processes of heat and mass transfer. Chemical reactions are so fast compared with flow, heat, and mass transfer processes that these processes are the main controllers of combustion. In fact, this is summarised in the 'proverb' of numerical modelling of combustion as 'Mixed is Burnt'.

In any combustion processes both fuel and oxygen take part. In most engineering combustion processes the oxygen is obtained by mixing the fuel with air.

Any engineering combustion process considered as a whole is an open system undergoing steady-flow processes with air and fuel entering the system and the products of combustion leaving it at a steady rate. This is illustrated in Figure 13.

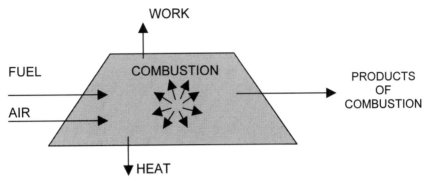

Figure 13: Combustion in a control volume (open system).

The energy produced in a combustion process can be transferred to the surroundings in the forms of heat and work. Most common fossil fuels consist mainly of hydrogen and carbon. The fuel state can be solid (e.g. coal), liquid (e.g. petroleum), or gaseous (e.g. natural gas). For solid and liquid fuels components are usually given as a percentage by mass of each in the fuel. In the case of gaseous fuels components are usually given as percentages by volume.

In typical combustion calculations we usually neglect the effect of minor components such as sulfur in the fuel. For the purpose of formulation of mass and energy balances this assumption is permissible. However, it should not be thought that the presence of a small quantity of a substance is always unimportant. We know that sulphur in fossil fuels is responsible for 'acid rain'. These trace substances produce corrosive compounds, which attack engine components.

10.1 Conservation of mass

The first step in the analysis of combustion processes is the formulation of the chemical stoichiometric equations. A chemical equation expresses the principle of the conservation of mass.

We consider the following simple chemical equation, which expresses the complete combustion of carbon and oxygen to carbon dioxide

$$C + O_2 \rightarrow CO_2 \tag{9}$$

This equation states that one atom of carbon will combine with one molecule of oxygen to produce one molecule of carbon dioxide.

The number of atoms of each element is the same on either side of the equation and the arrow represents the direction of the reaction.

The equation can also represent the amount of substance involved in the reaction and can be written in molar terms as:

$$1 \text{ kmol C} + 1 \text{ kmol O}_2 \rightarrow 1 \text{ kmol CO}_2 \tag{10}$$

(where the word 'kmol' means 1 kilo mol of substance), or in mass terms as:

$$12 \text{ kg C} + 32 \text{ kg O}_2 \rightarrow 44 \text{ kg CO}_2 \tag{11}$$

Now, all gases occupy equal volumes per kmol, when reduced to the same pressure and temperature. If any of the reactants or products is in the solid or liquid phase, the volume occupied by them can be neglected in comparison with the volume of the gases present. It follows that eqn (9) is also equivalent to:

$$0 \text{ vol C} + 1 \text{ vol O}_2 \rightarrow 1 \text{ vol CO}_2 \tag{12}$$

It follows that the volume of the products is approximately equal to the volume of the reactants.

Because oxygen is obtained from air, in any analyses we take into account that

Volumetrically O_2 is 21% in air mixture

and

Gravimetrically O_2 is 23% in air mixture

The chemical equations may be used to determine the stoichiometric air/fuel ratio. A stoichiometric mixture is one in which there is theoretically just sufficient oxygen to burn all the combustible elements in the fuel completely.

10.2 Conservation of energy – the First Law of Thermodynamics

The First Law of Thermodynamics applies to any system and to combustion processes. We will use for constant volume (non-flow) combustion processes the 'internal energy of combustion' concept and for constant pressure combustion processes (steady flow) the 'enthalpy of combustion' concept.

Let us first consider a non-flow process. Since the internal energy is a function of state, a change of internal energy is independent of the path of the process. A non-flow combustion process starts with a stoichiometric mixture of the fuel and air at an arbitrary state (p_1, T_1) and ends with the products in another arbitrary state (p_2, T_2). Each process can be regarded as consisting of three successive processes: going first, 'down to' in temperature, then 'through', then 'up from' a so-called standard state:

$$U_{P2} - U_{R1} = (U_{P2} - U_{P0}) + (U_{P0} - U_{R0}) + (U_{R0} - U_{R1}) \tag{13}$$

The first and third terms on the right hand side of the equation represent changes of internal energy in processes not involving chemical reactions and can be calculated in a conventional way, and the second term involves a chemical reaction at the standard state (p_0, T_0). The standard pressure is chosen as equal to 1 bar. The standard temperature is chosen as 298.15 K or 25°C. In engineering terms, then, the state p_0, T_0 represents ambient conditions:

$$\Delta U_0 = U_{P0} - U_{R0} \qquad (14)$$

For one kmol or kilogram of fuel ΔU_0 is called the 'internal energy of combustion' and is usually determined experimentally, especially for solid and liquid fuels.

We may similarly consider the change of enthalpy between reactants in state 1 and products in state 2 for flow combustion processes. It may be seen how the definition of enthalpy results in an identical equation format for the two sets of processes:

$$H_{P2} - H_{R1} = (H_{P2} - H_{P0}) + (H_{P0} - H_{R0}) + (H_{R0} - H_{R1}) \qquad (15)$$

Similarly to eqn (5):

$$\Delta H_0 = H_{P0} - H_{R0} \qquad (16)$$

For one kmol or kilogram of fuel ΔH_0 is called the enthalpy of combustion, and again is usually determined experimentally. Direct experimental determination of enthalpy of combustion is carried out in a steady-flow gas calorimeter.

In any application of combustion data, it is important to decide whether the H_2O in the products should be taken into account as a liquid or vapour and also whether ΔU_0 or ΔH_0 data are appropriate. During any combustion process in a closed system both work and heat may be transferred across the boundary in accordance with the First Law of Thermodynamics:

$$Q - W = U_{P2} - U_{R1} \qquad (17)$$

There are two combustion processes of special interest. One is carried out at constant volume and with $T_1 = T_2$. In this process the work is zero and the heat transferred according to eqn (17) is equal to the internal energy of combustion at temperature T_1:

$$Q = U_{P2} - U_{R1} \qquad (18)$$

The second process of special interest is adiabatic combustion ($Q = 0$) at constant volume. Since both the heat and work are not transferred the internal energy remains constant. There is a rapid rise of temperature resulting in the maximum possible temperature change for any given air/fuel ratio.

A similar analysis concerns flow processes with respect to enthalpy. Both work and heat may be transferred across the boundary of the control volume in accordance with the First Law of Thermodynamics formulated for open systems:

$$Q - W_{0P} = H_{P2} - H_{R1} \qquad (19)$$

Again, an adiabatic flow process ($Q = 0$) yields the maximum possible temperature rise for steady flow combustion. When the work is zero then the heat transferred according to eqn (15) is equal to the enthalpy of combustion at temperature T_1.

When dealing with energy transfers in combustion processes, engineers commonly relate these transfers to some calorific value of the fuel rather than to the properties ΔU_0 or ΔH_0. Calorific values refer directly to quantities of heat liberated when unit mass or volume of fuel is burnt completely in a calorimeter. These four calorific values have been in common use:

gross (or higher) calorific value, net (or lower) calorific value at constant volume corresponding to ΔU_0

and gross (or higher) calorific value, net (or lower) calorific value at constant pressure corresponding to ΔH_0.

Gross (or higher) and net (or lower) refer to the phase of H_2O in the products, the former to liquid and the latter to vapour.

In practice combustion tends to be incomplete. One reason is the dissociation process at high temperatures, which absorbs energy. Most fuels reach an adiabatic combustion temperature between 2200 and 2700 K. In such cases calculated temperatures are too high compared with practice. This fact can be explained only if we take into consideration the endothermic reactions of dissociation.

We may view dissociation as a kind of partial reversal of combustion, increasing with increase in temperature. In fact, then, the combustion is a 'net' dynamic process.

The presence of CO and H_2 in the products of combustion indicates that not all the chemical energy in the fuel has been released. In other words, the combustion process has not been completed. The question arises as to how in this case a state of chemical equilibrium can be achieved. The answer to this question is that conditions for equilibrium can be deduced from the Second Law of Thermodynamics.

11 The concept of maximum work

We will consider a calculation of the maximum work which a system can perform as it changes from one equilibrium state to another. Because we shall be interested in the maximum work output of a process we shall take the work as positive when it is done by the system.

Consider a closed system, Fig. 1, which receives heat from its surroundings at the constant temperature T_0. The system produces an amount of work dW, part of which may be $p \cdot dV$ work.

From the First Law of Thermodynamics we have:

$$dW = dQ - dU \qquad (20)$$

and from the Second Law of Thermodynamics we have for the system and surroundings:

$$dS + dS_0 \geq 0 \qquad (21)$$

The change in entropy for the surroundings is given by

$$dS_0 = -\frac{dQ}{T_0} \qquad (22)$$

Inserting eqn (22) in (21) and then in (20) gives

$$dW = dQ - dU \leq T_0 \cdot dS - dU \qquad (23)$$

where dS pertains to the system. The inequality holds for irreversible processes and the equality for reversible processes. If the overall process involving the system and its surroundings is to be reversible (external reversibility) the system temperature must be equal to the temperature of the surroundings.

Thus $T = T_0$ and

$$dW \leq T \cdot dS - dU \qquad (24)$$

where the inequality is retained to account for possible internal irreversibility within the system. Clearly dW is maximised when the equality sign holds, so it is a reversible process which produces the maximum work.

Hence

$$dW_{max.} = T \cdot dS - dU \qquad (25)$$

For a process at constant temperature we can write:

$$dW_{max.T} = -d(U - T \cdot S) \qquad (26)$$

Introducing a new thermodynamics function of state

$$A = U - T \cdot S \qquad (27)$$

we can write eqn (27) as

$$dW_{max.T} = -dA \qquad (28)$$

Function A is known as the 'free energy' or 'Helmholtz function'. This function is a potential function for work at constant temperature processes, because decrease of this function between two equilibrium states gives maximum work for these processes. The above is for <u>non-flow</u> processes.

A similar procedure can be applied for constant temperature and constant pressure <u>flow</u> processes.

$$(dW_{max} - p \cdot dV)_{T,p} = -dG_{T,p} \qquad (29)$$

where $G = H - TS$ is the 'free enthalpy' or 'Gibbs function'.

Eqns (28) and (29) may be used to set an upper limit to work done by the system. The more reversible the process the more work is obtained. If only heat is realised the process is irreversible. Combustion processes producing only heat are irreversible.

12 Analogy between combustion and energy release in cells

We are now in a position to make an explicit comparison between (engineering) combustion and (biological) energy release using foods. For further information to what follows, the reader may especially care to study two comprehensive undergraduate texts in engineering thermodynamics (Rogers and Mayhew [4]) and biology (Purves et al. [6]). Later, in our Design in Nature Series, a specialised Volume on Thermodynamics will cover these issues in much more detail.

13 Fuels and food

'Most common fossil fuels consist mainly of hydrogen and carbon, whether the fuel be solid (e.g. coal), liquid (e.g. petroleum) or gaseous (e.g. natural gas)' (Rogers and Mayhew [4], p. 321). 'All living cells contain ... glucose $C_6H_{12}O_6$. Green plants produce it by photosynthesis. Cells metabolise it to yield energy during cellular respiration' (Purves et al. [6], p. 45).

The above quotes show how close the analogy is between fuels and food. We describe fuels as <u>hydrocarbons,</u> in other words, the carbon and hydrogen in the fuel, together with oxygen in the atmosphere, provide the essential components for combustion. Similarly, glucose is an example of a <u>carbohydrate</u>, the carbon, hydrogen and oxygen in the fuel, again with oxygen from the atmosphere, providing the principal ingredients for all energy release. More specifically, the latter process is termed 'aerobic' (oxygen-containing environment). There is also fermentation, termed 'anaerobic' (oxygen-absent environment) with a much smaller cell energy release.

The other similarity is the quantitative description of the energy potential. By completely burning a sample of fuel in a calorimeter, the loosely-termed 'calorific value' is obtained. We have already seen that there are four alternative definitions.

As an example, with the combustion products in the gaseous phase, the calorific values at 25°C of the fuel C_2H_6 are about 47,600 kJ/kg. Now this may also be done with foods. A typical breakfast meal (Kelloggs Special K) has an 'energy content' of 370 kcal or 1600 kJ/100 g, giving 16,000 kJ/kg. The two calorific values, then, are of the same order of magnitude.

We will now compare the chemical equations for (a) combustion of a typical fossil fuel, (b) combustion of glucose and (c) aerobic metabolism involving glucose.

For the fossil fuel we will take hexane C_6H_{14}, as it has a fairly close hydrocarbon content to glucose.

13.1 Combustion of hexane

$$C_6H_{14} + 9\tfrac{1}{2}\, O_2 \rightarrow 6\, CO_2 + 7\, H_2O + \text{(energy release)} \tag{30}$$

If glucose is combusted in a conventional manner we will have:

13.2 Combustion of glucose

$$C_6H_{12}O_6 + 6\, O_2 \rightarrow 6\, CO_2 + 6\, H_2O + \text{(energy release)} \tag{31}$$

In both cases (a) and (b) there is an energy release, which for glucose in a biological context is expressed in molar terms.

Hence, energy release (b) is 686 kcal/mol.

13.3 Aerobic metabolism with glucose

This consists of two main pathways, glycolysis and cellular respiration, which we will consider in more detail later on. However, the <u>overall</u> consequence of aerobic metabolism is to store the greater part of the energy by way of transforming ADP (adenosine diphosphate) to the 'energy-storage compound' ATP (adenosine triphosphate) ([6], p. 137). Now what happens is that the change in Gibbs function (see eqn (29)) is −12 kcal/mol for the set of conditions typical of a living cell.

In aerobic metabolism, for each mol of glucose input, instead of all the energy release being in the form of heat and light, 36 mols of ATP are formed. This means that of the energy release of 686 kcal/mol, (36 × 12 ≡) 432 kcal/mol are used to store energy. Hence we may write:

(energy release from 1 mol of $C_6H_{12}O_6$) → (36 mol of ATP) + (254 kcal) (32)

This formation of ATP, then, is in the form of short-term, easily recoverable energy, analogous to the use of batteries. Long-term energy storage by the organism as a whole is by way of fat and carbohydrates.

We may finally make a comparison in terms of thermodynamic machines.

In Fig. 14 we show a comparison of engineering combustion (in a combustion chamber) with cellular aerobic metabolism, along the lines of [4], p. 436.

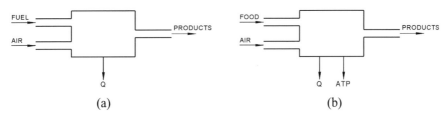

Figure 14: Comparison of (a) engineering combustion and (b) aerobic metabolism in a cell.

In Fig. 15 we extend this to a comparison of a notional power-plant with a biological organism. Part of the power-plant output provides input to a pumped storage scheme which provides an analogue to biological energy storage in fat.

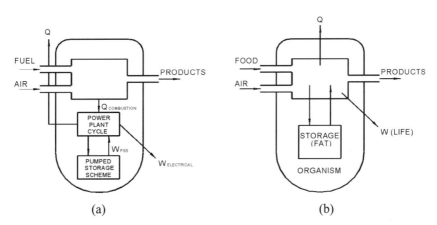

Figure 15: Comparison of energy transfers in (a) power plant using fossil fuel and (b) biological organism.

In Fig. 15(a) W and Q are respectively the electricity supplied to the grid, and the overall heat rejection, primarily from the condensers.

In Fig. 15(b) W and Q are respectively the (metabolic load + external activity) and the nett summation of cellular heat outputs. For a human (organism) the 'guideline daily amounts' (UK Government figures) of food energy intakes are 2000 kcal and 2500 kcal for women and men respectively. Such figures are intended to avoid weight increase via fat storage, of course, and should balance (W + Q), therefore, by the First Law of Thermodynamics.

14 Cell energy release – general

We may summarise our previous discussion by way of Fig. 16. This shows the relationship between production of ATP (the energy-storage compound) and the four processes of glycolysis, cellular respiration, fermentation and photosynthesis. We will consider these in detail below, but it should be pointed out that it is <u>cellular respiration</u> in particular that has a large ATP production.

Also shown in Fig. 16 is the engineering analogue of the solar car. This is a very significant design in terms of the above processes, as it bridges the entire sequence between solar radiation and energy production. The evolution of the design of solar cars is given in another chapter of this volume.

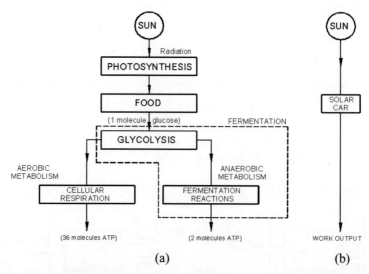

Figure 16: (a) Line diagram of cell energy release for aerobic and anaerobic metabolism.
(b) Analogue line diagram for solar car.

14.1 Cell energy release – glycolysis

Glycolysis is a complex 10 reaction pathway-type process, each reaction having changes in the Free Energy (Gibbs Function) (Ref. [6], pp. 144–147). For our purposes, we will consider glycolysis as a 'Black Box' with certain chemical inputs and outputs.

The strategic effect is to transform the food glucose into pyruvate:

$C_6H_{12}O_6$ (glucose)$_{IN}$ [TO] $C_3H_4O_3$ (pyruvate)$_{OUT}$

In addition, there are two major energy changes. Firstly, for each molecule of glucose, 2 molecules of ATP are produced.

2 × [ADP]$_{IN}$ [TO]   2 × [ADP]$_{OUT}$

(where $\Delta G = 2 \times 12 = 24$ kcal)

Secondly, the compound NAD (nicotinamide adenine dinucleotide) forms what Purves *et al.* term 'the main electron banking system'. NAD commutes between oxidising and reduction processes by having an oxidised (NAD$^+$) and a reduced form (NADH + H$^+$). The + sign indicates a free electron, which for H is of high-energy.
So the reduction of NAD$^+$ occurs as:

$$NAD^+ + 2H \rightarrow NADH + H^+ \qquad (33a)$$

and results in a free energy increase of − 52 · 4 kcal/mol if O_2 then oxidises the product, giving

$$NADH + H^+ + \tfrac{1}{2}O_2 \rightarrow NAD^+ + H_2O \qquad (33b)$$

We are now in a position to construct our glycolysis 'Black Box' as given in Fig. 17.

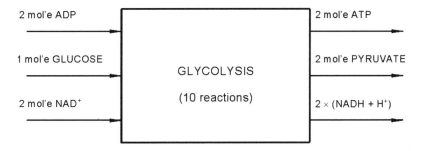

Figure 17: Black Box type diagram for glycolysis ($\Delta G \approx -140$ kcal, mol'e means molecule, rather than mol).

14.2 Cell energy release – citric acid cycle

It is in the citric acid cycle that the (organic) carbohydrate pyruvate is 'combusted' down to CO_2.

The overall equation is:

$$C_3H_4O_3 + 3\ H_2O + 5 \times \begin{bmatrix} \text{OXIDIZED} \\ \text{ELECTRON} \\ \text{CARRIERS} \end{bmatrix} \rightarrow 3\ CO_2 + 5 \cdot \begin{bmatrix} \text{REDUCED} \\ \text{CARRIERS} \end{bmatrix} \cdot 2\ H$$

For our Black Box, we use $2 \times C_3H_4O_3$ as input, to be consistent with the glycolysis output, as given in Fig. 18.

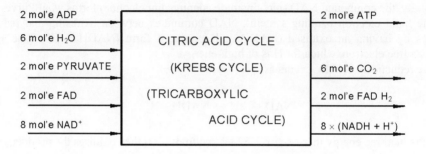

Figure 18: Black Box type diagram for the citric acid cycle ($\Delta G \approx -530$ kcal).

The new carrier appearing above is FAD (flavin adenine dinucleotide).

14.3 Cell energy release – respiratory chain

The effect of the respiratory chain is to restore the oxidised forms of NAD and FAD to be continuously used in glycolysis and the citric acid cycle. To this end the chain takes the reduced forms as input. The chain itself consists of six connected oxidation-reduction loops, with three input–output 'semi-loops'. One of these results in the eventual formation of further molecules of ATP. This process is called oxidative phosphorylation.

Again, we give a Black Box representation as in Fig. 19.

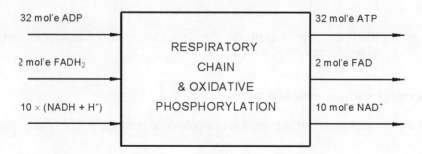

Figure 19: Black Box diagram for respiratory chain.

14.4 Aerobic metabolism

If the three processes are put together we have an overall summary of aerobic metabolism. Inspection of the three Black Box diagrams in Figs 17–19 shows that apart from CO_2, the principal product is 36 mols of ATP. Hence, per mol of glucose consumed, a free energy release of some 432 kcal is represented by the ATP.

14.5 Anaerobic metabolism and fermentation

When a cell operating by aerobic metabolism loses its supply of oxygen, the respiratory chain breaks down and no NAD^+ or FAD are available (see Fig. 19), thus causing the citric acid cycle to cease operation (see Fig. 18). Under these circumstances the cells also cease operation. An example is the human nerve cell, meaning that in the absence of oxygen, our nervous system, especially the brain, becomes the first part to die.

However, other cells, such as human muscle cells, contain the enzymes which permit glycolysis to continue by recycling the necessary NAD^+ (see Fig. 17). We will term such a process 'second stage fermentation', and this, when combined with glycolysis, is, overall, called 'fermentation'. Figure 20 gives a Black Box diagram for the second stage fermentation which occurs in human muscle cells, producing lactate or lactic acid.

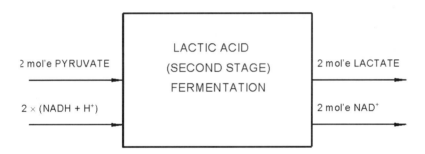

Figure 20: Black Box diagram for second stage fermentation (lactic acid fermentation as in human muscle cells).

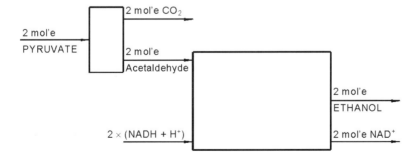

Figure 21: Black Box diagram for second stage of alcoholic fermentation.

An alternative important fermentation is alcoholic fermentation, which is the basic process for the brewing industry. In this case the second stage fermentation consists of two parts. Firstly, the pyruvate is converted to acetaldehyde, accompanied by the release of CO_2. Then there is a process similar to that in Fig. 20, whereby ethanol (or ethyl alcohol) is produced. This is shown in Fig. 21.

We conclude this section with a comparison of fermentation with aerobic metabolism in energy terms. The absence of ATP may be noted in Figs. 20 and 21. In fact, the only ATP production in fermentation is the 2 mol in glycolysis. This compares with the 36 mol of aerobic metabolism. The latter is therefore much more efficient, although Purves *et al.* [6], p. 155 point out that in cells capable of both metabolisms, there is a substantial increase in <u>rate</u> of metabolism. This tends, therefore, to 'balance up' the absolute energy supply, while retaining the inefficiency.

14.6 Photosynthesis – general

We have already noted in Fig. 16(a) that in using the sun's radiation the overall effect of photosynthesis is to produce food, notably glucose. In fact, the chemical basis of photosynthesis is comparatively simple, namely for glucose:

$$6\ CO_2 + 12\ H_2O \rightarrow C_6H_{12}O_6 + 6\ O_2 + 6\ H_2O + \text{(energy input)} \tag{34}$$

Water appears on both sides, because while being used as a reactant, it is also a photosynthetic product. Eqn (34) is the reverse of the glucose (food) combustion equation (31), where here the input energy is solar radiation.

Essentially, photosynthesis is composed of two pathways – photophosphorylation and the Calvin-Benson cycle. The first pathway converts ADP to ATP, and the second uses ATP to synthesise, say, glucose as above. Between these two pathways, also, two oxidising-reducing compounds are recycled, namely:

$$NADP^+ + 2\ H \rightarrow NADPH + H^+ \tag{35}$$

which may be compared with eqn (33a).

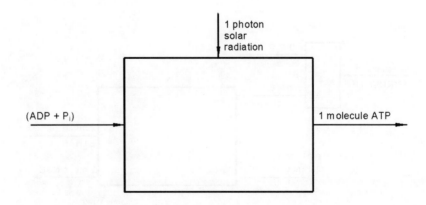

Figure 22: Black Box diagram for cyclic photophosphorylation (primitive form) (P_i means phosphorus-ion).

14.7 Photosynthesis – photophosphorylation

In photophosphorylation, then, ATP is produced, together with NADPH, and for one molecule of each being involved, the overall pathway requires one photon of solar radiation. This defines the evolutionarily ancient form called 'cyclic photophosphorylation' as given by the Black Box diagram, Fig. 22. This occurred before the earth's atmosphere contained much oxygen.

Certain primitive photosynthetic bacteria still use cyclic photophosphorylation only, but other bacteria can use both this form and 'noncyclic photophosphorylation' in which electrons are obtained from water. This allows the cycling of eqn (35) to take place, and also produces oxygen. These processes, requiring a further photon of radiation, constitute what we could term the developed form, for which the Black Box diagram is given in Fig. 23.

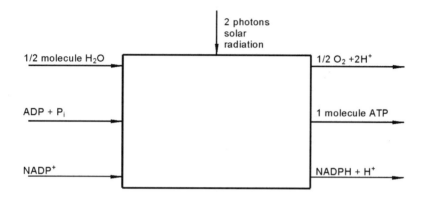

Figure 23: Black Box diagram for noncyclic photophosphorylation (developed form).

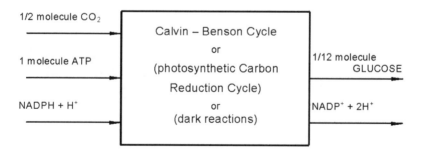

Figure 24: Black Box diagram for Calvin-Benson cycle to produce glucose.

14.8 Photosynthesis – the Calvin-Benson cycle

Proper understanding of this second pathway which produces glucose, other sugars and other foods is only some 50 years old. Basically it is a reflection of eqn (34), which to make it balance for a single molecule of (say) ATP, has to be divided by 12. Again, the Black Box diagram of Fig. 24 comprises some significant reactions. The alternative name for the cycle

58 *Nature and Design*

'dark reactions' is because unlike the other pathway, the Calvin-Benson cycle needs no light to operate.

14.9 The food chain – conclusion

We have explained the various sections of the food chain of Fig. 16 by a series of Black Box diagrams. This is partly for reasons of length of treatment, but mainly to systematise the thermodynamic aspects in terms of energy. By doing so, we have concentrated also on some of our world's great naturally occurring fluids, namely, water, carbon dioxide and oxygen. The second in particular is associated with the global warming environmental issue, to which we briefly turn.

15 Carbon dioxide and global warming

In the latter part of this chapter, the various processes we have considered (combustion, glycolysis, respiration and photosynthesis) all have carbon dioxide as a principal constituent. It is in the form of carbon dioxide that the main transfers take place in what is known as the global carbon cycle. Houghton [5] clearly explains (Chapter 3 – The Greenhouse Gases) how the enhanced greenhouse effect, or 'global warming', is associated with the great increase in carbon dioxide atmospheric concentrations since the industrial revolution. His Fig. 3.2. is reproduced here as Fig. 25. It rather neatly demonstrates the relative quantitative significance in global terms of photosynthesis, respiration and fossil-fuel combustion, namely 102 to 50 to 6 respectively. While the rationale of this chapter has sought to give a bottom-up synthesis of these natural and man-made processes in terms of the Laws of Thermodynamics, Fig. 25 epitomises a top-down synthesis in terms of their global consequences.

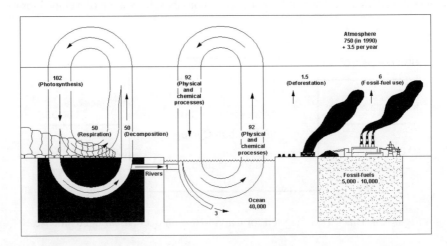

Figure 25: The reservoirs of carbon in the earth, the biosphere, the ocean and the atmosphere and the annual exchanges of carbon dioxide (expressed in terms of mass of carbon it contains) between the reservoirs. The units are thousand millions of tonnes or gigatonnes (Gt).

16 Conclusions

In this chapter, we have given a basic presentation of the Laws of Thermodynamics in engineering thermodynamics terms. Firstly, by this means it is hoped that a reader with no previous experience can understand the formalities of the definitions of work, heat, temperature and, above all, entropy. In the companion chapter in Volume 2, the interpretation of entropy is extended to encompass statistical thermodynamics and (Shannon) information theory. Again, therefore, such a reader should be able more fully to appreciate, for example, the range of statements presented by Paul Davies in his 'Fifth Miracle' [17]. We give just one (from p. 39): 'The conclusion we are led to is that the universe came stocked with information, or negative entropy, from the word go'. Davies leads on to consideration of Kauffman and his work on complexity ([17], p. 120). Now, it has already been explained in the Series Introduction in this volume, that fully to appreciate developments in evolutionary biology, it is necessary to give full consideration to a thermodynamic interpretation. Hence, we give as an Appendix a synthesis of the consideration (or its lack) of thermodynamics in some 30 typical texts relating to 'Design in Nature'.

Secondly, in this chapter we have given an outline treatment of cell energy processes, photosynthesis and respiration. The latter two also relate to the carbon (dioxide) cycle and global warming.

Appendix

Treatment of Thermodynamics in typical books on 'Design in Nature'

These have been systematised under whether the books refer to

    1st Law of Thermodynamics

    2nd Law of Thermodynamics

    Information/complexity.

Authors' speciality is given in parentheses, and the earliest locatable date of publication is also given. The books are fully referenced in the Reference list for this chapter, as [bracketed].

	First available date	1st Law?	2nd Law?	Information/Complexity?
1. Books on Engineering and Nature				
D'Arcy Thompson 'On Growth & Form' (Polymath) [7]	1917	×	×	×
J.E. Gordon 'Structures' [8]	1978	×	×	×
J.E. Gordon 'The New Science of Strong Materials' (Engineer) [9]	1968	×	×	×
M.J. French 'Invention and Evolution' (Engineer) [10]	1988	√	√	×
E.R. Laithwaite 'An Inventor in The Garden of Eden' (Engineer) [11]	1994	×	√	×
C. Mattheck 'Design in Nature' (Engineer) [12]	1997	×	×	×
S. Vogel 'Cats' Paws and Catapults' (Biologist) [13]	1998	√	√	×
2. Monographs on Entropy				
J.D. Fast 'Entropy' [14]	1962	(√)	(√)	√
J.S. Dugdale 'Entropy and its Physical Meaning' [2]	1996	(√)	(√)	×
3. Books on Molecular Biology				
Sungchal Ji (Ed.) 'Molecular Theories of Cell Life and Death' (Toxicologist etc) [15]	1991	√	√	√
M.J. Denton 'Nature's Destiny' (Molecular Biologist) [16]	1998	×	×	×
P. Davies 'The Fifth Miracle' (Theoretical Physicist) [17]	1998	√	√	√
4. Modern Expositions of Darwinism (with apologies to all other writers)				
N. Eldredge 'Reinventing Darwinism' [18]	1995	×	×	(see pp. 159, 192, 193)

R. Dawkins 'Climbing Mount Improbable' [19]	1996	×	×	×
5. Mainstream Biology Texts				
D.J. Futuyama 'Evolutionary Biology'	1986	×	×	×
W.K. Purves et al. 'Life', Volume I [6]	1995	√	√	×
6. Specialist Books				
E. Schrodinger 'What is life?' (origin of life) [21A]	1944	×	√	'order' (p. 91) 'orderliness' (p. 99)
I. Stuart 'Does God play dice?' (chaos) [22]	1987	√	×	√
J. Wicken 'Evolution, Thermodynamics & Information' (evolution & thermodyn.)[23]	1987	√	√	√
G. Nicolis & I. Prigogine 'Exploring Complexity' (complexity) [24]	1989	(√)	(√)	√
S. Goonatilake 'The evolution of information' (evolutionary history) [25]	1991	(√)	√	√
P. Davies 'The Mind of God' (Structure of the Universe) [26]	1992	√	√	√
S. Kauffman 'At Home in the Universe' (complexity) [27]	1995	×	√	√
J. Barrow 'The Artful Universe' (science and art) [28]	1995	×	√	√
M. Rees 'Just Six Numbers' (cosmic laws) [29]	1999	×	√	√
R. Penrose 'Introduction to E. Schrodinger's 'What is Life?' [21B]	2000	×	√	×
S. Hawking 'A Brief History of Time' (history of Universe) [30]	1988	×	√	×
S. Hawking 'The Universe in a Nutshell' (sequel to above)	2001	×	×	√

References

[1] Spalding, D.B. & Cole, E.H., *Engineering Thermodynamics*, (3rd Edition), SI Units, Arnold, 1973.
[2] Dugdale, T.S., *Entropy and its Physical Meaning*, Taylor and Francis, 1996.
[3] Kondepudi, D. & Prigogine, I., *Modern Thermodynamics. From Heat Engines to Dissipative Structures*, John Wiley, 1998.
[4] Rogers, G.F.C. & Mayhew, Y.R., *Engineering Thermodynamics. Work and Heat Transfer*, 4th Edition, Prentice Hall, 1992.
[5] Houghton, J., *Global Warming – The Complete Briefing*, Lion, 1994.
[6] Purves, W.K., Orians, G.H. & Heller, H.C., *Life*, (4th Edition), Volume I, Sinauer Associates, 1995. (Note: athough the 4th Edition has been used, in fact this publication is regularly re-edited – currently 6th Edition).
[7] Thompson, D., *On Growth and Form*, (Abridged. Ed. J.T. Bonner), Canto Edition Cambridge University Press, 1992.
[8] Gordon, J.E., *Structures*, Penguin, 1978.

[9]   Gordon, J.E., *The New Science of Strong Materials*, Penguin, 1968.
[10] French, M.J., *Invention and Evolution*, (2nd Edition), Cambridge University Press, 1994.
[11] Laithwaite, E.R., *An Inventor in the Garden of Eden*, Cambridge University Press, 1994.
[12] Mattheck, C., *Design in Nature*, (English Edition), Springer, 1998.
[13] Vogel, S., *Cats' Paws and Catapults*, Penguin, 1998.
[14] Fast, J.D., *Entropy*, (2nd Edition and further revised), Macmillan, 1970.
[15] Ji, Sungchal (Ed.) *Molecular Theories of Cell Life and Death*, Rutgers University Press, 1991. (Note: a number of chapters are relevant – by Ji, Rothstein and Prigogine – together with Sungchal Ji's Introduction).
[16] Denton, M.J., *Nature's Destiny*, The Free Press, 1998.
[17] Davies, P., *The Fifth Miracle*, Penguin, 1999.
[18] Eldredge, N., *Reinventing Darwin*, Phoenix Orion, London, 1995.
[19] Dawkins, R., *Climbing Mount Improbable*, Penguin, 1996.
[20] Futuyama, D.J., *Evolutionary Biology*, (2nd Edition), Sinauer Associates, 1986.
[21] Schrodinger, E., *What is Life?* A *What is Life?* with 'Mind and Matter' & 'Autobiographical Sketches', Canto Edition, Cambridge University Press, 1992. B *What is Life?* with Introduction by Penrose, R., The Folio Society, London, 2000.
[22] Stuart, I., *Does God Play Dice'*, Penguin, 1997.
[23] Wicken, J.S., *Evolution, Thermodynamics and Information*, Oxford University Press, 1987.
[24] Nicolis, G. & Prigogine, I., *Exploring Complexity*, Freeman, New York, 1989.
[25] Goonatilake, S., *The Evolution of Information*, Pinter, 1991.
[26] Davies, P., *The Mind of God*, Penguin, 1992.
[27] Kauffman, S., *At Home in the Universe*, Penguin, 1995.
[28] Barrow, J.D., *The Artful Universe*, Oxford University Press, 1995.
[29] Rees, M., *Just Six Numbers*, Weidenfeld and Nicolson, London, 1999.
[30] Hawking, S., *A Brief History of Time*, Bantam Press, London, 1988.
[31] Hawking, S., *The Universe in a Nutshell*, Bantam Press, London, 2001.

CHAPTER 4

Robustness and complexity

M. Atherton[1] & R. Bates[2]
[1]*Department of Engineering Systems, London South Bank University, UK.*
[2]*Department of Statistics, London School of Economics, UK.*

Abstract

A robust system is defined as one that maintains stable behaviour when subjected to disturbance or noise. In this chapter, the behaviour of dynamic systems is investigated in relation to natural structures, revealing the roles played by perturbations, imperfections, random protrusions, energy and interactions in the formation of pattern. The relationships between noise and deterministic, self-governing processes in nature are illustrated. From these examples emerges the important role of mathematical transformations in engineering design, highlighted through phase space reconstruction in tracking performance despite the blurring effects of noise. Robustness can be achieved through system design that deflects noise away from where it has greatest effect and also through feedback mechanisms that compensate for the effects of noise.

1 Introduction: Robustness of function

Natural systems have to contend with disturbances to their vital functions. For example, consider the leaves on a tree offering their surface area to the sun for photosynthesis. This is fine until the tree is ruffled by a strong wind; the drag caused by the leaves is now a potential threat to the tree's survival. However, in high winds leaves have been observed to reconfigure into cones and cylinders (Figure 1) that get tighter as wind velocity increases thereby reducing their drag [1]. Hence shape is a critical 'design factor' in the function of a leaf with respect to a 'noise factor' in the form of high winds.

Leaves conform to the wind; i.e. they change mode, whereas current human technology commonly maintains shape through the use of stiff structures when faced with changing flows. The design of rigid structures can benefit from interpretations of nature, for example through using methods such as *Evolutionary Structural Optimisation* (ESO) [2], which optimises the shape of mechanical structures under load through a simulation analogy of the biological growth mechanism observed in trees and bones. The *uniform stress axiom* is the driving principle behind adding (or removing) material at an overstressed (or understressed) locality. Structures determined in this way are not hefty or 'overengineered' but sturdy and relatively lean. These structures could be described as *robust* if they are more tolerant to damage compared with alternative designs made from the same amount of material. However, ESO is limited to a single class of design parameter, namely shape parameters, whereas the design of most technical systems is concerned with achieving a robust configuration involving different materials and architectures, in addition to shape. We hope to learn more from the natural world about the ingenuity of the operating principles of natural devices. In particular, how classes of design parameters relate to each other in order to maintain a successful function under changing conditions.

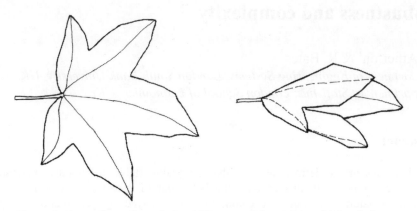

Figure 1: Reconfiguration of a leaf (*Hamamelidaceae*) in a high wind.

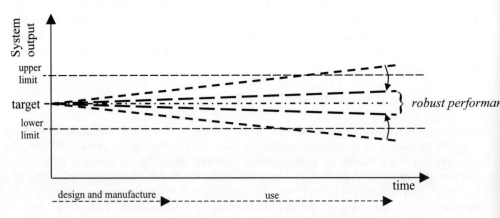

Figure 2: Limits of performance over lifecycle.

In engineering, a *robust system* or product is a design that is insensitive to variations in context, such as manufacturing or operating conditions. Ideally, a robust design is insensitive to the effects of these sources of variability over its entire lifecycle, even though the sources themselves may not be eliminated [3]. In other words, robustness of function is that which remains close to target (Figure 2).

This leads us to consider two classes of factors, design and noise. Sources of variability or *noise factors* are often not controllable by the design engineer. Noise factors include those that are external (e.g. ambient temperature, applied load), internal (e.g. deterioration through ageing or wear) and unit-to-unit (e.g. supplied quality, process variability). It is possible for the designer to improve the robustness of function through changing the nominal values of *design factors* - features of the system specified and controlled by the designer. Figure 3 is an abstract general representation of a system, where the system is treated as if the constituent parameters are mathematically related to each other such that $y = f(x, z)$.

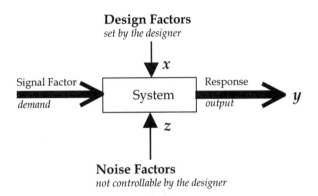

Figure 3: Basic parameters of a system.

The output, y, can remain robust to the noise factors, z, if the system is designed to operate through non-linear relationships between the output, y, and the design factors, x (Figure 4).

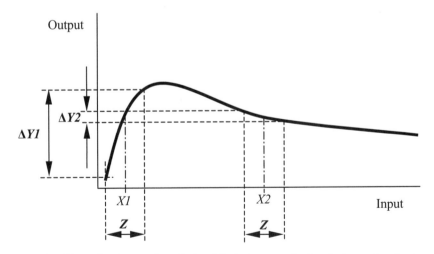

Figure 4: Effect of noise on the relationship between a design factor and output.

In this illustration Z could be the variation in manufacturing two nominal values $X1$ and $X2$ of a design factor, i.e. noise factor $Z = \Delta X1 = \Delta X2$. Observe that the character of the non-linear function means that Z has less effect on Y at $X2$ than its effect at $X1$, i.e. $\Delta Y2 < \Delta Y1$.

Non-linear relationships can be straightforward to identify and exploit in systems with discrete components of known characteristics. But in natural systems this can be much more difficult. However, we shall look at some examples of natural phenomena and endeavour to identify relationships that exhibit robustness. These examples reveal *complexity* in that they consist of a large number of interacting components, often with many things going on simultaneously to organise ornate structures. For such systems a holistic view is more vital to our understanding of robustness than the pervasive reductionist view would yield.

2 Robustness through noise interactions

When attempting to understand the behaviour of a system, it is often useful to build a mathematical model that mimics the system, and then to see what happens when the model parameters are modified. This physics-based view of design and noise factors operating on systems can be observed in natural systems, and can be effective in solving engineering problems. However, when looking at natural systems, the propagation of noise through the system can be convoluted. For example if a natural system such as sand dune formation is sensitive to small changes, the response of the system, in this case the shape and position of the dune, may not be simply related to the input factors (wind, sea etc.). Therefore we need to think more carefully about how noise affects complex systems.

In the following sections, we will discuss more broadly the concepts of noise, equilibrium, near-equilibrium, bifurcations and self-enhancing phenomena. The intention is that the examples given will provide the basis for a more natural understanding of noise and robustness than the deterministic view that dominates current engineering practice.

2.1 Noise in pattern formation

A vortex travelling down a river and sand ripples on a seashore are 'structures' that demonstrate a notable robustness to their circumstances. That is, under certain conditions their formation is inevitable and for a while curiously maintained amidst damaging forces. Noise also plays a vital role in the regimes in which such patterns are generated.

In dynamic natural systems pattern is generally produced spontaneously. This happens by reducing a very high state of symmetry; namely randomness or homogeneity, to a level at which pattern can be discerned, sometimes as structure or order. In other words total randomness, uniform and apparently featureless, promotes perfect symmetry, which must be broken to yield the lower symmetry of perceptible pattern. The most distinct patterns tend not to have the highest symmetry and symmetry breaking tends to produce complexity rather than reduce it.

Spontaneous patterns represent a compromise between competing forces. For example, a vortex (Figure 5) can be formed when, under certain conditions, uniform flow of a homogeneous fluid is sufficiently perturbed from its course such that the driving shear force overcomes the viscous drag.

The conditions for this state of flow are characterised by the ratio of inertial to viscous forces (known as the *Reynolds number, Re*) rising above a critical value. At low Reynolds numbers ($Re<1$), the dominant viscous forces tend to smooth out any irregularities in the flow unless energy is supplied. At high Reynolds numbers, representing dominant inertial forces ($Re>1$), regions of shear activity in the form of vortices become viable. Indeed vortices are quite organised for laminar flows of approximately $Re = 10$ to 150 and continue to exist, and increase, well into turbulence, up to approximately $Re = 100,000$. Thus, a vortex pattern emerges from a featureless laminar flow as the inertial forces due to disturbance shear the fluid layers held together by viscosity. Eventually at *Re* >200,000 the vortices are consumed by the overwhelming disorganisation of strong turbulence.

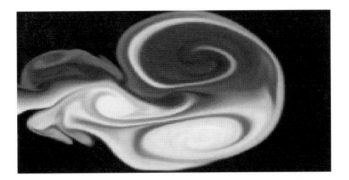

Figure 5: Vortex.

Symmetry tends to break a little at a time as such systems are driven harder and harder, but if the competition becomes too one-sided all pattern disappears. Thus patterns in nature are said to live at the edge of chaos where competing forces banish uniformity without quite inducing chaos. Waves lapping on a sandy shore impose a periodic perturbation (noise) when viewed along one dimension. This creates bands in the sand, whilst parallel to the bands, symmetry is not broken and homogeneity remains (Figure 6).

Figure 6: Pattern produced by 1-D periodic perturbations of the sea on sand.

Does the symmetry of the effect differ from the symmetry of the cause? An understanding might emerge from looking at the root causes of abrupt changes, for example, depending on whether the system is in equilibrium or non-equilibrium, by considering phase transitions and critical phenomena respectively.

2.1.1 Equilibrium: instability thresholds at phase transitions

In nature 'equilibrium is death', as it is often a temporary, precarious state that cannot persist amidst change.

Phase transitions are generally jumps from one equilibrium state to another. Just like spontaneously formed patterns, the whole system is changed at once; that is to say the effect is global. Consider the magnetisation of iron. Above the Curie point temperature (T_C), magnetisation is zero due to the random orientation of the atomic magnetic poles. Below T_C it *bifurcates* into North or South as the poles align one way or the other. The bifurcation is where the stable state of the system splits into two and this change is not a step, it is continuous. There cannot be coexistence of both states - it's either North or South - and there is no hysteresis. In practice, the perfect bifurcation is unlikely in real systems, as there will be preferred solutions due to imperfections. For example, consider the mid-point deflection, d, under a load, P, of a classic *Euler* elastic strut, as shown in Figure 7.

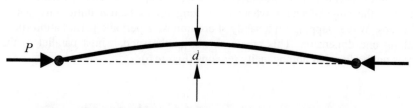

Figure 7: Idealised Euler elastic strut buckling due to an end load P.

Employing a simple mathematical model of the strut, under very small loads the deflection will be zero, but above a critical load, P_c, the strut will adopt either of two deflection states with equal probability (Figure 8).

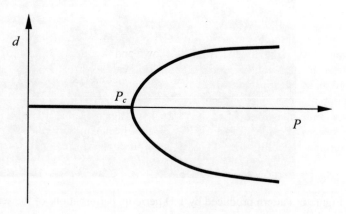

Figure 8: Mid-point deflection of strut against load for an idealised Euler strut.

However, it will be evident in basic testing of a plastic ruler that it will bend in a preferred direction. The imperfections of real systems bias the results and they have the effect of disconnecting the theoretical bifurcation (Figure 9).

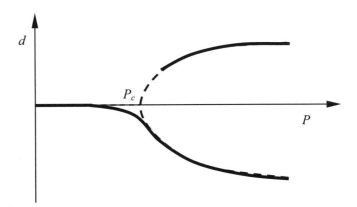

Figure 9: Mid-point deflection of strut against load for an imperfect Euler strut.

The preferred state will develop smoothly with increasing loads and the disconnected state materialises from a discontinuous jump in the load or variability in its application.

To summarise, a characteristic of continuous phase transitions is extreme sensitivity to small perturbations (noise), the slightest bias will swing the result between the options. However, when the options are unequal, which is usually the case in reality, a system at equilibrium will generally tend to adopt the configuration with the lowest free energy. The main point for us here is that noise operates most dramatically at a point of instability as it determines the outcome.

2.1.2 Non-equilibrium: dissipative structures and critical instability

Many pattern-forming systems are out of equilibrium, i.e. not in the most thermodynamically stable state. Directionality in an irreversible process is associated with increased entropy, and this entropy production ceases when a system reaches equilibrium. It has been suggested that in the near-equilibrium (linear) regime, non-equilibrium systems tend to minimise the rate of entropy production [4], thus providing a criterion for determining the most stable state out of, but close to, equilibrium. Further from equilibrium the state of minimal entropy production reaches some crisis point at which it breaks down and becomes transformed to another state.

Therefore in non-equilibrium systems bifurcations can develop into a cascade of bifurcations until an apparently irregular or chaotic region is reached where the bifurcation zones overlap. For example, mathematically a bifurcation cascade emerges when plotting long-term values of the logistic equation, $f(x) = kx(1 - x)$ against values of k (Figure 10).

As the system of bifurcations is driven further from equilibrium (k increased), the patterns further from the instability threshold are laced through with defects, sometimes such that all appearance of pattern is lost. It has been argued that noise is central to the shift between these different states [7].

Dissipative structures are dynamically stable patterns supported away from equilibrium by the generation of entropy. The vortex is a good example and demonstrates how a dissipative structure can be maintained in the midst of a constant throughput of energy and matter.

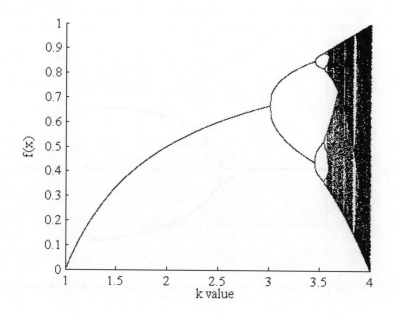

Figure 10: Bifurcation cascade (after Stewart & Golubitsky [6]).

A simple pendulum illustrates the point. As it swings under gravity, energy will be dissipated through friction and the system will eventually come to rest unless there is a driving spring, as in a clock. Then the amplitude and period of pendulum oscillation will be quite consistent soaking up any small disturbances such as a momentary exaggeration of the swing (Figure 11).

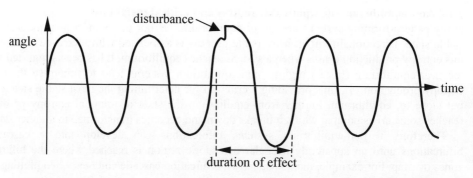

Figure 11: Time series plot of a clock pendulum oscillation with a small disturbance.

Trees, snowflakes, crystal dendrites, vascular networks, bacterial colonies and other branching growth structures are generically different in their pattern formation. They have non-equilibrium growth rules known universally as *Laplacian branching instability* (Figure 12), which includes:

(a) *Diffusion-Limited Aggregation* (DLA) (e.g. bacterial colonies):

Where diffusing particles undergo a random walk and attachment to a surface is irreversible such that there is no chance of reshuffling into a regular stacked arrangement. Growth develops from random bumps due to their greater chance of encountering the diffusing particles, which is in effect a self-promoting instability.

(b) *Viscous fingering* (e.g. vascular networks):

A process in which a fluid sprouts branching fingers as it forces its way through a more viscous surrounding fluid. Fingers emerge where a pressure gradient at the tip of a random initial protrusion is in a self-amplifying alliance - known as the *Saffman-Taylor instability*.

(c) *Dendrite formation* (e.g. snowflakes):

Regular branched structures formed under rapid cooling and therefore in non-equilibrium circumstances, where latent heat is conducted away from an advancing edge. This solidification front proceeds more rapidly where the temperature gradient is steepest, which again is self-enhancing at a random protrusion - known in this context as the *Mullins-Sekerka instability*.

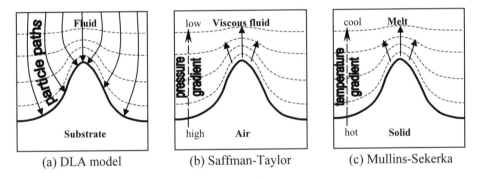

Figure 12: Laplacian branching instabilities (after Ball [5]).

DLA clusters in particular exhibit self-similarity, a form of scale invariance, which looks more or less the same on all size scales. Noise can significantly influence the branching growth patterns in all cases. Investigators have shown that viscous fingering can be driven to behave like DLA by increasing the noise [8]. Again we can deduce that noise affects a system at a point of instability.

Instabilities are not limited to growth structures. In fluids the shear flow between two fluid layers travelling in opposite directions can reach a point, at which, if the boundary is displaced, it spontaneously develops into vortices. This event defined by a particular Reynolds number value is known as the *Kelvin-Helmholtz instability*. Undulations in the boundary creates faster flow over the peaks and slower flow through the troughs thereby setting up a pressure imbalance, according to the Bernoulli principle, which exaggerates the undulations further still (Figure 13), and so on to vorticity, and beyond, to turbulence.

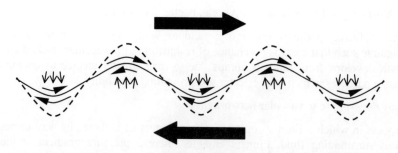

Figure 13: Kelvin-Holmholtz instability between opposing flows.

A critical state in a non-equilibrium system is a least stable state; e.g. as can be 'sought' by sand piles as they are deposited by the wind or the sea. The sand pile is on the brink of instability and susceptible to fluctuations (e.g. avalanches) *on all scales* at the slightest perturbation (noise). Noise is again a determining feature of critical states. The universal patterns and forms of such noise-dominated systems are commonly 'hidden', and they become apparent only in 'mathematical space' i.e. a transformation of some kind such as a power law. This is termed *Self-Organised Criticality* (SOC), which is an apparently self-organised precarious state and often exhibits a power law relationship when plotted on appropriate axes [9]. For example, a simple power law, $y = kx^{-1}$, emerges when sand avalanche frequency versus size is plotted on log–log scales (Figure 14).

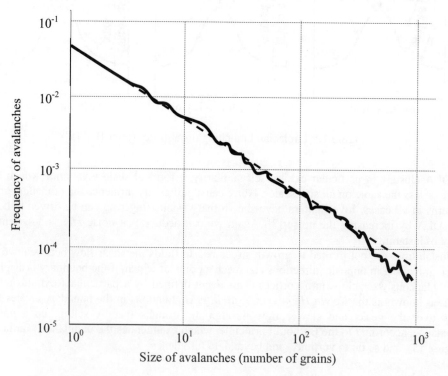

Figure 14: Frequency of sand pile avalanches of different sizes (after Bak [9]).

Figure 14 illustrates that in contrast to most equilibrium structures, the spatial scale of a non-equilibrium pattern typically bears no relation to the size of its constituents and this scale is robust in the face of perturbation. However, SOC is not universal and is a limited metaphor for nature's complexity. A problem in observing SOC is that it needs a lot of data obtained by careful measurement, which can be difficult to obtain.

The behaviour of sand is converse to that of water in respect of flow, as water of course slumps to its most stable state (least free energy). However, all liquids achieve a critical state of another kind at a well-defined temperature and pressure called the critical point, a point at which the distinction between two phases disappears and certain properties vary sharply. At the critical point domains of both states exist. As a system approaches its critical point, the variables that describe its behaviour also start to obey power laws, thus apparently different systems can exhibit the same pattern-forming sequences. It has therefore been suggested that noise, power-law behaviour, scale invariance, SOC and fractal forms are intimately connected in some profound way [5].

2.2 Interactions and systematic organisation

There is more to note about the humble vortex, in particular the fact that under certain circumstances any two points within the flow are correlated, that is they are not independent but regarded as connected parts of a system. Such co-ordinated behaviour is robust. Furthermore, notions of interaction, correlation, coupling and the like are linked with a self-governing growth process that is demonstrably robust to the starting conditions.

2.2.1 Correlation across length scales

The characteristic length scale acquired by patterns formed in non-equilibrium is often far removed from the size of the constituents, i.e. interactions on unimaginably small scales give rise to patterns millions of times larger. This implies that the components of a system at non-equilibrium must be co-ordinated over much longer distances than when it is at equilibrium.

For example, in rolling convection flow above the critical point at which buoyancy forces of the fluid overcome the frictional forces, molecular motions (measured by the *Rayleigh number, R_a*) become correlated over roll wavelength distances, called *Rayleigh-Bérnard convection cells* (Figure 15). At this critical point ($R_a = 1708$), the convection rolls that appear are almost as wide as the depth of the fluid. Thus, virtually in an instant, high symmetry (randomness) is broken, as the system is driven away from equilibrium and fluid molecules suddenly begin to cooperate across distances millions of times greater than their atomic influence.

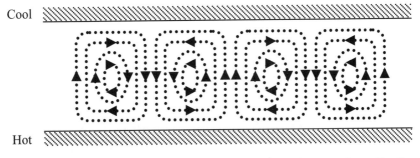

Figure 15: Coordinated molecules in Rayleigh-Bérnard convection roll cells.

Indeed, even in turbulence, there is long-range correlation. After all, turbulence has an innate structure that distinguishes it from a totally random flow and therefore there will be some correlation between velocities at different points. The vortex is a fundamental structure of turbulence and, unlike those in laminar flow, these vortices are formed over a wide range of length scales. These length scales echo an energy cascade where large vortices are transferring their energy to smaller vortices and so down before dissipating through viscous drag into heat.

In both cases it is as if the system loses all sense of scale, i.e. the system becomes scale-invariant and so is able to support fluctuations over a range of scales. Thus a link is established between scale invariance and correlation. Patterns such as vortices should be regarded as global *emergent* properties of the system and are likely to remain hidden to a highly reductionist analysis.

2.2.2 Non-linearity in morphogenesis

Do interactions play a part in organisation and pattern formation in living organisms? The *neo-Darwinian* stance maintains that organisms generate highly improbable structures only because they are functionally useful. However, Goodwin [10, 11, 12] has reconciled random variation of genomes with systematic morphological order through proposing a model for the growth of whorls in *Acetabularia acetabulum*, known as *Mermaid's Cap* (Figure 16). It is a giant unicellular green alga found in the Mediterranean Sea, which is 30–50 mm long with a cap diameter of 5–10 mm.

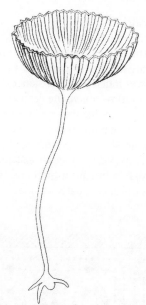

Figure 16: Mermaid's Cap (*Acetabularia acetabulum*).

As it grows the alga forms branching structures called whorls that eventually drop off until finally the process produces a cap instead of whorls. The study centred on the interaction between calcium and cytoplasm in the cell as the alga regenerates its whorls and cap after its stem is cut (Figure 17).

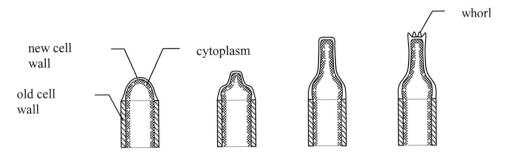

Figure 17: Whorl regeneration of Mermaid's Cap (after Goodwin [10]).

The cytoplasm has a gel-like property and is an excitable medium in that from an initially uniform state it spontaneously generates non-uniform states of calcium concentration and mechanical strain under perturbation.

Two principal mechanisms are at work in the cytoplasm of a cut stem:

(i) Deformation of the cytoplasm results in the release of calcium from a bound or stored state.

(ii) Elevated calcium levels increase breakdown and liquefaction of the cytoplasm gel.

These mechanisms cause changes in the elastic modulus of the cytoplasm as a function of calcium (see Figure 18):

(a) An increase in the regulated concentration of calcium arises due to, say, the strain of internal cell pressure on the new wall where the gradient of strain and calcium concentration has its maximum at the convex tip.

(b) Beyond approx. 0.1 micromole of calcium concentration, the cytoplasm gel begins to break down and liquefies thus reducing its elasticity and allowing more strain. This increased strain causes a greater release of calcium and a potentially destructive positive feedback loop.

(c) However, beyond about 1 micromole of calcium concentration, opposing processes of calcium diffusion reduction and filament contraction effectively overwhelm the positive feedback and thus increase elasticity.

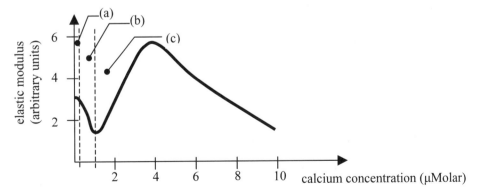

Figure 18: Elasticity of cytoplasm versus cell free calcium concentration (after Goodwin [11]).

Robustness can be observed here to operate through a non-linear feedback function. This growth process exhibits *reaction-diffusion dynamics*, where calcium is the short-range activator and mechanical strain is the long-range inhibitor. Whilst biologically this is a simple concept, mathematically more than 26 parameters were initially used to model it [11], and thus finding the parameter values that would successfully simulate the correct morphogenesis was expected to be a time-consuming process. Actually, the model was only sensitive to six parameters and it showed that a whorl is the natural consequence of the coupling between cytoplasmic strain and wall elasticity. In other words, this coupling is a morphogenetic attractor in the design space of all possible forms where whorl formation apparently occurs over quite a wide range of parameter values. Note that the parameterisation process defines the design space. It was also found that environmental conditions, in particular changes in the calcium concentration of seawater (normally 10–12 micromoles), determine whether or not certain features are formed.

Similarly, it has been pointed out that 80% of higher plants have spiral phyllotaxis (spiral leaf arrangement) the rest have one of two other arrangements, which might be a reflection of the greater domain of attraction of this pattern in the parameter space [11, 12]. There is also a hint here that interactions are an intrinsic aspect of attractors. Insistence on ubiquitous random variation and natural selection omits such dynamic possibilities, and the above supports the case for a revision that reconciles random variation of genomes versus systematic morphological order and complex regulatory feedback.

2.2.3 Highly Optimised Tolerance

The importance of feedback dynamics to robustness is easier to see in the sophisticated process of whorl regeneration of Mermaid's Cap than it is in the relatively primitive action of sand pile avalanches (section 2.1.2). In the former case, the coupling of cytoplasmic strain and wall elasticity and also the number of degrees of freedom of the morphogenesis design space takes us beyond the realm of Self-Organised Criticality (SOC) and towards what has been termed Highly Optimised Tolerance (HOT) [13], which is a different theoretical framework for understanding complexity. We have seen that SOC explores generic states associated with bifurcations and phase transitions. However, HOT states are considered to be rare and specialised to the environment, history and build of the system, namely its design.

In HOT complexity is determined by extreme heterogeneity of component parts typified by forests, river systems, organisms, modern aircraft and the Internet. *Highly Optimised* relates to highly structured interacting networks of parts, i.e. non-generic designs. *Tolerance* emphasises that robustness in complex systems is limited to expected conditions of loading and disturbance; therefore it is fragile to design faults and also rare or unanticipated conditions. The core principal is that complex natural systems achieve robustness through highly structured mechanisms that create barriers to cascading failure events or noise effects. Power laws are their signatures. These barriers are not easy to spot but usually involve complex feedback dynamics [13], as in the case of the Mermaid's Cap.

3 Robustness out of transformation

Looking at natural systems may give us clues as to how to improve our treatment of robustness. Many natural phenomena are clearly observable, but may not be transferable to a mathematical framework. Transformation of what is observable may be necessary.

We need to start building a bridge between nature's methods of dealing with noise and a mathematical framework useful for engineering design. We need also to think about observation, measurement and mathematical treatment of data. Choosing factors and responses carefully may make it easier to construct mathematical models. Patterns in data may then be revealed more clearly and the relationship to patterns in nature may be revealed.

3.1 Ideal function and noise

If noise is a key to the emergence of pattern in nature and transformations in mathematical space are a means of revealing this pattern, then what is their relevance to engineering systems?

In general, engineered systems transform an input energy into specific intended output energy. That is, they are usually dynamic systems. The intent of the *ideal function* is the state in which all input energy is transformed into creating the desired output energy [14]. However, in reality energy is transformed with less than 100% efficiency, and the energy lost can produce an unintended function (Figure 19). Normally this means that whilst noise performs no useful role it inevitably manifests itself in the output performance of the system as variability and inefficiency due to the effects of heat, vibration etc.

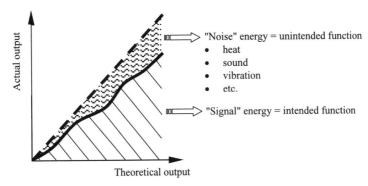

Figure 19: Ideal function versus actual function of an engineering system.

This ideal relationship between input and output provides a means to evaluate actual system performance, using a *Signal-to-Noise Ratio* (SNR) [14] as an index of robustness based upon the quality of energy transformation within the system (equation 1). An SNR applied in this way and expressed as a ratio of squared mean level to variance of performance, provides a measure (in decibels) akin to the power of the intended function in relation to the power of the unintended function.

$$SNR_{NB}(dB) = 10 \log_{10}(\frac{\mu^2}{\sigma^2}) \qquad (1)$$

In effect the SNR is a mathematical transformation to a domain where the structure of the data is viewed under a different light. While specifying a straight-line relationship may be an oversimplification, there may be other mathematical transformations that could offer a more useful insight into robustness.

3.2 Phase space in low-dimensional non-linear dynamics

3.2.1 Attractors in phase space

The simple pendulum plotted as a time series in Figure 11 can be plotted in *phase space*, in which the co-ordinates of a point represent the pendulum's position and velocity at a particular point in time (Figure 20). The phase space captures the system's long-term behaviour as a repeating elliptical path where any disturbance is conspicuous by comparison.

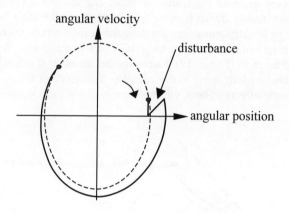

Figure 20: Phase space of clock pendulum oscillation with a small disturbance.

We have observed that highly structured complex behaviour, such as a vortex, can emerge out of straightforward settings. General fluid flow is described mathematically by the highly non-linear *Navier-Stokes* partial differential equations, whose solutions for turbulent flow almost certainly exist in an infinite-dimensional phase space [15]. Therefore in order to investigate chaos analytically, it is necessary to focus on non-trivial low-dimensional non-linear dynamic behaviour. Indeed, the long-term behaviour of dissipative systems no more complex than the pendulum can resolve into an *attractor* (Figure 21) as long as non-linear phenomena such as friction are included. A transient *perturbation* may temporarily disrupt the structure of these systems, but the disturbance will pass as long as it has not knocked the structure into another state.

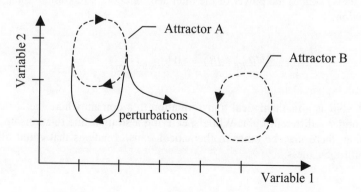

Figure 21: Two attractors in phase space (after Ball [5]).

3.2.2 Phase space reconstruction with real data

There is a distinct difference between the patterns generated from a precise mathematical model and that of associated data from a validation experiment. Viewed as a time series, the resultant disorderly pattern of test data can mask the underlying structure, whereas a phase space representation provides the means for separating the determinate parts of such behaviour from the noise. Artificially high derivatives are likely when using a co-ordinate axis based upon a value derived from data finitely sampled in the presence of noise. In Figure 20 angular velocity derived from the rate of change of angular position could be such a case. Therefore *phase space reconstruction* techniques are used for producing topologically equivalent portraits of attractors from noisy sampled data.

The *method of delays* is the most primitive approach, in which co-ordinates are selected from pairs of data separated by a delay period T. For example, amplitude values in a pendulum time series with sampling intervals of 0.125 seconds are superimposed with 10% maximum random error (shown over 10 seconds in Figure 22).

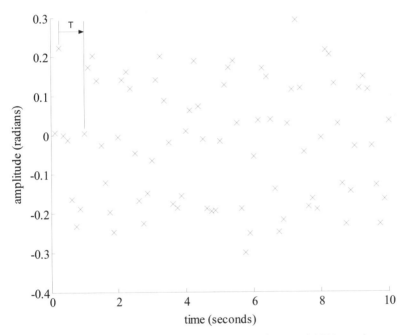

Figure 22: Pendulum amplitude time series with superimposed 10% maximum random error.

The underlying sine wave has an amplitude of 0.2 radians and frequency of 1 Hz. The elliptic attractor is readily apparent in phase space reconstructed using an arbitrary delay, T, of 0.75 seconds on 800 values (Figure 23).

A limitation of this method in practice is that high sampling rates are needed for complicated time series. *Singular Value Decomposition* [15] is an improved phase space reconstruction method that systematically takes account of the information content of the data in finding a projection of the time series onto co-ordinate axes of optimum trajectories. Pairs of axes using different values of T are compared to find which axes best reveal the attractor. This highlights again the principle that there are transformations (trajectories in this case) that maximise deterministic understanding of complex systems and conversely those that contain mostly noise. Therefore robustness in chaos can be addressed along principal axes.

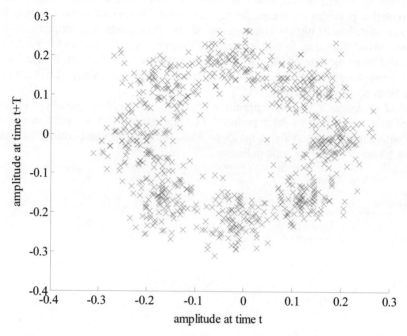

Figure 23: Phase space reconstruction of pendulum time series using method of delays.

Phase space reconstruction has been exploited in the manufacture of small coil springs [16, 17]. It is briefly summarised below as an example of how an appropriate transformation can be used to characterise complex systems and thereby improve their robustness.

The wire from which the coil springs are made undergoes a series of twenty or more drawing operations potentially causing variations in the wire properties and geometry that reflect in the quality of the springs. For a compression spring, the coiling machine is set to produce an open coil by forming a batch of springs and making adjustments to the process accordingly until acceptable results are obtained. Thus *Coilability* of the wire is determined by a large number of factors and eludes detection via any standard quality control tests until after the springs are made - accompanied by a 10% to 25% scrap rate.

Springs made from a wire batch of poor coilability quality will differ significantly in their finished length, expressed by the difference in successive coil pitch distances along the spring length. In other words, good coilability wire produces regular pitch sequence and poor coilability wire produces erratic pitch sequences (shown schematically in Figure 24) as determined from special evaluation coils of 500 pitch lengths [17].

In treating the sequential pitch as a 'time series', phase space can be reconstructed in order to characterise the hidden complex behaviour. Ideally the attractor would be a 45-degree line for T = 1, corresponding to successive pitch measurements that are almost the same value (Figure 25).

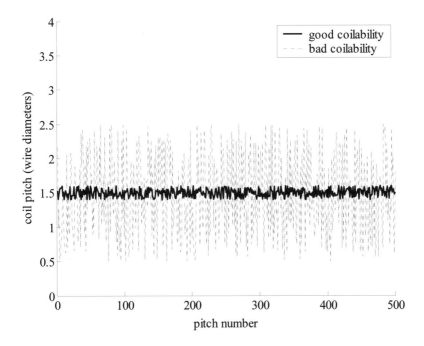

Figure 24: Schematic of sequential wire pitch measurements for two samples.

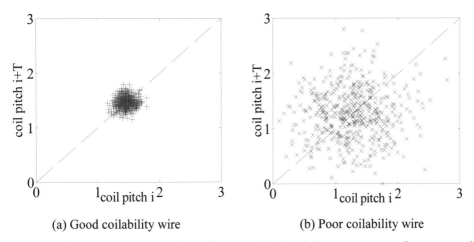

(a) Good coilability wire (b) Poor coilability wire

Figure 25: Phase space reconstruction of sequential wire pitch measurements for two samples.

By using principal component analysis [18] the point clouds of Figures 25(a) and 25(b) can be equated to ellipses whose axes lengths, σ_1 and σ_2, are used to characterise the coil wire quality.

The mean square, $\left(\sqrt{\sigma_1^2 + \sigma_2^2}\right)$ is a *statistical descriptor* and relates to the overall pitch variability.

The ratio $\left(\dfrac{\sigma_2^2}{\sigma_1^2}\right)$ is a *chaotic descriptor* and a value greater than one indicates increasingly inconsistent variation, which means the properties of the wire are changing along the time series.

4 Summary

Robust design is intended function maintained in the face of disturbances. In the light of attractors, we could describe robustness generally as at its most deterministic when function is projected onto principal axes.

Noise is inevitable in any. Indeed, noise is seen to be central to the formation of pattern and structures in nature. It sometimes acts through portals of instability or initiates conflict between competing forces from which complexity or high symmetry emerge out of randomness. Many of these complex patterns are attractors and as such their occurrence is inevitable when the conditions are right. Some organisms use self-governing mechanisms fashioned out of non-linear relationships that render noise factors ineffectual within certain bounds.

Operating at a point of equilibrium is a common strategy in engineering design, and this promotes attempts to keep systems rigid against noise. We have considered how noise operates at equilibrium points to swing performance one way or the other, whereas in some non-equilibrium systems noise can act to shift between multi-stable states as seen with attractors. Thus critical states are noise dominated and there are conflicting examples in nature. Whilst the critical states of fluids are infinitely unstable, the critical state of a sand pile is robust as it constantly seeks to return to this state.

Dynamic engineering systems often aim for pattern in that there is often a desired output characteristic rather than a randomness that is sought. The correct pattern may well be hidden to direct measurements but an appropriate mathematical transformation of this output will reveal it for characterisation. The designer can then attempt to modify the system design accordingly. Studies of fluid flows [15] indicate that this is practicable at the higher symmetries of well-controlled systems, but turbulence, for example, is almost certainly an infinite-dimensional phenomenon and so only vaguely understood through the drastic restrictions of small-scale experiments.

Traditionally, designing robust products concentrated on preventing noise entering the system but today reducing the *effect* of noise is technological best practice. Indeed, in recent years the idea has emerged of *adding* noise to the input of a non-linear system in order to enhance the performance of devices such as A/D converters. Two such approaches are *Stochastic Resonance* (SR) and *dithering* [19]. Basically, both improve system sensitivity to signals with amplitudes lower than the threshold. Whilst dithering is an artificial technique, SR is observed in the natural world, such as in variations in the volume of glaciers over time, the eccentricity of the Earth's orbit, human hearing and the use of electro-receptors by paddlefish in order to detect the electrical signals from their plankton prey.

It is clearly evident that in nature robust function is a common and necessary property. Usually the robustness is a feature of the whole principle of operation, emerging from interactions and feedback within the system. Increased robustness requires greater internal complexity and so advanced technologies have the propensity for complicated architecture that have component parts with much wider tolerances than the precision parts used in old technology.

From this brief investigation it is apparent that there are numerous examples in nature of mechanisms by which robustness can be improved in engineering systems.

References

[1] Vogel, S., *Life in Moving Fluids: the Physical Biology of Flow* (2nd edition), Princeton University Press: Princeton, pp. 120–124, 1994.
[2] Xie, Y.M. & Stevens, G.P., *Evolutionary Structural Optimisation*, Springer-Verlag, 1997.
[3] Fowlkes, W.Y. & Creveling, C.M., *Engineering Methods for Robust Product Design*, Addison-Wesley, pp. 2–3, 193–196, 1995.
[4] Prigogine, I., *The End of Certainty: Time, Chaos, and the New Laws of Nature*, Free Press: New York, pp. 57–72, 1996.
[5] Ball, P., *The Self-Made Tapestry: Pattern Formation in Nature*, Oxford University Press, Oxford, pp. 115, 119, 124, 256, 1999.
[6] Stewart, I. & Golubitsky, M., *Fearful Symmetry: is God a Geometer?* Penguin, pp. 224–226, 1992.
[7] Landauer, R., Stability in the dissipative steady state. *Physics Today*, pp. 23–29, 1978.
[8] Nittmann, J. & Stanley, H.E., Non-deterministic approach to anisotropic growth patterns with continuously tunable morphology: the fractal properties of some real snowflakes. *J. of Phys. A: Math Gen.*, **20**, pp. L1185–L1191, 1987.
[9] Bak, P., *How Nature Works: the Science of Self-Organised Criticality*, Oxford University Press: Oxford, pp. 118–121, 1997.
[10] Goodwin, B., *How The Leopard Changed Its Spots*, Phoenix: London, pp. 92–96, 1997.
[11] Goodwin, B., Development as a Robust Natural Process. *Thinking about Biology*, eds. W.D. Stein & F.J. Varela, SFI Studies in the Sciences of Complexity, Lect. Note Vol. III, Addison-Wesley, pp. 123–148, 1993.
[12] Goodwin, B.C., Kauffman, S. & Murray, J.D., Is morphogenesis an intrinsically robust process? *J. Theor. Biol.*, **163**, pp. 135–144, 1993.
[13] Carlson, J.M. & Doyle, J., Complexity and robustness. *Proc. Natl. Acad. Sci. USA*, **99(1)**, pp. 2538–2545, 2002.
[14] Taguchi, G., Chowdhury, S. & Taguchi, S., *Robust Engineering*, McGraw-Hill: New York, pp. 1–9, 1999.
[15] Mullin, T.J., Finite-dimensional dynamics and chaos. *Plasma Physics: An Introductory Course*, ed. R.O. Dendy, CUP, pp. 129–165, 1995.
[16] Stewart, I.N., Bayliss, M., Reynolds, L.F., Saynor, D., Muldoon, M. & Nicol, M., Chaos theory and wire coilability: The FRACMAT Project. Part 1. Data analysis. *Mathematics Today*, pp. 104–110, August 1997.
[17] Stewart, I.N., Bayliss, M., Reynolds, L.F., Saynor, D., Muldoon, M. & Nicol, M., Chaos theory and wire coilability: The FRACMAT Project. Part 2. The test machine. *Mathematics Today*, pp. 176–180, December 1997.
[18] Broomhead, D.S. & King, G.P., On the qualitative analysis of experimental dynamical systems. *Nonlinear Phenomena and Chaos.* ed. S. Sarkar, Adam Hilger Ltd, pp. 336, 1986.
[19] Andò, B. & Graziani, S., *Stochastic Resonance: Theory and Applications*, Kluwer, pp. 1–73, 2000.

CHAPTER 5

D'Arcy Thompson: nature and design through growth and form

M.A.J. Chaplain
The SIMBIOS Centre, Division of Mathematics, University of Dundee, Scotland.

Abstract

The application of mathematics to biology may be viewed by some as a relatively recent phenomenon. However this was being done in a systematic fashion over a century ago by the great polymath Sir D'Arcy Wentworth Thompson. In this chapter we examine the themes of his great work "On Growth and Form" and show how they are still relevant to and underpin the modern mathematical biology theories of pattern formation and morphogenesis.

1 Introduction

The influence of Sir D'Arcy Wentworth Thompson and his *magnum opus* "On Growth and Form" are still with us today. First published in 1917 it was his great *tour de force*, an attempt to offer an alternative to Darwinism and to synthesise (applied) mathematics and biology long before it became the discipline of mathematical biology or biomathematics that we know today. D'Arcy was indeed a man ahead of his time, the first mathematical biologist or biomathematician. Although his book may be seen as "out of date" in some respects (the advancement of knowledge in modern biology, especially biochemistry, and the life sciences in general shows that many of his proposals are incorrect) there is still much that is refreshingly contemporary about it. The opening lines of the Introductory set the scene for the whole book:

> "Of the chemistry of his day and generation, Kant declared that it was a science, but not Science – *eine Wissenschaft aber nicht Wissenschaft* – for that the criterion of true science lay in its relation to mathematics. This was an old story: for Roger Bacon had called mathematics *porta et clavis scientiarum* and Leonardo da Vinci had said much the same. [*Nessuna humana investigazione si può dimandare vera scienzia s'essa non passa per le matematiche dimostrazione*]".

However, in spite of the "old story" that mathematics is both the key to and the door of science D'Arcy notes:

> "But the zoologist or morphologist has been slow, where the physiologist has been eager, to invoke the aid of the physical or mathematical sciences".

Nonetheless, it is D'Arcy's firm conviction that just as mathematics and physics can provide rational explanations for the growth and form of waves, hills, clouds and their dynamics:

> "Nor is it otherwise with the material of living things. Cell and tissue, shell and bone, leaf and flower, are so many portions of matter, and it is in obedience to the laws of physics that their particles have been moved, moulded and conformed. They are no exception to the rule that Θεὸς ἀεὶ γεωμετρει. Their problems of form are in the first instance mathematical problems, their problems of growth are essentially physical problems, and the morphologist is, *ipso facto*, a student of physical science."

Although modern biology has demonstrated that certain propositions and proposals forwarded in *On Growth and Form* are completely wrong, nevertheless, the book as a whole still stands as an unsurpassed masterpiece of scientific prose. D'Arcy Thompson was a very fine writer and although his opinion of the quality of his written work was perhaps rather modest:

> "The little gift I have for writing English which I possess, and try and cultivate and use, is, speaking honestly and seriously, the one thing I am a bit proud and vain of - the one and only thing."

P. Medawar (Thompson [17]) puts it rather more objectively and offers the opinion that *On Growth and Form* is

> "…beyond comparison the finest work of literature in all the annals of science that have been recorded in the English tongue".

This opinion is seconded by Hutchinson [8] who notes that:

> "*On Growth and Form* is a great book because it shows us science as a traditional activity and tells us that the tradition is one of daring and of imagination… To foster the values of civilised men takes many talents, and few possess a great share of them. D'Arcy Wentworth Thompson had more than most of us and he used them superbly. What he wrote brings home to the scientific mind, perhaps better than the work of any other writer, what it means to be civilised; the splendour of the panorama that he has presented gives value to what we try to do, and in giving us this he achieved reward given to few men."

Given this (rightly) glowing appraisal of D'Arcy's own not inconsiderable literary talent, this short chapter will make good use of D'Arcy's own words (who better to describe his work?) in detailing his work, scientific career and his finest achievement and scientific legacy, *On Growth and Form*.

2 A brief biography

D'Arcy Wentworth Thompson was born in Edinburgh in 1860, the son of a classics professor. His father was appointed Professor of Greek at Queen's College (now University College), Galway, when D'Arcy was 3 years old but he returned to Edinburgh at the age of 10 to attend Edinburgh Academy. There he won the prize for Classics, Greek Testament, Mathematics and Modern Languages in his final year at school. These talents were to remain with him throughout his long and distinguished scientific career and the influence of each can be felt on every page of *On Growth and Form*. After starting a medicine course at Edinburgh University he changed to study natural science at Cambridge.

He graduated with a BA in Zoology in 1883 and the following year, 1884, D'Arcy Thompson was appointed Professor of Biology at the newly founded University College Dundee. The letters of his recommendation for this post are kept in the archives of the library of Dundee University. Finally, in 1917 he was appointed to the Chair of Natural History in St. Andrews University. He was to hold a professorial chair for a total of 64 years (33 years in Dundee, 31 years in St. Andrews), a record which will not now be broken. He was a rather colourful individual, often seen in the streets of St. Andrews and in the university lecture theatres with his pet parrot upon his shoulder.

D'Arcy combined skills in a way that made him unique. He was a Greek scholar, a naturalist and a mathematician. Indeed, he remained a sufficiently talented classicist to be appointed as the President of the Classical Associations of England, Scotland and Wales. In addition to his scientific articles and books, he translated Aristotle's *Historia Animalium* and published a *Glossary of Greek Birds* and a *Glossary of Greek Fishes*.

He was the first biomathematician, although he followed in the tradition of another great natural historian with mathematical skills, namely Buffon. His understanding of mathematics was of the modern subject but based on the firm foundations of an understanding of Greek mathematics.

In terms of scientific honours and awards, D'Arcy Thompson was elected a Fellow of the Royal Society of London in 1916 and was later awarded the Darwin Medal of the Society in 1946:

> "... *in recognition of his outstanding contributions to the development of biology.*"

He also received recognition for his mathematical endeavours, being made an honorary member of the Edinburgh Mathematical Society in 1933. A knighthood followed in 1937, as did an honorary degree from Oxford in 1945. Yet despite these many accolades he "...*suffered for his iconoclasm during his life*" and "...*marginality was his fate throughout the central years of his professional life.*"

A concise and eloquent picture of the man, his work and his style is given by Medawar (Thompson [17]):

> "An aristocrat of learning whose intellectual endowments are not likely ever again to be combined within one man. He was a classicist of sufficient distinction to have become President of the Classical Associations of England and Wales and of Scotland; a mathematician good enough to have had an entirely mathematical paper accepted for publication by the Royal Society; and a naturalist who held important chairs for sixty-four years, that is for all but the length of time into which we must nowadays squeeze the whole of our lives from birth until professional retirement. He was a famous conversationalist and lecturer (the two are often thought to go

together, but seldom do), and the author of a work which, considered as literature, is the equal of anything of Pater's or Logan Pearsall Smith's in its complete mastery of the *bel canto* style. Add to this that he was over six feet tall, with the build and carriage of a Viking and with the pride of bearing that comes from good looks known to be possessed.

D'Arcy Thompson (he was always called that, or D'Arcy) had not merely the makings but the actual accomplishments of three scholars. All three were eminent, even if, judged by the standards which he himself would have applied to them, none could strictly be called great. If the three scholars had merely been added together in D'Arcy Thompson, each working independently of the others, then I think we would find it hard to repudiate the idea that he was an amateur, though a patrician among amateurs; we should say, perhaps, that great as were his accomplishments, he lacked that deep sense of engagement that marks the professional scholar of the present day. But they were not merely added together; they were integrally – Clifford Dobell said chemically – combined. I am trying to say that he was not one of those who have made two or more separate and somewhat incongruous reputations, like a composer-chemist or politician-novelist, or like one man who has both ridden in the Grand National and become an F.R.S.; but that he was a man who comprehended many things with an undivided mind. In the range and quality of his learning, the uses to which he put it, and the style in which he made it known, I see not an amateur, but, in the proper sense of that term, a natural philosopher."

Although his scientific legacy through *On Growth and Form* is here to stay, since as G.E. Hutchison [8] puts it this is *"one of the very few books on scientific matters written in this century which will, one may be confident, last as long as our too fragile culture"*, its importance and influence has, as Medawar notes, *"been intangible and indirect."*

3 On growth and form, style and substance

The first edition of *On Growth and Form* appeared in 1917. This was followed by a second (expanded) edition in 1942 (this is currently available from Dover publications). The contents of the book are as follows:

- I. Introductory
- II. On Magnitude
- III. The Rate of Growth
- IV. On the Internal Form and Structure of the Cell
- V. The Forms of Cells
- VI. A Note on Adsorption
- VII. The Forms of Tissues or Cell-Aggregates
- VIII. The Same (continued)
- IX. On Concretions, Spicules and Spicular Skeletons
- X. A Parenthetic Note on Geodesics
- XI. The Equiangular Spiral
- XII. The Spiral Shells of the Foraminifera
- XIII. The Shapes of Horns, and of Teeth or Tusks: With a Note on Torsion

XIV.   On Leaf Rearrangements or Phyllotaxis

XV.   On the Shape of Eggs and Other Hollow Structures

XVI.   On Form and Mechanical Efficiency

XVII. On the Theory of Transformations or the Comparison of Related Forms

An abridged edition has now appeared edited by J. Bonner (Thompson [16]). In this shortened version, chapters III, IV, VI, VIII, X, XII, XIV and XV of the original book have been omitted. This at once highlights the strengths and weaknesses of the book and D'Arcy's lasting influence. To take one omitted example - internal structures of the cell - this chapter of the original book has been removed because it is now simply out of date and incorrect. Advances in microscopy, cell biology and biochemistry since 1942 have surpassed D'Arcy's speculations. And yet, although incorrect in the precise detail, D'Arcy was incredibly prescient with his choice of subject matter – biochemistry is one of the largest areas of research in Life Sciences – as we shall see in the next section.

4 D'Arcy's legacy and influence today

Even glancing at the titles of the original chapters of the 1917 [15] edition one can see immediately how contemporary *On Growth and Form* is – intra-cellular structures, cells, cell-aggregates, tissues and organs, bones and skeletons. D'Arcy deals with how these interact and the causes of their forms (teleology) and patterns. Indeed one might say that the whole book is concerned with self-organisation and pattern formation, key fields of research in mathematical biology today. D'Arcy's legacy is thus alive and well in today's mathematical biology research, over one hundred years later as can be seen by the current activity in the field (cf. Chaplain *et al.* [3]). A few specific areas of research are worth mentioning in particular:

1) *Turing's theory of morphogenesis or pre-pattern theory*: this theory was first proposed in a seminal paper of 1952 by Alan Turing (also famous for his work on cracking the Enigma Code during World War II and his theoretical computer the "Turing machine"). This theory states that (two) chemicals can *react* together and *diffuse* and set up spatially heterogeneous patterns which arise from an original spatially homogeneous state (Turing [18]). The radical counter-intuitive concept introduced by Turing was that it was the introduction of diffusion which *destabilised* the system and caused the patterns to form. Prior to this, diffusion had always been thought of as a mechanism for stabilising the system and "smoothing out" any potential heterogeneities. A general (2-variable) Turing "reaction-diffusion" system may be written in the form:

$$\frac{\partial u}{\partial t} = \overset{\text{diffusion}}{\nabla^2 u} + \overset{\text{reaction}}{f(u,v)}$$

$$\frac{\partial u}{\partial t} = d\nabla^2 v + g(u,v)$$

where u and v are the concentrations of the two morphogens which diffuse and react together with f and g the two functions describing their reaction kinetics.

2) *Gierer-Meinhardt activator-inhibitor theory*: 20 years after Turing, Gierer and Meinhardt developed the pre-pattern theory further and more precisely, identifying the chemicals (Turing's morphogens) as *activators* (i.e. chemicals which stimulate cells to divide) and *inhibitors* (ie. chemicals which stop cells from dividing) (Gierer and Meinhardt [7]). This theory is also a *pre-pattern theory* – since chemicals normally react and diffuse on a much faster timescale than cells, for instance, the initial (spatially heterogeneous) pre-pattern is laid down by the reacting and diffusing chemicals. Any cells which exist in the same spatial domain as the chemicals then "interpret" this pre-pattern, react accordingly (e.g. migrate or differentiate according to the chemical pre-pattern) and provide the newly organised structures in response to the underlying chemical pre-pattern. Meinhardt [11, 12] applied this activator-inhibitor theory to many aspects of developmental biology and also to patterns of leaves (phyllotaxis), and sea shells. Murray [13] also applied this theory to the coat markings of animals.

Recent work by Crampin *et al.* [4] and Lolas [10] have combined growing domains with pre-pattern theory to move the theory forward still further. A true combination of "growth" and "form"! Figures 1 to 3 show some of the patterns produced on growing domains.

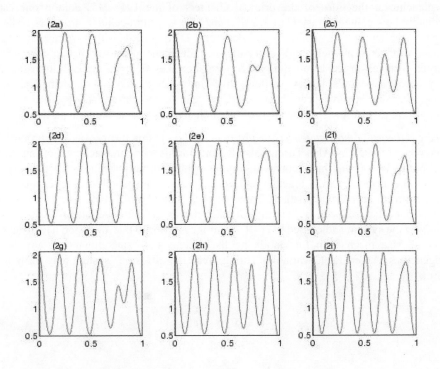

Figure 1: Plots of the evolution of a spatially-heterogeneous pattern in a one-dimensional growing domain.

D'Arcy Thompson: nature and design through growth and form 91

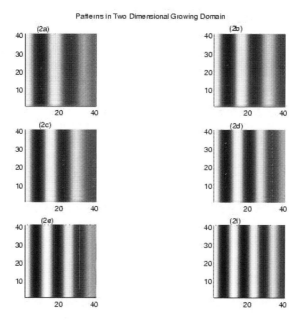

Figure 2: Plots of the evolution of a spatially-heterogeneous "stripy" pattern in a two-dimensional growing domain.

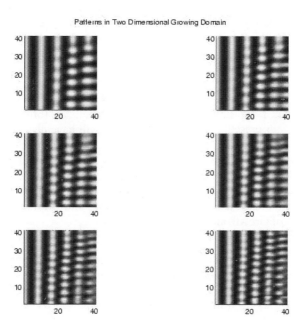

Figure 3: Plots showing the evolution of a spatially-heterogeneous pattern in a two-dimensional growing domain. The plots show the evolution of "spots" from "stripes".

3) *Murray-Oster mechanochemical pattern formation theory*: an alternative pattern formation theory to Turing and Gierer-Meinhardt was proposed by Murray and Oster [14]. In this theory chemicals and cells act not sequentially but **together** to produce a spatial pattern. The Murray-Oster approach considers cells first secreting chemicals and then moving in response to these chemicals. In moving, the cells deform their underlying substrate (e.g. the tissue they are migrating through) and this in turn affects the cells, the chemical distributions within the tissue and so on (each feeding back to the other). This theory combines the previous reaction-diffusion modelling of the chemicals with solid mechanics of the cells (linear visco-elasticity theory and small deformation theory). The theory incorporates explicitly the physical concepts of stress, strain, pressure and deformation and has been taken up by many subsequent researchers to model the dynamics and biophysics of individual cell movement and deformation (see, for example, the excellent book of Alt *et al.* [2]).

In terms of areas of biological research activity today, cell biology (Alberts et al. [1]), biomechanics (Fung [6, 5]) and tissue engineering (Lanza *et al.* [9]) are all key areas of research in the Life Sciences.

5 Conclusions

D'Arcy Wentworth Thompson was a man ahead of his time who has left us the lasting legacy of *On Growth and Form*, a book which, although incorrect in some of its finer detail, is highly contemporary in terms of its overall scientific vision. This vision is the product of a unique individual, a scholar of breadth and depth, a man of learning and wisdom. His attempt to apply the rigour of mathematics (and physics and dynamics) to biology, to provide a synthesis of the natural and life sciences ("a unification theory"?) has left us with not only much food for thought in how we can describe today's biological processes, but, perhaps, just as importantly, a beautifully crafted literary masterpiece. Indeed, long before Stephen Hawking caught the public imagination with his talk of "knowing the mind of God", D'Arcy had said this and more (infinitely more eloquently) almost a century before. It is perhaps fitting to end this piece with the words of the great man himself from the Epilogue of *On Growth and Form*:

> "The fact that I set little store by certain postulates (often deemed to be fundamental) of our present-day biology the reader will have discovered and I have not endeavoured to conceal. But it is not for polemical argument that I have written, and the doctrines which I do not subscribe to I have only spoken of by the way. My task is finished if I have been able to show that a certain mathematical aspect of morphology, to which as yet the morphologist gives little heed, is interwoven with his problems, complementary to his descriptive task, and helpful, nay essential to his proper study of Growth and Form.
>
> ………
>
> "And while I have sought to show the naturalist how a few mathematical concepts and dynamical principles may help and guide him, I have tried to show the mathematician a field for his labour – a field which few have entered and no man has explored".
>
> ………
>
> "I know that in the study of material things, number, order and position are the threefold clue to exact knowledge; that these three, in the mathematician's hands, furnish 'the first outlines for a sketch of the Universe'; …. He hath measured the

waters in the hollow of his hand, and meted out heaven with the span, and comprehended the dust of the earth in a measure".

....

"So the living and the dead, things animate and inanimate, we dwellers in the world and this world wherein we dwell - παντα γα μαν τα γιγνωσκομενα - are bound alike by physical and mathematical law'. Conterminous with space and coeval with time is the kingdom of Mathematics; within this range her dominion is supreme; otherwise than according to her order nothing can exist, and nothing takes place in contradiction to her laws'. So said, some sixty years ago, a certain mathematician, … and who with Plato and Pythagoras, saw in Number *le comment et le pourquoi des choses*, and found in it *la clef de voûte de l'Univers*".

References

[1] Alberts, B., Bray, D., Lewis, J, Raff, M., Roberts, K. & Watson, J.D., *Molecular Biology of the Cell*. 3rd edn. Garland Publishing: New York, 1994.
[2] Alt, W., Deutsch, A. & Dunn, G.A. (eds.), *Dynamics of Cell and Tisuue Motion*. Birkhäuser: Basel, 1997.
[3] Chaplain, M.A.J., McLachlan, J. & Singh, G.D. (eds.), *On Growth and Form: Spatio-temporal pattern formation in biology.* Wiley: Chichester, 1999.
[4] Crampin, E.J., Gaffney, E.A. & Maini, P.K., Reaction and diffusion on growing domains: Scenarios for robust pattern formation. *Bull. Math. Biol.*, **61**, pp. 1093–1120, 1999.
[5] Fung, Y.C., *Biomechanics: Mechanical Properties of Living Tissues.* 2nd edn., Springer: New York, 1993.
[6] Fung, Y.C., *Biomechanics: Motion, Flow, Stress and Growth.* Springer: New York, 1990.
[7] Gierer, A. & Meinhardt, H., A theory of biological pattern formation. *Kybernetic*, **12**, pp. 30–39, 1972.
[8] Hutchinson, G.E., In Memoriam, D'Arcy Wentworth Thompson. *American Scientist*, **36**, p. 577, 1948.
[9] Lanza, R.P., Langer, R. & Vacanti, J. (eds.), *Principles of Tissue Engineering.* 2nd edn., Academic Press: San Diego, 2000.
[10] Lolas, G., *Spatio-temporal pattern formation and reaction diffusion equations.* MSc Thesis, University of Dundee, 1999.
[11] Meinhardt, H., *Models of Biological Pattern Formation.* Academic Press: London, 1982.
[12] Meinhardt, H., *The Algorithmic Beauty of Sea Shells.* 2nd edn., Springer: New York, 1998.
[13] Murray, J.D., A pre-pattern formation mechanism for animal coat markings. *J. theor. Biol.*, **88**, pp. 161–199, 1981.
[14] Murray, J.D. & Oster, G.F., Cell traction models for generating pattern and form in morphogenesis. *J. Math. Biol.*, **19**, pp. 265–279, 1985.
[15] Thompson, D.W., *On Growth and Form.* Cambridge University Press: Cambridge, 1942.
[16] Thompson, D.W., *On Growth and Form.* Cambridge University Press: Cambridge (abridged and edited by J.T. Bonner), 1961.
[17] Thompson, R.D., *D'Arcy Wentworth Thompson, the Scholar-Naturalist, 1860–1948, by His Daughter. With a Postscript by P.B. Medawar.* Oxford University Press: Oxford, 1958.
[18] Turing, A.M., The chemical basis of morphogenesis. *Phil. Trans. Roy. Soc. Lond.*, **B237**, pp. 37–72, 1952.

CHAPTER 6

Design in plants

D.F. Cutler
Royal Botanic Gardens, Kew, UK.

Abstract

Examples are given of the parallels between structures occurring naturally in plants, and their engineering counterparts. The mathematics of boundary layer physics is explored.

1 Introduction

In this chapter, we shall look at the anatomical structure of a wide range of plants to see if parallels can be found in engineering or other technological 'inventions'. This introduction can be passed over by those confident in their knowledge of the main features of plants. It provides those who have little or no botany with a very compressed and oversimplified account of the basics. Wood is the focus of another chapter in this volume, and only the cell types that constitute wood (xylem) will be considered in any detail. Here leaves, stems and roots provide most of the examples. General texts pertinent to the anatomical parts of this paper are [1] Fahn and [2] Cutler.

Many factors have played and continue to play a part in the wide range of cell arrangements we can see in multicellular land plants. Plant evolution started in the water. This led from single cells to large but relatively simple structures, such as *Macrocystis* from tropical seas, and kelps which have species growing in a wide range of water temperatures. The holdfast secures these seaweeds to a firm substrate, but does not take up nutrients, and the large fronds are supported by the salt water in which they are immersed. They need to exhibit flexibility and strength to withstand normal movement of the sea water, but have no need of internal mechanical systems. Neither do they need formal 'plumbing' systems, because they are bathed in water and nutrients. Nor do they need a waterproof skin for water retention, since water and mineral nutrients are absorbed through the whole plant body.

Fertilisation occurs through motile male gametes swimming through the water, to reach the egg cells. Dispersal of sexually produced propagules is also through the medium of water.

1.1 Move to land from water

Early land plants were relatively simple, often assuming a dichotomously branched form. After a number of 'experimental' forms, several lines continued to evolve into more complex structures. Some early ones became extinct, but some lines remain, for example equisetums are related to some of the very large plants making up a proportion of the coal measures, as are clubmosses. Also among the groups of ancient origin that are still extant, some of the thallose liverworts have thin, flattened bodies. The spore containing capsules are raised on stalks. Mosses and leafy liverworts have a more complex 'stem and leaf' arrangement in their gamete bearing structures, from which arise after fertilisation, a short stalked spore containing capsule. Fertilisation is still effected by a self motile male gamete, that has to swim through a water film to reach the female egg cell.

These early lines of land plants and their living representatives had to conquer a number of obstacles to life out of the water. Among these were the need to retain water and the development of specialised cells for mechanical support. The tough skin (epidermis) evolved which prevents water loss but permits gas exchange. Sometimes it represents the mechanical system as well, containing the main leaf or stem material under hydraulic pressure. In many plants, the strength of the skin is supplemented by tough mechanical cells arranged in mechanically appropriate places. With the development of the epidermis and a primitive plumbing (vascular) system came the development of regulated pores in the epidermis, the stomata (Fig. 1).

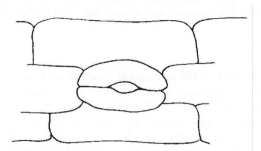

Figure 1: A leaf epidermis (skin) in surface view. The pair of kidney-shaped cells with a pore between them form a stoma. Changes in the shape of these cells can regulate the size of the pore. Line drawing.

When these are open, the flow of water (transpiration stream) through the plant is encouraged. Development of stomata and the evolution of complex vascular systems may have occurred hand in hand.

The gymnosperms, and the flowering plants which evolved more recently, among other, much smaller groups, have developed pollen which carries the male gamete in a form protected from dehydration to special receptive structures in the female part of the flower (pollination). After its arrival on a receptive surface of the same species (and often on another plant), each pollen grain develops a tube which grows through special tissue to deliver the male gamete (here just a nucleus) right to the egg cell where fertilisation can occur. This mechanism cuts out the need for a continuous film of water. In the more ancient group of cycads, the last stage following pollination still involves a swimming male gamete, but it is delivered close to the egg cell. During evolution the ability to overcome water dependent reproductive systems and dispersal mechanisms has influenced the capacity of plant life to disperse and conquer the wide range of seasonally dry niches on earth.

In higher plants a 'double fertilisation' occurs. In addition to the fertilisation of the egg cell, a second nucleus (actually with a double set of chromosomes) fuses with the second male nucleus transported in the pollen tube, to form the cell which can go on to divide and form endosperm, giving the seed a convenient source of food reserve. Seeds that are adapted to a period of dormancy can call on this reserve when they germinate. These, and another seed type in which the seed leaves store reserve food material, need not germinate at once or even in the next season. They have added survival value by being able to germinate when conditions are favourable.

In the flowering plants, the dicotyledons normally have two seed leaves, and the monocotyledons one. They also have other distinguishing structural features that will be mentioned below.

Plants submerged in water are afforded some protection from damaging UV light. Land plants need other mechanisms to prevent UV damage. The green pigment, chlorophyll, is readily damaged by UV. Since this pigment and its cohort of specialised enzymes is responsible for transforming the energy of sunlight through its action on CO_2 and H_2O into sugars, the starting point for nearly all stored organic energy on earth, it is vitally important that the UV screening methods developed are effective! These are discussed further.

1.2 Competition for light

All green plants need light for photosynthesis. In a mixed community of plants, a range of different strategies can often be observed by which plants obtain exposure to light.

Some put out leaves before others, complete their annual cycle and either form seed (annuals) or retire to a dormant form (some perennials and biennials). Others develop long stems or trunks and overshadow the competition (some annuals and many perennials). Yet others do not invest material in mechanically strong stems, but use the support provided by those which do, climbing or scrambling over them (both annuals and perennials). Biennials, noted above, are plants with a two year life cycle. They build up a plant body and food reserves in the first year, and then flower and fruit in the second.

Plants growing in shady habitats often show special adaptations to maximise the use of the small quantity of light that falls on them. Those adapted to very exposed, bright, dry or salty habitats show another range of characteristics. Leaf form – the shape, size and thickness – varies greatly, and generally gives a good indication of the type of environment to which the species is adapted. This will be discussed more fully below, but as a couple of examples, reduced, needle-like leaves are common on plants exposed to long periods of drought (surviving through periods of lack of water brought about by climates that are hot and dry, cold and dry or through soil salinity, for example). Very large, soft leaves are common on ground layer plants in wet tropical forests, where they have a continuous, adequate water supply and are sheltered from the wind.

Anatomy and morphology of plants are often strongly associated with ecology, but sometimes plants which show no obvious physical adaptation to their conditions may survive through having modified physiology, or simply have developed a strategy of leafing and growing when the conditions are good.

1.3 Economy

Underlying all of the adaptations in shape and size is the need for economy of use of materials in most of the habitats available to plants. It goes without saying that a larger plant body needs more minerals, water and sunlight to reach its maturity than a smaller plant body. In extreme conditions of competition, with very low water supply, mineral resources and a restricted growing season, such as in deserts, on sea cliffs and at high altitude, and in places where strong winds are frequent, few plants grow tall. In the majority of other conditions, however, some plants in most environments are tall perennials, whether as bushes or trees, or columnar stems, as in some cacti. The development of the 'tree' habit in many different plant families must reflect a high degree of competitive success for this life form. The expenditure of materials in short supply in the production of long-lived, mechanically robust forms must confer survival benefits to such plants. Synthesis of materials for mechanical support of the plant uses resources that otherwise might have been directed towards reproduction. We see an elegant use of strengthening tissues that parallels engineering solutions. Although expensive in mechanical tissues, the tree habit prolongs the period over which an individual may produce seed; over a long period successful seed formation and germination is more likely.

When we consider the anatomy and morphology of plants, then, there is a multitude of factors that have had a bearing on the end product. To begin to understand the significance of what we see, as a minimum, some knowledge of the evolutionary history, habitat preference and general biology of a species is necessary. Even after more than three centuries of recorded, structured and increasingly scientifically based study of plants, there are very few examples for which we have adequate data to reach this level of understanding. Consequently examples given here are flawed. They can only serve as a challenge! Where parallels are seen between plant structure and engineering or technology, these are often conceptual. There are some elegant pieces of experimental work and mathematical calculation which go beyond the philosophical, but these are rather few. Some of these are listed for further reading. Here, some of the engineering problems overcome by plants are described, in the full knowledge that a rather simple view has to be taken at present, through lack of proof. Perhaps a few postgraduate students might be stimulated by them, and do further research.

2 What holds a plant up?

2.1 Water

Plant cells need to be fully hydrated to work properly (except in periods of dormancy, as for example in many seeds). Individual vegetative cells in plants, unlike those in animals, are encased in a cellulose cell wall. The cellulose cell wall may be very thin, in cells that are actively dividing, as for example, in growing shoot or root tips. However, once developed into their mature form, the cell walls may become thicker, and additional substances, mainly lignins, incorporated into their structure. The cells themselves, then, contribute to the mechanical strength of the plant. Thin-walled cells when fully hydrated, are like small, pressurised containers. Mature cells, especially those with thick walls, have mechanical strength of their own, even without watery contents. Indeed, many fibres lack living contents when mature.

Young plants that are short of water wilt visibly. Temporary wilt does not cause death, and fortunately for those of us who may have spent a lot on a pot plant, we get a warning before it is too late. Permanent wilt brings with it cell damage, and death of the severely dehydrated parts.

2.2 Strong skin

In order to support a plant, the turgid cells need to be contained within a strong skin to be mechanically effective. A succulent tomato is mechanically coherent until its skin is split. Then the pressurised contents spill out. The technical term for the skin is the epidermis (Fig. 2). In succulent plants, such as cacti, the epidermal cells have thickened walls. Often the next couple of cell layers below the epidermis are also composed of thick-walled cells.

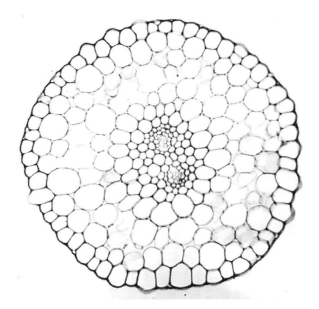

Figure 2: Stem with circular outline. This small plant (200 mm tall) from moist parts of sand dunes relies on the outer skin and water pressure in the thin walled cells to hold it up. There are no thick-walled cells. Cross section, light micrograph.

2.3 Xylem

In young plants, often in addition to the epidermis, the cells specialised for conducting water from root to leaves and shoots, have a mechanical function. The xylem vessels and tracheids are elongated cells (in the case of vessels, the vessel itself is composed of a series of shorter 'vessel elements' forming an axially elongated structure). These cells have thickened walls which help prevent their collapse when water in them is under tension through the pull of the transpiration stream (Fig. 3). The drying effect at the leaf surface promotes water movement from the roots through the plant body. The first formed conducting cells of the xylem consist of rather thin-walled, elongated cells that have to extend with the growth in length of the stem. Their collapse during the time they are needed to function is prevented by specialised thickening in their walls. This takes on the form of a series of annuli, or of a spiral (helical) winding. In this respect some hose pipes and extending exhaust tubes for tumble-dryers have been designed in the same way. The tracheids and vessels formed after extension growth is complete tend to have thick, rigid walls with either thin areas (pits), as in both tracheids and vessel elements, or clear openings between cells in line, as in vessel elements alone. These facilitate water movement from cell to cell. Even here, some of these cells in a range of species have an additional helical thickening on the inner side of their walls.

The tracheids in conifers have a specialised arrangement of fail-safe valves in the thin-walled pit areas. The pit pairs themselves are more or less circular in surface view. The cell wall bulges out on either side of the thin membrane between the cells (a laminate – cell wall, middle lamella, cell wall). The aperture in the out-bulging part of the wall is narrower than the diameter of the central membrane.

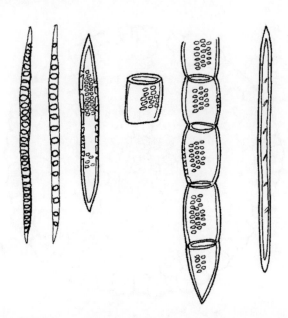

Figure 3: A selection of cells with mechanical strength; that on the far right is a fibre, a strengthening cell, the others are concerned with both water transport and providing strength. The two on the left are tracheids from the first stage of development of the vascular tissue. The spirals or annuli help keep the water pathway open as the cells grow in length. The next cell is a tracheid from the mature second stage of growth. The short cell is a vessel element, and represents one of the series to its right, shown joined end to end as part of a vessel. Only the end cells in a vessel are closed off. The elements in the line have openings in their end walls. Longitudinal view, line drawings.

The central membrane has an annular rim which is not thickened, but the centre of the membrane has a thickened area. This gives a bordered appearance to the pit pair (Fig. 4).

If there is a failure of the water column in a tracheid, the thickened portion of the pit membrane moves across under the differential pressure, blocking the pit aperture, and helping prevent the spread of an embolism.

The arrangement of the xylem in formalised bundles, distributed in a softer matrix of ground tissue, disperses the strengthened columns regularly around the young stem towards its periphery (Fig. 5).

Design in plants 101

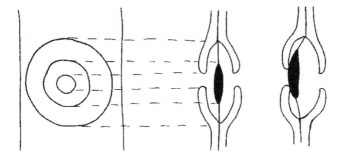

Figure 4: Pits which act like valves in the tracheid of a conifer. The left hand drawing shows one in surface view; those to the right, in longitudinal section. The far right shows the condition when the valve is closed, and the thickened central part set in a porous diaphragm has been forced over to occlude the pore. Line drawings.

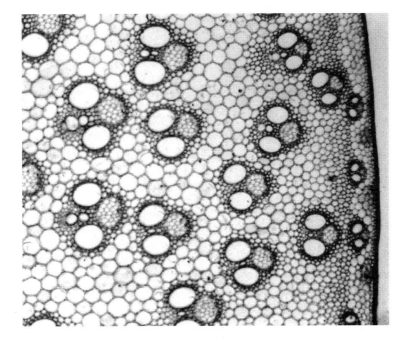

Figure 5: Part of a stem of a robust grass, in cross section. Here mechanical strength of the stem is provided by the vascular bundles set in a matrix of thinner-walled cells, rather like rod reinforcements. Each vascular bundle has an outer sheath of fibres, forming a strong tube in which the two wide vessels can conduct water, and the strand of thin-walled, narrow cells (phloem) can transport sugar solutions with little risk of damage. Just to the inner side of the outer ring of smaller vessels the several layers of narrow cells eventually become thick-walled and provide additional strength in the form of a cylinder to the whole stem. Light photomicrograph.

In monocotyledonous vascular bundles there is no increase in xylem cell number after maturity. In dicotyledons and gymnosperms as the stems age new xylem is added. Fibre strands are often present in the outer cortex (Fig. 6). Eventually, in dicotyledons and gymnosperms the individual bundles become united in a cylinder, conferring more strength on the stem structure. Further consideration will be given to the secondary xylem, known familiarly as wood, below. Each strand is accompanied by cells that retain thinner walls, the phloem, whose function is to conduct synthesised food materials around the plant body. The conducting phloem cells provide little mechanical strength, but their associated fibres may.

In the leaf, the veins are composed of xylem and phloem strands (Figs 7, 8). These are often reinforced with fibres and collenchyma, and will be discussed below.

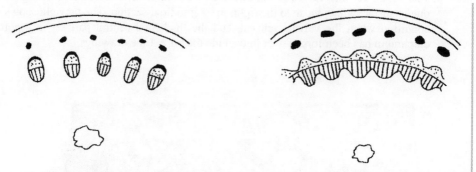

Figure 6: Diagrams of sectors of stem cross sections. In the left hand picture the vascular bundles are arranged in an outer cylinder, like rod reinforcement. They are supplemented by fibre rods, shown solid black. This type of arrangement, where the vascular bundles remain separate, prevails in the monocotyledons, a group including ranging from herbs to tree-like palms. As in Fig. 5, there can be many more than just one cylinder of such bundles. In the right hand picture, the single ring of vascular bundles has developed into a closed cylinder. This is done through a zone of cells called the cambium. Cambial cells continue cell division to form new xylem and new phloem throughout the life of the plant, causing large increases in stem diameter. The much thickened xylem forms the wood of trees. This type of structure is found in the dicotyledons and gymnosperms. *The shading in this Figure is used where appropriate in all the line drawings. Solid black represents fibres, dots represent phloem and parallel hatching represents xylem.*

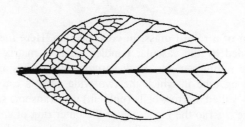

Figure 7: Diagram of the veins strengthening the laminar structure of a simple leaf. The primary and some secondary level veins are indicated.

Design in plants 103

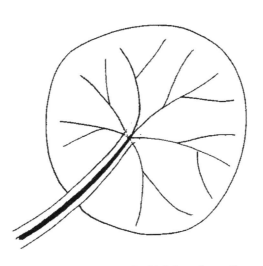

Figure 8: Diagram of the veins of a peltate leaf which has the stalk attached to the centre of a disk-like lamina. The main veins act rather like the ribs in a cantilevered structure.

2.4 Fibres

Plant fibres occur in the wood of many plants, and because of their association with the xylem, are called xylary fibres. They are also often found in the outer part of young stems, bark and leaves, where they are called extraxylary fibres. Their main function is in strengthening. The common feature of fibre cells is that they are elongated and thick-walled, with lignin permeating the cellulose of the cell wall. Fibre cells normally have pointed ends (Fig. 3). They often extend in length during development, growing between cells that may not be lengthening at the same rate. Fibres may be only about 10 times longer than wide, but many are 20-30 and even up to and exceeding 100 times longer than wide. They may remain flexible, as in many extraxylary fibres, or have more limited flexibility, as in xylary fibres.

2.5 Lignified parenchyma

Sometimes axially elongated cells of the 'packing' tissue, parenchyma, become thick-walled and lignified. These have similar functions to fibres, but their ends tend not to be pointed. Often no distinction is made between this cell type and true fibres. Cells of this type make up the bulk of the strengthening tissue in bamboos. They are arranged towards the periphery of the stem, the centre of which is often hollow, with transverse septa at intervals.

2.6 Fibres in composite structures

Fibres are typically found as components of xylem of angiosperms, where the individual cells may be grouped together, making up a significant proportion of the mass of the wood. This is further described in the chapter on wood. Here the extraxylary fibres will be considered. They infrequently occur singly in an organ. Generally they are grouped together as strands several to many cells wide, oriented axially in relation to stem or bark, and following vein direction in leaves. They may be associated closely with veins or their equivalent in stems, the vascular bundles, or grouped strategically, strengthening leaf margins, for example, or forming rod-like reinforcement in amongst softer tissues.

As mentioned earlier, forming thick-walled cells make considerable demands on the energy balance of plants. During evolution there has been a refinement of the proportion of strengthening tissue produced, and its disposition in the plant to a high level of efficiency. Consequently close parallels can often be seen between the proportion and disposition of strengthening material in man-made structures, and those with similar form in plants. In man-made structures the solution has been derived through experiment and calculation, in plants through a natural selection process. Selected examples are described in the section on Micro-engineering, below.

2.7 Sclereids

Sclereids are also cells with thick, lignified walls. They are grouped with fibres under the general term sclerenchyma. They differ from fibres in generally being shorter in relation to their length, but there is some overlap in the range of cells. They may be branched, sinuous or short – often more or less isodiametric. The longer ones commonly feature in the sheaths to veins, particularly near the ends of the finer branches. They can be pit-prop-like when they extend between the upper and lower surfaces of leaves, and appear to help prevent collapse of softer tissues at times of water stress, as in olive leaves and the leaves of many mangrove plants. These plants, and many of the hard-leaved plants found in arid habitats, often have abundant elongated or branched sclereids. Fig. 9 shows a shorter prop-like cell in a leaf of an *Osmanthus* species, extending from the epidermis through the thinner walled cells that in life contain the chloroplasts.

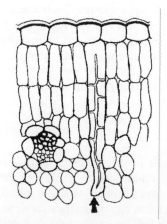

Figure 9: Diagram of part of the cross section of an *Osmanthus* leaf showing a pit-prop-like sclereid (arrowed) set among thin walled cells of the photosynthetic tissue. Note also the thick cuticle over the epidermal cells (top). This is part of the system that helps reduce water loss.

Most people are more familiar with the isodiametric sclereids that make up the shell of hazel nuts (*Avellana* sp.), or the gritty component of the skins and flesh of pear (*Pyrus*) fruits. Sclereids along with fibres are very common in bark. As a plant stem ages, its outer part, the cortex, may become progressively sclerified. This is an important consideration for those wishing to propagate by cuttings. Since many new 'adventitious' roots develop from inside the cutting, and grow out through the cortex, any strong mechanical barrier can hinder the process. Here young, softwood cuttings would be chosen, with little sclerenchyma. Some roots will develop directly from the cut end wound tissue; these are not so affected by woodiness.

In trees with flaky bark, short fibres and sclereids may predominate, but in stringy bark trees longer fibres, organised into long bundles or strands are more abundant. Early string and cloths were made from bark, relying on the fibrous nature of the material for their suitability.

2.8 Collenchyma

In addition to the 'mechanical' cells - fibres and lignified parenchyma – a third cell type has mechanical functions. This is collenchyma. Collenchyma cells have walls which during their development and extension are mainly cellulosic. They grow with the surrounding tissue as it expands or lengthens. They are more flexible than fibres, and if they remain unlignified, as they might in association with leaf veins and midribs, or in leaf stalks (petioles), they allow for a high degree of flexibility in the organ itself. Often, after growth in length of stems has occurred, and more mechanical rigidity is an advantage, we find that the collenchyma cells become lignified, and function much as fibres.

3 Micro-engineering

3.1 Fan vaulting and leaf architecture

Veins in leaves are composed of xylem and phloem cells, concerned principally with the movement of water and solutes, and the soluble products of photosynthesis within the leaf. They are directly connected with the vascular system of the shoots. In general they import water for photosynthesis and cooling, and export the synthesised products of photosynthesis. When leaves are developing they are net users of photosynthetic products.

In most land plants, the xylem itself has relatively little mechanical strength, and in many species, the vascular bundles are accompanied by sheaths or rod-like arrangements of fibres, sclerified parenchyma or collenchyma. The mechanical tissue may be accommodated within the thickness of the leaf, so that both surfaces are smooth, or it may produce prominent ridges above or below (or both). The arrangement of veins in leaves is very varied. In the monocotyledons and some dicotyledons, many species have strap-shaped leaves, with the main veins parallel to one another. Grasses are typical examples. The centrally placed vein may be the largest. The axial veins are connected at intervals by transverse veins, which are in general narrower than most of those of the axial system. This produces a net-like arrangement, with the softer, non-load-bearing green tissue suspended between the veins. This type of arrangement is reminiscent of glazing bars in glasshouses, where the axial bars are thicker than the transverse bars. Joseph Paxton, designer of the Crystal Palace, was inspired by the structure of leaves of *Victoria amazonica*.

Some monocotyledons, such as many aroids, are more like the broad leaved dicotyledons in leaf form. They have a definable stalk (petiole) and an expanded flattened blade (lamina). The veins in the lamina generally consist of a midrib, which follows directly in line with the veins of the petiole, and a series of lateral veins. The first order lateral veins may depart from the midrib in a regular pinnate manner up its length, fairly evenly spaced, and extending towards the leaf margins on either side. They might all originate near the base of the lamina, and fan out towards the margins, as three, five or more branches. In some leaves, for example *Gunnera,* and rhubarb, the ribbing is very pronounced, and parallels can be seen with fan vaulting in buildings.

Cantilevers are common in nature. Some leaves are very large, and their strengthened, ribbed venation provides mechanical support with economy of materials, cantilevering the relatively delicate green tissue out into the light.

3.2 Guttering or thin tubes: the petiole

The arrangement of mechanical tissue in petioles is of considerable interest. In most plants, the petiole has at least two major functions. It provides a mechanically strong extension piece, complete with plumbing, holding the leaf out from the stem in a suitable position to intercept light for photosynthesis. Leaf arrangement on stems usually helps minimise self-shading, but the fine tuning is carried out by the petiole. In some plants, the petiole may also assist the lamina in tracking the sun. The petiole also enables the lamina to twist and partly rotate in the wind, minimising the damaging effects of wind on the thin, sail-like structure. The design is such that the elasticity in the system allows the lamina to return to its preferred orientation for light interception, when the wind falls.

This remarkable self-righting flexibility comes about through the structure and properties of the mechanical tissue. The cross section can be U-shaped, with some variants, or it may consist of a cylinder, with thicker and thinner parts, or a cylinder with additional rods either internally, externally, or both (Fig. 10).

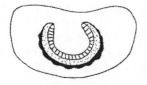 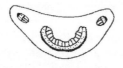

Figure 10: Diagrams of leaf stalks (petioles) in cross section. Each of these either has the vascular bundles united into a U-shaped structure, or into a cylinder composed of discrete strands. Both types allow torsional rotation without loss of load-bearing strength. The U-shaped sections are reminiscent of plastic guttering profiles. Shading as in Figure 6.

The mechanical tissue is set in a matrix of compressible parenchyma cells, and the cells of the epidermis are often thick-walled, and can have a few layers of thick-walled cells to their inner side. Those mechanical tissues with a U-shaped cross section are reminiscent of plastic guttering, which when horizontal can withstand a degree of vertical loading without undue deflection, but twists readily (too readily when it is being fitted!) around its short axis and recovers when released from the torsional force.

It is not so clear how those with closed cylinders function. Such petioles will be able to withstand vertical loading well. Clearly they also withstand torsional forces, because they work!

In a number of species, the base of the petiole, where it is attached to the stem (pulvinus) is a specialised structure. It has hydraulic properties. When fully turgid, the leaves are held in their position for effective photosynthesis. At times of drought, pressure is lost and the leaves hang down. They are then not as exposed to the sun, and the rate of further dehydration is reduced. Sensitive plants, like *Mimosa pudica*, can move water rapidly from the pulvinus using ion pumps. This causes the leaf to droop rapidly. In this mimosa, each leaflet has a miniature pulvinus that can also respond to the leaf being touched. The rapid collapse of the leaves is thought to deter animals from eating the leaves. Interestingly, light winds do not often trigger the mechanism. Leaves recover more slowly than they droop.

Of course, in feathery, pinnate leaves, and other forms with several leaflets, wind can be spilled by movement of the individual components. Many palms have large leaves, which when immature are entire. As they develop, predetermined lines of weakness in the lamina split, and the mature leaf gives the appearance of being pinnate. Images of palms in tropical storms show how flexible, yet strong, these leaves are. In addition to strength provided by reinforced veins, they often contain fibre strands, and possess thick-walled epidermal cells. Coconut palm is a good example. This passive approach to survival is effective. The leaves may become more tattered, but they generally survive to function satisfactorily. Pre-determined lines of weakness are built into cars, so that air bags may deploy readily when needed, and some potentially dangerous parts protruding into the passenger space snap off or collapse on impact.

3.3 Girders, I beams

Mention has already been made of the vascular bundles in leaves, and the mechanical tissues that accompany the weaker conducting cells. Thin leaves, with one row of vascular bundles arranged parallel to one another, extending axially, often have a configuration of strengthening cells similar to that found in I beams or girders. The conducting tissue may be situated equidistant from the upper and lower surfaces. Above (adaxial) and beneath (abaxial) this, fibres, collenchyma or axially elongated parenchyma cells provide the upper and lower flanges of the girder. The illustrations show some of the configurations that occur (Fig. 11).

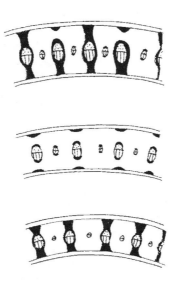

Figure 11: Diagrams of leaf cross sections showing a few of the options providing mechanical strength. These are reminiscent of I beams and girders. Shading as in Fig. 6.

108 *Nature and Design*

In manufactured I beams, the upper and lower flanges are more robust than the upright part between them. Indeed, it is common to find circular holes to have been cut in this, or for it to take on a lattice-like form, saving materials where they play only a small part in the structural rigidity of the beam. Not uncommonly in nature, the real strength is supplied in the 'flanges', and the tissues between them serve mechanically only to hold the flanges in their respective spatial arrangement. Commonly soft, turgid parenchyma cells may be found in such parts of the beam, but sometimes there may even be air spaces. Here we see an excellent parallel in the economy of use of materials in plants, and calculated efficiency in engineering.

3.4 Reinforced concrete, rod reinforcement and the peripheral placement of stem vascular bundles

Many stems in their earliest stages of development consist of a skin, a soft matrix of thin-walled parenchyma cells and a series of axial vascular bundles with some mechanical strength, situated near the periphery. In the herbaceous monocotyledons, the early stages of development of stems involve such axial strands in more than one ring. The outermost are often narrow in diameter, but those of the next layer in are often the widest and most robust. Later, in most dicotyledons, the vascular bundles become united in a cylinder by cambial development. This cylinder becomes in effect a single rod in the wood of trees, as the stem expands in diameter. Certain species of woody climbers retain separate vascular bundles (see later). In many monocotyledons the outer bundles may become embedded in mechanically thickened cells, thus developing a mechanical cylinder by a different route (Fig. 12).

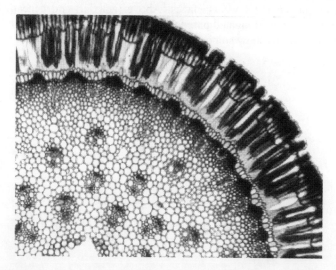

Figure 12: Cross section of part of a monocotyledonous stem showing concentration of strength in the outer skin (epidermis and cuticle) and an inner cylinder of fibres. Rod reinforcement is also provided. The dark, thick-walled cells projecting inwards from the epidermis add an extra pit-prop-like protection to the thin-walled cells of the photosynthetic tissue. The stomata are deeply sunken. The pit-prop cells also help reduce water loss, because their exposed walls are cuticle-covered. This plant grows in dry areas. Light photomicrograph.

There are some monocotyledons in which the vascular bundles remain separate from one another during the life of the stem. It is common for leaves that are held upright, and are modified to have a more or less circular or elliptical outline, to have separate vascular bundles.

This arrangement is reminiscent of concrete columns with metal rod reinforcement, and the mechanics of the two are probably similar. In stems, at each leaf insertion (node), the column may have a mechanically strong transverse partition, but this is by no means always the case. In a large number of plants the leaf base surrounds the stem at the node, forming a strong sleeve.

Climbers like *Vitis vinifera*, the grape vine and *Clematis* species, often retain separate vascular strands, even though these become radially elongated as the stem thickens with age. Between these are radial bands of thin-walled cells. When the stem becomes compressed as it climbs, the thin-walled regions are deformed, but the main conducting cells in the vascular tissue do not.

3.5 Tubes

Tubes or cylinders have been described above for stems, as a developmental stage as the stem ages or increases in diameter. It is common for the mechanical tissue to be concentrated towards the periphery, where it can provide considerable mechanical support, with an economy of material. People are often concerned about the possible reduction in strength of hollow trees, but providing the outer tissues are alive and flourishing, and sufficient wood remains at the points of branch insertion to hold them up, there may be little to be concerned about. Tubes are commonly applied in engineering because of their particularly efficient use of material in relation to strength. Fluted tubes and columns are also reflected in nature. The additional mechanical strength concomitant with fluting comes with the use of little additional material. Many columnar cacti are fluted columns.

3.6 Ropes and roots; central location of strengthening material

In their early stages of development, roots differ from stems, in that their vascular tissues are concentrated near the centre and not at the periphery (Fig. 13).

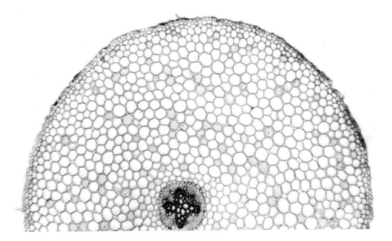

Figure 13: Part of a root cross section, showing the concentration of strength in the centre, like a rope. Light photomicrograph.

The main stresses on roots are not bending stresses, as stems, but pulling stresses provided by leverage of the stems. The mechanical strengthening of roots is, then, reminiscent of ropes, concentrated in the centre. Water plants may have roots growing in mud that lacks sufficient oxygen. In such cases it is common to find air-containing compartments that have gas exchange with other cavities formed in the stem and leaf, thus supplying oxygen for respiration. Is this an early form of air conditioning?

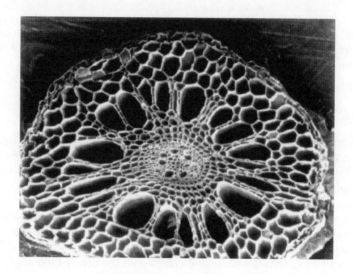

Figure 14: A root cross section, showing the concentration of strength in the centre. This is a water plant; the root also shows a ring of air chambers. Scanning electron microscope image (×60).

3.7 Gelatinous fibre, tracheids and inclined trunks

These were discussed in Section 2.3, on xylem.

3.8 Self shading – keeping cool

Temperature gain and the problems that can bring to a plant in hot sunlight are considered in the sections on surface features. However, the overall shape of plants, their gross morphology, can have a marked effect on heat regulation. It is no coincidence that the tall columnar form in many cacti, which are of South American origin, is paralleled in the tall, columnar form in some African euphorbias. As mentioned above, fluting of such columns confers additional strength. It also leads to self-shading, and hence less heat uptake. As the sun tracks, flutes cast shadows on the stem. Without them, more of the surface of the cylinder would be exposed to insolation at any one time. Close spiralling of leaves on columnar plants, and plants having closely packed opposite leaves set in pairs with those next above at 90° to them, also achieve a high degree of self-shading. Such forms are common in hot dry areas, and at high altitude in the tropics.

4 Surface features

4.1 Introduction

The surfaces of leaves and stems are subjected to all that the environment can throw at them. Plants growing in different environments show surface features that appear to be adaptations for survival and efficient functioning. Since there are great differences between conditions in dry deserts, exposed wet mountain tops and sheltered, dimly lit floors of tropical forests, for example, it is not surprising that the plants living in these habitats show a range of very different leaf surface features. Often a correlation can be made between environment and surface adaptation.

In this section leaf surfaces are concentrated on, but it should be remarked that stems, pollen grains, petals, and seeds, among other plant parts also have highly adapted surfaces, relating to their function. Petals, for example, may have cells adapted to assist pollinating insects to get a foothold and may be shaped in such a way that they produce colour by refraction.

Considering leaf surfaces: they must be mechanically adapted to meet environmental stresses, but translucent, to allow photosynthetically active radiation to pass through them to reach the pigment chlorophyll in cells beneath. In all but the wettest environments, leaf surfaces must also be capable of regulating water loss. This is brought about in most cases by the presence of a transparent outer layer, the cuticle, which retards water loss. This tends to be thinnest in species not normally subject to water stress, and thickest in those that are. The cuticle may also give added mechanical strength. It helps resist abrasion by blown sand particles, or in the case of some conifers, blown ice crystals. Cutin, the main component of cuticle, may permeate the walls of epidermal cells, or just the outer walls, in species confined to extremely arid habitats. Low availability of water to the plant can be induced by saline soils, so plants growing on these often show adaptations similar to those from dry habitats.

The main control of water movement is provided by stomata. These consist of a pair of guard cells (often kidney-shaped) with a pore between them (Fig. 1). The size of the pore is regulated by changes in shape of the guard cells. As hydraulic pressure is altered, the cells deform in a regulated way, aided by specialised, uneven wall thickening. The pressure in guard cells is normally under the control of the plant. This mechanism could be pursued in development of pressure sensitive valves.

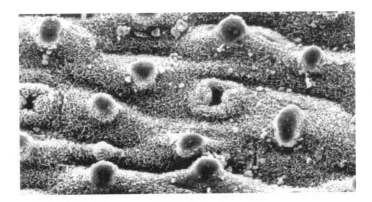

Figure 15: A leaf surface showing waxy particles, raised papillae on the epidermal cells, giving surface roughness, and deeply sunken stomata. Scanning electron microscope image (×300).

Often there is an additional layer of material on the surface of the cuticle, of a waxy nature, although other chemicals such as flavenoids are sometimes involved (Fig. 15). Surface wax may be smooth, or may show varying degrees of roughness. It may function in helping reduce water loss, but it has reflecting and other properties that will be considered later.

4.2 Boundary layer, wind speed and turbulence: morphological considerations

The outer surface of the epidermis, with its cuticle and usually, its surface wax, is the functional interface between the plant and the environment (Fig. 16).

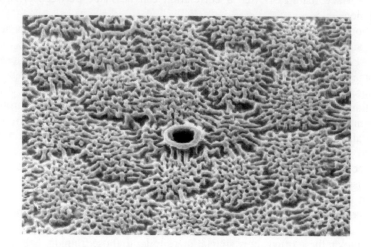

Figure 16: A leaf surface showing intricate raised ridges providing surface roughness. The deeply sunken stoma is surrounded by a raised cylindrical rim. Scanning electron microscope image (×300).

The texture of the surface can have marked effects on the flow pattern of air above it. The layer of air immediately next to the surface is termed the boundary layer. It is thinner over a smooth than a rough surface. The boundary layer consists of relatively still air, but surface roughness can lead to turbulent air flow. Water vapour loss through stomata tends to be more rapid when the boundary layer is thin. This is reflected in the observation that in plants growing in humid environments, sheltered from the wind, surfaces tend to be smooth. In the more exposed environments, for example hot, rocky hillsides, it is common to find leaves with quite rough surfaces.

Increased surface roughness can be brought about at the macro-level by leaf form, prominent ribbing and corrugated laminas, but often it is contributed to by micro-characters as well (Fig. 17). But it is at the micro-level that less obvious adaptations occur. The epidermis is composed of numerous small cells. In smooth-leaved plants, these cells each have a flat outer wall. Even a slight doming of this wall can increase surface roughness, but if the walls are produced into papillae, they have a marked effect on the boundary layer. The outer walls themselves may also be 'rough'. This fine roughness can be the product of fine striations and micropapillae. The scale of these features, and their apparent correlation with plants exposed to wind and heat stress, suggests that they play a part in modifying the boundary layer. They may also increase light and heat scatter (Fig. 17).

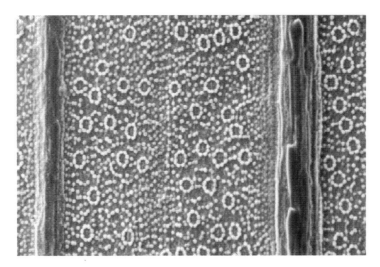

Figure 17: A leaf surface with micropapillae and raised veins, producing surface roughness. Each stoma is surrounded by a ring of micropapillae. The raised cells are over the veins. Scanning electron microscope image (×100).

Extreme disturbance of the boundary layer occurs when hairs are present. These often trap a layer of still air, which can provide insulation, both against heat and cold. Extreme hairiness is found in the plants of high tropical mountains, subjected to extreme diurnal temperature fluctuation. Some plants, such as mesembryanthemums, have balloon-like hairs, which when fully turgid allow gas exchange to occur between the air and the stomata beneath them. If water is short, the hairs partly collapse, press against one-another and effectively block gas exchange and water vapour loss. Hairs can have thick walls, and with effective waterproofing inserts (Caspian thickenings, a waxy band within the cell wall) near their bases, provide protection against water loss. In the other extreme, they might have thin walls, and thus be vulnerable to water loss – but equally capable of taking up water from mist or moist air. Hairs like these are common on plants growing high in the canopy of trees (epiphytes) in cool mist forests. The roots of epiphytes serve principally for anchorage; water is taken in by the leaves, aided by hairs like these, when present.

4.3 Heat and mass transfer, turbulence, and the boundary layer

4.3.1 Heat and mass transfer

As indicated above, the outer surface of the epidermis, with its cuticle and often, its surface wax, is the functional interface between the plant and its environment. Across this interface two major transport processes take place, critical to the well-being of the plant. These are heat transfer and transpiration, or mass transfer. While solar radiation is the heat transfer mode of immediate significance to the plant's metabolism, in fact conduction and convection also occur. These latter, together with transpiration, are intimately affected by the air movement across the plant surface, or using terms as in physics, by the fluid mechanics.

Now the laws of fluid mechanics, the Navier-Stokes equations, were established some one hundred and fifty years ago, and the phenomenon of fluid turbulence defined by the British mathematician/engineer Reynolds well before 1900. However, the highly non-linear partial differential form of the equations (let alone the further complication of involving turbulence) has meant that analytical solutions have been very few. It was not until the advent of high-speed digital computers post World War II, that numerical solutions of any practicality were feasible. In fact, because turbulence is fundamentally 3-dimensional and time dependent in character, it has only been over about the last fifteen years that commercial computer codes, exploiting the latest high-capacity computers, have been able to generate heat transfer and fluid mechanics data of *engineering significance*. In its turn, this has future relevance for plant studies, as complex leaf and plant geometries may be modelled using the same methods as for engineering design problems. The technology and terminology used by biologists in surface transport plant studies (for example, Foster and Smith [3] and Gurevitch and Schuepp [4]) are borrowed from the semi-analytical dimensional analysis traditionally used in engineering heat transfer. For example, Van Gardingen and Grace [5] point out: 'Firstly, g_{ca} may be calculated from the theoretical consideration of heat transfer, developed principally for the engineering sciences.' This quote, moreover, illustrates the 'one-world' rationale of this Chapter, Volume and Series, namely that universal scientific laws apply to both engineering and natural structures.

There are two pairs of regimes associated with fluid mechanics, namely laminar or turbulent flow, and forced or free convection. The boundaries between these regimes are not absolute: a 'transition region' occurs between laminar and turbulent flow, and there is 'combined free and forced convection' where both forced and free modes are of comparable significance.

The convective heat transfer alternatives arise from the cause of the fluid flow over the solid surface. If that flow is caused by some external means (in plant heat transfer a *wind* is the best example) the convection is said to be forced. If the flow is caused by density differences due to heating, and hence by buoyancy changes (gross air movement known as 'thermals' is a good example) then the convection is free (or natural). It can be readily appreciated that for low local wind speeds, and high local heating, combined convection is a distinct possibility in plant heat transfer.

The above discussion is paralleled by Jones (in [6] p. 58), with a detailed quantitative description given by Foster and Smith [3]. The latter includes consideration of combined convection, the alternative expression 'mixed convection' being used. There is not space here to discuss explicitly the various dimensionless groups used to correlate convection data, that is mainly the Nusselt, Prandtl Grashof and Graetz Numbers, but the Reynolds Number is addressed below in the context of turbulence and boundary layers. However, the almost extraordinary complexity of the various issues involved in synthesising engineering convection data may be seen, for example, from the review by Piva *et al.* [7] of combined convection in horizontal flows.

Heat and mass transfer are analogous transport processes, and in mass transfer, dimensional analysis leads to the Sherwood Number in place of the Nusselt Number for heat transfer. Again this is explained by Foster and Smith [3], who also give some correlation data for both processes taken from the engineering heat transfer text of Kreith [8].

4.3.2 Electrical network analogues

Another engineering heat transfer technique, less popular now that commercial Computational Fluid Dynamics and heat transfer codes are used in design problems, is that of the electrical network analogy. Conduction heat transfer in particular is completely analogous to electrical conduction (that is, 'heat flow' to 'current', and 'temperature difference' to 'voltage drop') and this leads to the concept of 'thermal resistance' (or its reciprocal 'thermal conductance') for layers of material through which heat passes. The various scientific molecular transport laws (of which Fourier's and Ohm's Laws for heat transfer and electricity form part) are quite elegantly and thoroughly presented by Jones ([6], pp. 48-57), being described by him (p. 52) in terms of a *general transport equation*. Now the analogy between thermal and electrical conduction extends to complete electrical networks, with a similar analogy for mass transfer, so we see a comprehensive water transport electrical network given by Jones ([6] p. 96) and thermal and gas transport 'networks' by Jones ([6] p.109) and in Van Gardingen and Grace ([5], p. 220, Fig. 13).

4.3.3 Turbulence

The above gives a brief overview of the application of engineering-related methods to plant surface transport processes. The question of laminar and turbulent flow is now addressed. A fluid (liquid or gas) deforms continuously when it is exposed to a shearing force. Up to a certain (critical) level the shear stress may be accommodated by the fluid sliding smoothly by molecular means only. This is termed laminar flow. The shear stress (τ) is related to the local velocity gradient (dV/dx) by a constant of proportionality (μ) which is a property of the fluid:

$$\tau = \mu \, (dV/dx).$$

Such is called a Newtonian fluid: if μ varies with 'shear rate' the fluid is termed non-Newtonian, and in addition μ can be temperature-dependent.

Beyond this critical level the fluid oscillates, and then forms fluctuations, the intensity of which increases with the imposed shear stress. This we call turbulent flow.

Reynolds, by carrying out experiments for liquids in a tube, found that the critical state could be expressed non-dimensionally for all fluids by (what we now call) the Reynolds number (Re) This is given by:

$$Re = \frac{\rho V d}{\mu}$$

where ρ is the fluid density, flowing with a mean velocity V in a tube of diameter d, with:

$$Re_{crit} \simeq 2,300.$$

Above and below this 'critical Reynolds Number' the flow is turbulent or laminar respectively, with an intervening transition regime between them. Transition to turbulence is affected by the surface character of the tube (its roughness) and by whether small fluctuations of a 'triggering' nature occur in the flow.

In his landmark Royal Society paper [9] Reynolds showed that by using the simple concept of an instantaneous velocity (u, say) he could express the 'locally averaged' velocity ($\bar{u}$) in terms of u, and a turbulent fluctuation u'

$$u = \bar{u} + u'.$$

This allowed the original form of the Navier-Stokes equations for fluid-flow to be re-expressed in a laminar-like form, but with extra stress-like terms (now called Reynolds stresses). In conventional turbulence modelling these terms are subsumed in a so-called turbulent viscosity (μ_{turb}) with the overall effective viscosity (μ_{eff}) expressed as:

$$\mu_{eff} = \mu + \mu_{turb}.$$

116 *Nature and Design*

In its turn, this concept allows a fresh set of transport equations to be formulated for turbulent flow, similar to the laminar flow equations in Jones's Table 3.1 ([6], p. 54). This set, together with parallel equations enabling μ$_{turb}$ to be quantified, has also been described in terms of a 'general transport equation' (compare [10], p. 10, bottom).

4.3.4 Eddy structure of turbulence

In the above, turbulence has been described in terms of time-dependent fluctuations about a mean, and a sensitive monitoring experiment can readily register the changes in the 'instantaneous' value, and deduce the mean turbulent quantity. These are u and ū respectively, as above, ū being termed more precisely as the 'ensemble average' of many experiments. However, more macroscopically, turbulence may be viewed as the behaviour of three-dimensional time-dependent eddies, whose sizes range (for internal flows) from that of the confining duct to those where extinguishing dissipation takes place, via fluid friction, to heat energy.

This is often presented in terms of an energy spectrum-wave number graph, as do Finnigan and Brunet ([11], p. 15) and Van Gardingen and Grace ([5], Fig. 8) in the context of the boundary layer above a forest canopy. Fig. 20 is re-presented from Finnigan and Brunet's Fig. 14.

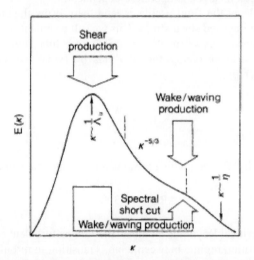

Figure 18: Schematic diagram of the energy spectrum E as a function of wavenumber k in a plant canopy. Extra processes that are peculiar to canopies are marked, as is the inertial subrange ($k^{-5/3}$): the dissipation range occupies wavenumbers greater than η, η being the Kolmogoroff microscale.

Figure 18, apart from the features 'peculiar to canopies' shows three *general* features of the eddy description of turbulence. Firstly, the high-energy-containing eddies are the largest and are where the turbulence energy is produced (termed 'shear production'). Secondly, this energy 'cascades down' a succession of smaller eddies, the cascade having a distinctly linear slope of $k^{-5/3}$. However, the cascade is described in turbulence terms as 'leaky' – in other words much of the energy disappears to dissipation before the end of the linear range. Thirdly, the whole process 'vanishes' by virtue of extinguishing of the smallest, fastest eddies by fluid friction to thermal energy (dissipation). This last dissipation region is characterised by eddy scales whose wavenumbers are close to the Kolmogoroff microscale, a theoretical description of dissipation.

This eddy description of turbulence also provides a convincing interpretation of transition from laminar to turbulent flow. Fig. 21 reproduces Fig. 2 of Voke and Collins [12]. (Turbulent Re is defined differently from channel Re, as in Fig. 1 of [12].)

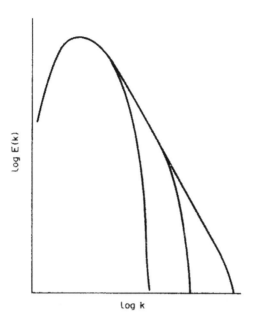

Figure 19: Typical mean kinetic energy spectra at low, medium and high turbulent Reynolds numbers. Spectra are normalised to the production peak level.

The key difference in shape is the extent of the inertial subrange (see title, Fig. 19). At low turbulent Re (around transition) it is absent, and with increasing turbulent Re, the length extends to be more like Fig. 19. What this means is that near transition, the eddy size which directly dissipates to thermal energy is of the same order as the 'production' eddy range, and there is no energy cascade. Also, as the Re increases so the dissipation eddy range becomes increasingly small (η, Kolmogoroff microscale, as in Fig. 19).

In fact, for biological boundary layers, this aspect is quite relevant, as the 'low turbulent Re' shape of Fig. 19 would correspond to low wind speed cases, and the developing inertial subrange to higher wind speeds.

4.3.5 The turbulent boundary layer

What is termed the boundary layer arises because of the zero velocity of fluid immediately adjacent to a solid surface; this is known as the no-slip condition. There is a consequent region where the fluid flow velocity varies between zero and its free-stream value: this region is termed the boundary layer. Most plant, leaf or canopy surfaces may be viewed as living world extensions of the engineering standard case of a boundary layer development of external flow over a flat plate. In discussing this, it becomes clear that almost all the preceding technology is involved. Fig. 20 shows boundary layer development for a flat surface over which a wind blows, and is reconstructed from Fig. 2.2 of the engineering text by Rogers and Mayhew [13] and Fig. 3.3 of Jones [6].

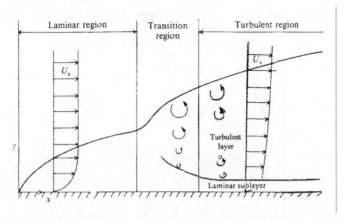

Figure 20: Development of turbulent boundary layer for a flat plate. U_s is free stream, or wind, velocity.

Several features are distinguishable. Firstly, irrespective of wind speed, turbulence energy levels (expressed as kinetic energy percentage ratios to the kinetic energy of the main stream velocity) and exact surface configuration, there is an initial laminar boundary development. This then undergoes transition to a turbulent boundary layer, encompassing a laminar sublayer. In this instance the Reynolds Number, defined with a free-stream velocity V_s instead of mean tube velocity V, and a characteristic dimension 'x' (distance from 'flat plate' entry instead of tube diameter 'd') gives a critical value of around 500,000 (Rogers and Mayhew [13], p. 537) for standard controlled conditions. Within the turbulent boundary layer there is a range of eddy sizes reflecting the energy spectrum of Fig. 20.

4.4 Overall assessment

In this section, the engineering knowledge relevant to typical biological studies has been briefly reviewed. It might be queried whether this knowledge is strictly necessary to understand the biological implications. However, because often air movement is locally rather slow for plants, the situation can well be transitional (as opposed to definitely laminar or turbulent), can be intermittent (time-dependent variation of laminarity and turbulence), with heat transfer in combined free and forced mode (as opposed to definitely free or forced). The biological case then, represents the most general engineering problem. Of equal importance is the nature of the plant or leaf surface. Whereas traditionally engineering surfaces are rigid and smooth, with designs made on these bases, leaf surfaces as mentioned above have irregularities such as hairs, veins and transpiration, and move in the wind. In fact, with the advent of more sophisticated engineering materials, it is becoming possible to mimic the kinds of devices used in biology.

Further, space has not allowed for the question of *particle* transport to be addressed. In fact, this can be an integral part of the engineering design problems, especially in chemical engineering situations. The importance of understanding biological particle and gaseous transport can hardly be overstressed, as it represents a crucial contribution to environmental studies, including global warming. A number of contributions to the 1979 Edinburgh

Conference of the British Ecological Society (reference [15]) illustrate this. Jones ([6], pp. 313-324) gives a quite comprehensive account of the global warming issue, providing many quantitative data in his Tables 11.3 and 11.4. In fact, the whole section may usefully be compared with a more engineering-oriented treatment of the same topic given by Houghton [16].

As a concluding comment, it should now be possible to use realistic models of symmetric plant shapes such as cacti (Fahn and Cutler [14], p. 95) within an overall computational fluid dynamics, heat and mass transfer treatment. This, in conjunction with validating experimental data, should lead to a more thorough understanding of the transport processes involved in leaf surface studies.

4.5 Leaf margins - dentate and air flow

The edges of leaves, their margins, can be entire, or variously toothed (dentate). Dentations have marked effects on reducing turbulent air flow, and are thought to help maintain the integrity of the leaf margins in strong winds.

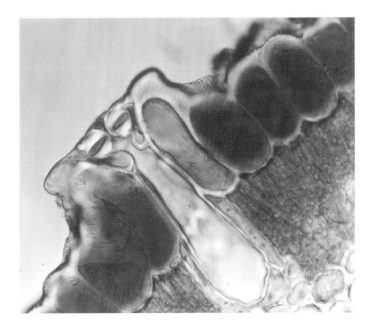

Figure 21: Raised stoma in stem epidermis of a plant from a dry habitat. The epidermal cells are thick-walled. They contain tannins that in life are translucent and serve as a UV screen. The tall narrow cells below them have chloroplasts which are essential to photosynthesis. Cross section, light micrograph.

4.6 Chimneys and prairie dogs

Prairie dogs along with some other burrow-dwelling rodents commonly construct mud chimneys at the entrances to their burrows. When of different heights, these influence the air flow and ventilation through the burrows, using the venturi effect. Some plants, exposed to extreme periods of high temperatures and low water supply have similar chimneys above their stomata. Naturally there are several orders of magnitude difference in size! In plants, the chimneys are developed by extension of epidermal cells surrounding the stomata (Fig. 21), or in a few examples, are formed from the surface wax. In plants that have extensive protection from water loss, it may be that even when water is in adequate supply, it is difficult to maintain enough evaporation from the leaves to drive the transpiration stream. In these circumstances, if the stomata at the bottom of the chimneys are open, external air flow above the chimneys could cause reduced pressure in them, enhancing the flow of water vapour from the plant.

4.7 Reflection: shiny cuticle, hairs, wax, self-cleaning

The heat load can be very high for plants growing in areas of high insolation, particularly at higher altitudes or in arid to semi-arid regions. Self-shading can only play a part in helping reduce heat load. Ground surface temperatures can exceed 40°C. The plant body in hydrated plants must be cooler than this, or problems occur in their internal chemistry, leading to death. There are species, like the resurrection plant, *Anastatica*, which dehydrate, and survive high temperatures during drought, only to unfurl and green up rapidly on wetting.

Some surface features appear to reduce the effect of insolation. A very shiny cuticle, accompanied by smooth wax can act as an effective reflector. This effect is often noticed in photographs taken of some species in bright sunlight, when the leaves appear white.

Hairs, as mentioned above, may trap an insulating air layer, but they may also, by their pale colour, reflect light and heat. A similar effect can be produced by a rough waxy coating. The wax bloom on some species can make the plant look glaucous, or even white.

It has also been shown that wax in certain crystalline configurations prevents wetting of leaves. The 'lotus' effect researched by Neinhuis & Barthlott [17] describes how the surface wax can cause rounding up of water droplets, which effectively wash the leaf surface and keep it clean. Some surface waxes of different crystallinity erode and wash off, taking dust and spores with them. They are replenished by formation of new wax from within the leaf. These observations have led to the successful commercial development of self-cleaning paints, that do not rely on toxins to prevent fouling.

5 Windows and light pipes

Some plants, of which *Lithops* 'living stones' and some *Haworthia* species are examples, grow in arid environments, with only the tips of the leaves above ground. They have clear, translucent leaf tips, like windows. The main blade of the leaf is thick and succulent and either entirely or mostly underground. Some other haworthias with windows may have leaves in a tightly packed whorl on a short stem, so that in effect only their tips are visible. Light passing through the window is conducted through axially elongated thin-walled, more or less translucent cells, but with sufficient scatter (by transverse walls) to illuminate 'underground' chloroplasts. These are natural fibre optics. The chlorophyll surrounds the central light pipe, and is illuminated, by this internal mechanism.

6 Light acceptors and internal mirrors

Plants living on forest floors, in dense shade, have to survive in conditions almost diametrically opposite to those described above. They have to be extremely efficient in light collection. The relatively humid, windless environment enables species with large, thin, relatively unprotected leaves to thrive. They are often very chlorophyll-rich, and appear dark green.

Shafts of sunlight may penetrate the leaf canopy above; their angle changes during the day. Light reaching a leaf surface at 90° is less likely to be reflected than oblique light. Many of the species adapted to this environment have means of presenting to light a high proportion of their surface at 90°. Some lower plants, such as *Selaginella* species, achieve this by having highly domed epidermal cells, containing chloroplasts. As the sun tracks, each cell can present part of its surface normal to the source. Other species, for example from among the begonias, have a surface in which the epidermal cells develop together to produce a series of multicellular domes, with much the same effect. Undulating leaf surfaces are also common.

In general the leaves themselves are thin. This enables the chloroplasts to be presented in one to three cell layers into which the light can penetrate. Each layer effectively shades the next one in. Most interestingly, when viewed from the underside, many of these leaves adapted to low light interception appear purple. This is because they have an inbuilt sheet of anthocyanin containing cells which act as reflectors. The anthocyanin may be in the lower epidermis, but more commonly, it is in an inner layer, immediately below the chlorenchyma. This means that light striking the upper surface of the leaf, which would otherwise pass through the leaf, is reflected back for a return visit to the chlorenchyma. This parallels the heat/light reflecting membranes applied to windows in buildings, but makes efficient use of the reflected energy.

7 UV absorption

On the mountain tops, species in thin air, and subject to high insolation, have to contend with very high levels of damaging UV light. The chloroplasts are bleached and rendered ineffective if they intercept too much UV light. It has been found that some species have evolved natural sun screens, often involving polyphenols, filtering the UV. These substances may be in the epidermal cells (Fig. 21), or in a layer or two of otherwise colourless cells beneath the epidermis but above the chlorenchyma. While in the cells the substances may be translucent and colourless, but sometimes they are yellow to brown, and could have a direct blocking effect on light, reducing its intensity. There is considerable commercial interest in finding out how these chemicals work, and their structure.

8 Fire retardants

Plants in many parts of the world are subjected to natural or 'wild' fires as part of their normal existence. Some, in desert regions, produce essential oils. In the heat of the day these form a haze over the vegetation. They burn very quickly, and may carry the fire over the plants, without it becoming locally hot enough to burn extensively or kill the woody stems above or below ground.

The bark of some species contains large quantities of crystals of calcium oxalate. At certain temperatures, this breaks down and in due course produces carbon dioxide, with its fire dampening effect. Provided the bark is thick enough to insulate the cambium from damage, the tree may survive to sprout again. Calcium oxalate crystals are a component of the better fuelwoods, which burn slowly and hot. The survival value here is difficult to understand, but the information is of value to those wishing to select species for good burning properties! Fire retardants for fabrics involve similar substances.

122 *Nature and Design*

9 Preservatives

Most of the chemicals used by plants to prevent the growth of bacteria or fungi, or make them distasteful or poisonous to grazing animals, are of little interest to engineers. They are, of course, of medicinal or agricultural interest, though. But there is one component of some leaves and woods that can be of importance in construction. This is opaline silica, in the form of silica bodies. These small abrasive particles in woods may reduce the value of the timber because of saw wear. However, their resistance to wood boring organisms, presumably by blunting their mouthparts, makes them ideal for use in submerged conditions, particularly in the tropics.

Silica bodies in leaves are common in parts of the world where vegetation co-evolved with grazing animals. The abrasive silica wears their teeth. Such animals have evolved continuously growing teeth, and the battle goes on.

10 Floating devices and air chambers

Many aquatic plants contain buoyancy tanks. The internal septa, multicellular walls bounding the air cavities are often very thin. Individual cells are star-like, efficient in the use of materials, and permeable to gas flow. In bulrush, *Typha* species, the upright leaves are divided into chambers very reminiscent of bulkheads in supertankers. They are the model of economy of use of materials in producing a tall, mechanically strong structure. The upright, flattened blades are twisted. This adds to their strength, and improves wind spillage, thus reducing the impact and potential damage of strong winds. Architects of tall buildings take note.

Buoyancy may be achieved by using trapped air, rather than by having air-filled chambers. Water lettuce, *Pistia*, for example, has a layer of closely arranged, hydrophobic hairs on its upper surface. It is difficult to wet or sink.

Figure 22: Drawings of the original Velcro-like fastener. The hairs in this plant are flat to the stem surface. Each has a pair of large, thin-walled transparent and window-like cells. These are surrounded by a frame of thick-walled cells, with complex, recurved micro-hooks, shown in detail in the small drawing. They zip the hairs up round their edges to form a flat sheet. The sample of four hairs shows this arrangement at a lower magnification. The single hair is approximately 100× life size.

11 Velcro

In closing, we often think that we are masters of technology. As has been indicated above, plants often got there first (before the animals, too!). The inspiration for Velcro came from the hooked hairs of burdock fruits. But there is an even more convincing example in nature of the original zip/Velcro. This is found in the stem hairs of a unique Australian plant, *Meeboldina*. These hairs are composed of two large transparent cells, through which light can pass. They have a short stalk, and their diamond-shaped outline is defined by a rim of thick-walled cells. These cells are equipped with numerous short, reflexed hooks. The hooks of adjacent hairs interlock, effectively zipping them together (Fig. 22).

12 Conclusion

Engineering and technology have gone a long way by using empirical methods. But there could still be many examples from nature that might prompt the development of new materials and new concepts.

Acknowledgement

The section on turbulence and the boundary layer was written in conjunction with MW Collins, the Series Editor. DFC and MWC hope to give a more comprehensive treatment of the issues in a future publication.

References

[1] Fahn, A., *Plant Anatomy*, Fourth Edition, Pergamon Press: Oxford, 1990.
[2] Cutler, D.F., *Applied Plant Anatomy*, Longman: London, 1978.
[3] Foster, J.R. & Smith, W.K., Influence of Stomatal distribution on transpiration in low-wind environments. *Plant, Cell and Environment*, **9**, pp. 751-759, 1986.
[4] Gurevitch, J. & Schuepp, P.H., Boundary layer properties of highly dissected leaves: an investigation using an electrochemical fluid tunnel. *Plant, Cell and Environment*, **13**, pp. 783-792, 1990.
[5] Van Gardingen, P. & Grace, J., In: *Advances in Botanical Research*, **18**, ed. J.A. Callow, Academic Press: London, 1991.
[6] Jones, H.G., *Plants and Microclimate*, Second edition, Cambridge University Press: Cambridge, 1992.
[7] Piva, S., Barozzi, G.S. & Collins, M.W., Combined free and forced convection in horizontal flows: A review (Chapter 8). *Computational Analysis of Convection Heat Transfer*, eds. B. Suncten & G. Comini., WIT Press: Southampton and Boston, pp. 279-359, 2000.
[8] Kreith, F., *Principles of Heat Transfer*, Third edition, Intext Educational Publishers: Scranton, Pa., 1973.
[9] Reynolds, O., On the dynamical theory of incompressible viscous fluids and the determination of the criterion. *Phil. Trans. Roy. Soc., A*, **186**, pp. 123-164, 1895.
[10] Collins, M.W. & Ciofalo, M., Computational fluid dynamics and its application to transport processes. *J. of Chemical Technology and Biotechnology*, **52(1)**, pp. 5-47, 1991.
[11] Finnigan, J.J. & Brunet, Y., Turbulent airflow in forests. (Chapter 1), *Wind and Trees*, eds M.P. Couts & J. Grace, Cambridge University Press: Cambridge, 1995.

[12] Voke, P.R. & Collins, M.W., Large-eddy simulation: retrospect and prospect. *Physico-Chemical Hydrodynamics*, **4.2**, pp. 119-163, 1983.
[13] Rogers, G.F.C. & Mayhew, Y.R., *Engineering Thermodynamics*, Fourth Edition, Prentice Hall, Pearson Education: Harlow, U.K., 1992.
[14] Fahn, A. & Cutler, D.F., *Xerophytes, Encyclopedia of Plant Anatomy*. Borntrager: Berlin and Stuttgart, Germany, 1992.
[15] Grace, J., Ford, E.D. & Jarvis, P.G. (eds), *Plants and Their Atmospheric Environment*. (See in particular Chapters 1, 7, 8 and 10.) Blackwell: Oxford, 1981.
[16] Houghton, J., *Global Warming*, Lion Publishing: Oxford, 1994.
[17] Neinhuis, C, & Barthlott, W., Characterization and distribution of water-repellent, self cleaning plant surfaces. *Ann. Bot.*, **79**, pp. 667-677, 1997.

CHAPTER 7

The tree as an engineering structure

D. Hunt
South Bank University, London, U.K.

Abstract

This paper discusses the tree from the engineering point of view, in particular showing how the tree retains a memory of its environment during its life. The paper lists the main requirements of the tree as a living organism. It is pointed out that all adjustments to allow for changes in the tree's environment must be made in new growth, most of which is made at the perimeter of the tree. The structure of the tree is explained, at the various levels of scale, and it is shown how its various features are beneficial from an engineering point of view, and in particular the avoidance of either total fracture or the production of minor fissures into which harmful fungi or other organisms can penetrate. By comparison with the tree as a structure, some man-made wooden structures are discussed from the point of view of the overall philosophy and some of the structural details.

1 Introduction

> *[The girl began playing the lute in the garden...] "At this touch the lute shivered and moaned, suddenly remembering its life and destiny: it recalled the earth of its planting, the waters of its refreshment, the places where its stem had silently grown, the birds of its hospitality, the woodcutters, the clever craftsman, the varnisher, the ship which carried it, and all the fair hands between which it had passed".* [From "The Thousand Nights and One Night" [1].

Whilst the above quotation from the Arab classic may be a poetic exaggeration of the properties of wood, it nevertheless reminds us that the tree is a living organism. As such, it reacts to various stimuli and so bears a 'memory' of its environment throughout its life. This environment is continuously changing, not only with changes in the weather and the seasons, but also with changes in its own size and shape and those of neighbouring trees, and also in damage that it might sustain from inclement weather, animals (including humans), insects and fungi. Moreover, evolution has equipped the tree to react in an optimum way to most of the environmental conditions it is likely to meet, so as to promote its strength and lengthen its life.

Thus, uniquely among plants because of its long life, the tree structure is a record of its history and that of its environment over a period of many years, decades, or even centuries.

What are these stimuli and environmental conditions that the tree must withstand? The list of basic requirements in the life of a tree must include ingestion of sufficient water and nutrients from the soil, absorption of sufficient light from the sun, sufficient anchorage at its base to retain its upright position, and sufficient strength and stiffness to resist the forces of gravity and the wind. These requirements must be met, regardless of any changes in the tree's environment.

2 Structure and growth of a tree

Before discussing how the tree has been designed to cope with these conditions or with changes in them, the reader should be reminded of some basic facts about a tree's growth mechanism and internal structure.

2.1 Growth

The mode of growth of the tree above ground level is most strongly governed by three, sometimes conflicting, controlling factors [2]. *Apical dominance* ensures the dominance of the main trunk or its replacement. The result is that side branches take a subsidiary role, generally growing away from the main trunk. *Phototropism* ensures that the leaves obtain the maximum amount of sunlight for photosynthesis, by arranging for the branches to grow towards the light, in whichever direction that happens to be. *Negative geotropism* ensures that the tree grows upward, against the pull of gravity. In the case of side branches, this means bending upwards, with the formation of special 'reaction wood'. Obviously these three controlling factors are usually working in concord with each other, although in some cases there may be conflict between them.

Although the tip of the tree grows longitudinally, as is well known, the existing trunk and limbs expand radially, with the new growth being added to the outer surface. This allows the outside of the tree to react to any changes in its environment. The actual growing part is known as the cambium, situated between the wood and the bark, in which growth occurs by radial division of the cells. As the diameter and perimeter dimensions increase, the number of cells also increases, by tangential division.

The effect of the growth mechanism is that internal stresses are developed in the wood. Generally the magnitudes of the growth stresses are lower in softwoods than in hardwoods. In the latter the longitudinal stresses may be as high as 10 MPa in large-diameter European hardwoods, although considerably lower in softwoods [3]. In the tangential direction of the surface these stresses are compressive, whilst in the longitudinal direction they are tensile. The surface compression stresses may be as high as 2.8 MPa in hardwoods, but lower than this on softwoods [3]. The condition of stress equilibrium means that these surface stresses are balanced by the core stresses: core tension in the radial and tangential directions and core compression in the longitudinal direction.

2.2 Chemistry, transport, storage

During the spring, in temperate climates, large amounts of sap rise from the roots through the outer layer of the tree towards the crown: this sap consisting of water and nutrients that have

been stored in the roots or in specialised storage cells, or that are newly absorbed from the soil. This sap is required to promote new spring growth. The leaves of the tree absorb energy in the form of sunlight, which is used together with the nutrients to build new plant material. Particularly during autumn, some of these nutrients are stored in specialised storage cells or returned through the outer layer of the tree to the roots for storage until the following spring. This process then defines some of the requirements of the living tree: a root system that can absorb water and nutrients from the soil, free passage of water and nutrients between the roots and the crown, and access to sunlight.

2.3 Wood macro- and microstructure

The previous two paragraphs describe the gross characteristics of a tree trunk or branch, having the characteristic annual growth rings, as shown for a softwood in Figure 1(a). The structure of one softwood growth ring, magnified, is shown in Figure 1(b). The microstructure of a broad-leaved tree, or *hardwood* is shown in Figure 2.

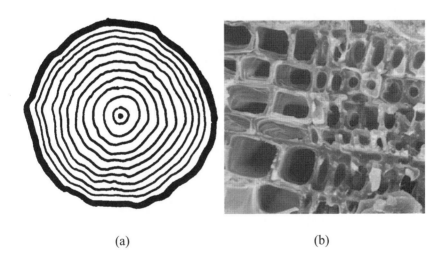

(a) (b)

Figure 1: Structure of a softwood. (a) Cross section of trunk showing growth rings; (b) Microstructure showing previous year's latewood on right, with new season's earlywood on left.

It can be seen that the wood cells (known as tracheids in softwoods) consist of hollow tubes. These are effectively the 'fibres' in a 'fibre-reinforced composite'. In the springwood (earlywood) of the softwood, the tubes are of larger diameter, about 40 pm, with thinner walls, about 8 pm, but in the summerwood (latewood) the cells are of smaller diameter but with much thicker walls. Typically the earlywood, when dried, has a density of 300 kg/m^3, and the dried latewood has a density of about 800 kg/m³. With wood substance having a dry density of about 1500 kg/m³ this means that dry earlywood contains about 80% of air volume, and dry latewood about 45%. In the living tree this space contains liquid.

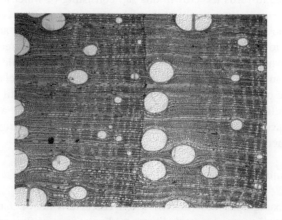

Figure 2: Microstructure of hickory, *Carya* spp. The radial direction is shown horizontal. Reproduced by permission of the Building Research Establishment Ltd.

The tubular structure has a large influence on the structural properties of the tree and of the dry wood. The lengths of the softwood tracheids vary from about 2 mm to 5 mm, giving an aspect ratio of approximately 100:1; and they have tapered ends, as shown in Figure 3. Transport of liquids from one cell to another is possible through pits in the cell wall. The woody material of hardwoods is different from softwoods, as can be seen in Figure 2, in that they contain larger-diameter open *vessels* for the transport of liquids, whilst the thick-walled fibres are specialised for their structural functions and are about 1 mm to 2 mm in length. Both types of woods contain specialised nutrient storage cells (*parenchyma*).

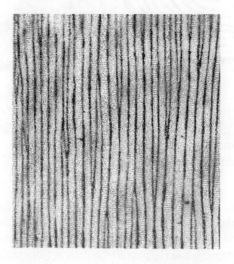

Figure 3: Longitudinal section showing softwood tracheids and some vertical parenchyma.

2.4 Cell walls

At a still smaller level, we must consider the structure of the walls of the cells. A typical cell-wall structure is indicated in Figure 4, where it can be seen that the wall consists of several layers of structural microfibrils, each layer being effectively a fibre-reinforced composite. The cell wall is therefore a kind of multi-layered laminate of fibre-reinforced sheets. From a structural point of view the thickest and most important layer is the S2 layer, whose main fibre direction spirals around the cell at a range of spiral angles, but generally between 10° and 20° to the cell axis.

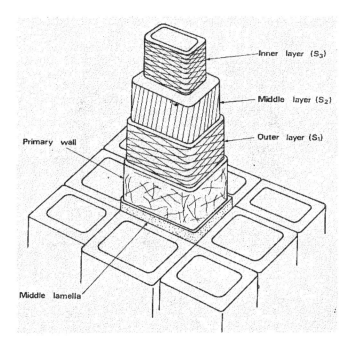

Figure 4: Simplified structure of the cell wall showing orientation of the microfibrils in each of the major wall layers. Reproduced by permission of the Building Research Establishment Ltd.

The structure of the cell wall, and especially the spiral angle of the S2 layer, has a strong influence on the properties of the wood. In young trees, up to an age of about ten years, the spiral angle of the 'juvenile wood' tends to be large, but the new 'mature wood' built up in later years tends to have a much smaller spiral angle. In situations where the living tree is subject to a large stress, the newly produced wood tends to have abnormal wall thickness and spiral angle. In softwoods, this is known as 'compression wood', which grows in areas of high compressive stresses. In compression wood the cell walls are thicker and shorter and have large spiral angles. In hardwoods, 'tension wood' is produced in regions of high tensile stresses. This has the microfibrils at a very low angle, and there are differences in chemical composition.

2.5 Ultra-structure

At a still lower level of scale, the individual microfibrils form a mat in which there is a predominant direction, but with some variation about the mean, as shown in Figure 5. The individual microfibrils consist of bundles of molecules of cellulose. These bundles are partly crystalline and partly amorphous, and a simplified view of the structure at this scale could be that the bulk of the cellulose, comprising the core, is held in place by a 'matrix' consisting of non-crystalline cellulose, semi-crystalline hydrophilic hemicellulose and amorphous lignin. The lignin also protects the hydrophilic cellulose and hemicellulose from the softening effects of water.

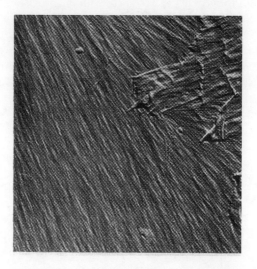

Figure 5: The S2-S3 transition lamellae: at the lower left is the S2 layer, with the S3 layer at the upper right [5]. By permission of Springer-Verlag. Photo: H. Harada.

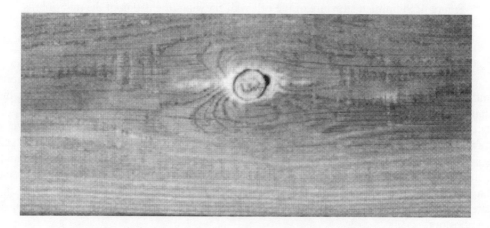

Figure 6: Knot in a softwood. Cracks indicate grain orientations.

2.6 Grain orientations at knots

Although the orientation of the cells is normally axial with respect to the trunk or branch, the orientation changes locally in the region of the joining of a branch, where a knot such as that shown in Figure 6 is formed.

3 The mechanical properties of the wood structure described above

When an engineer designs a structure, he normally considers all possible mechanisms of failure, and then ensures that the structure is sufficiently protected against each type of failure so that it is fit for the use for which it is designed, with the provision of a safety factor that is appropriate to the risk. Such an approach will be taken in this study.

The effects must be considered at the various levels of scale. To avoid extensive discussion, it will be mainly the trunks of *softwoods* that will be considered here.

When considered as a structure, the tree consists of a series of joined cantilever beams. The individual beams, and therefore the structure as a whole, are quite flexible. Such a structure therefore has good resistance to dynamic loadings, or in other words, wind loadings. The flexibility gives improved impact resistance and also allows smaller members to bend into the wind for reduced wind resistance.

On the macroscopic scale level, since the growth rings consist of alternative layers of high and low density, one may consider the growth rings as a series of stiff strong and dense concentric structural cylinders, separated and bonded together by low-density and relatively weak layers. Moreover the structural cylinders, of alternating latewood and earlywood, are also anisotropic, since their fibrous cells are aligned parallel to the trunk of the tree.

3.1 Strength

Most of the strength properties can be attributed to the structure at the microscopic scale level. As shown in Figure 1, the cells consist of tubes. In the earlywood these have thin walls and a large central space (lumen), and have a cross section of shape somewhere between hexagonal and rectangular. In latewood the tubular cells have thick walls and small lumens, with generally a rectangular shape of cross section. It is well known that under axial compression loads the walls of a tube will fail by buckling, and that the resistance to buckling increases strongly with the wall thickness. In wood the result is that collapse by buckling, and hence compression failure, takes place at a lower stress than tensile rupture. As would be expected, the large thin-walled earlywood cells buckle much more easily than the small thick-walled latewood cells. The axial tensile strength, which is not controlled by buckling, may be about five times the axial compression strength [4], depending on the conditions. In reality, neither wood in ordinary structural use nor wood within a tree is normally subjected to pure tension, but usually to a combination of bending and compression. The result is that when bending failure takes place, it occurs on the compression face of the 'beam'. In the case of a tree trunk, the natural loads are a combination of compression due to the weight of the tree, and bending due to wind and any leaning of the tree. In the case of branches, the bending component is a larger proportion of the total loading, since the weight is less. In fact, often the tree may *appear* to have failed in tension, even if the beginning of failure was in compression. Compression buckling tends to move the neutral axis of bending nearer to the tension face, thus increasing the tensile stresses until tensile failure occurs.

In addition to failure resulting from axial stresses, in other words tensile or compressive stresses running along the grain direction, wood can fail in other loading modes. The two other most important modes of stressing are normal stresses perpendicular to the grain, and shear stresses in which the shearing plane is parallel to the grain. In both cases there is much variability, and there are big differences between failure strengths according to whether the crack is in the radial plane or the tangential plane. Typical tensile strengths of softwood perpendicular to the grain are 1.5 to 2 MPa [5], which is about 1/50 of that parallel to the grain [4]. Especially in softwoods, the low-density earlywood provides a crack plane with low fracture resistance. On the other hand the elastic modulus in this direction is also low, typically 1/15 to 1/30 of that in the axial direction [4], so that dangerous concentrations of stress at discontinuities are also lowered. The compression strength perpendicular to the grain direction is not a relevant property since the tubular structure of the cells means that the first failure takes place by local buckling. This type of damage can be repaired by the tree with time. However, the local buckling of the cells, since they also act as axial load-bearing members, means that there is a weakness when loaded in axial compression.

The shear strength parallel to the grain direction is also quite small, the failure stress typically being about 1/14 of the axial tensile strength. Typical shear strengths of green softwood are 4 to 6 MPa [6]. However, the most important loading mode of tree members is usually bending, and it may be noted that the maximum shear stresses that result from bending occur at the neutral axis of a beam. In the case of a tree member of circular section this maximum shear stress is located at the maximum width. Moreover, on this section the growth rings are orientated perpendicular to the neutral plane, so that the weak earlywood is closely supported by a neighbouring latewood region. The shear stress distribution is shown in Figure 7.

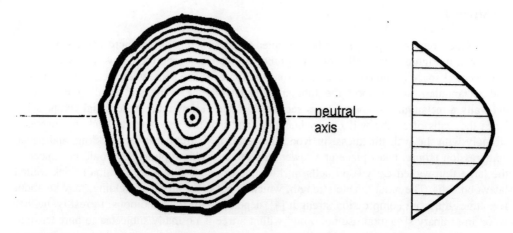

Figure 7: The maximum shear stress is located at the neutral axis of the member. The shear-stress distribution is shown on the right.

3.2 Crack avoidance in tension

Examination of axial tensile fractures shows that on the macroscopic scale there is gross tearing accompanied by considerable shear failure, giving a very jagged fracture appearance. If axial tensile failure is to occur, its mechanisms are largely controlled by the crack blunting and deflection mechanisms within the wood. Part of the reason for this is the **crack-deflection** mechanism, especially in 'green' wood (i.e. the living tree), whereby the tensile strength of the wood perpendicular to the grain is so much less than that parallel to the grain that the front of the crack which is running across the fibres tends to be deflected relatively harmlessly parallel to the grain, by fibre separation as shown in Figure 8. This process is further helped by the low shear strength parallel to the grain. On the microscopic scale there is a **crack-blunting** mechanism whereby the microfibrils can be separated as the result of the tensile component of stress perpendicular to their spiral alignment. The effect of microfibril separation is to change the microfibrillar orientation from the normal spiral arrangement to a straighter, axial arrangement within the lumen (central space), thus relieving the concentrated stress at the crack tip. This could be considered as a mechanism for increasing the 'work to fracture', which is well known as an indicator of toughness.

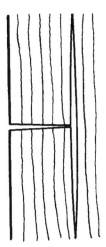

Figure 8: Crack blunting by fibre separation.

Resistance to dynamic loading can be increased by flexibility: it being well known that a flexible structure is better able to absorb energy than an otherwise equivalent stiff one. This particularly applies to members of small section such as young trees and thinner branches. This is discussed in Section 3.4.

3.3 The effects of growth stresses on bending failures

The overall effect of the much greater tensile strength than compression strength of the wood material, coupled with the generally greater compressive stresses caused by the weight of the standing tree, would mean that the first failure is likely to be by compression. This especially applies to the perimeter of the tree, where a bending load is added to the 'weight' loading. The tree guards against this by means of the growth stresses mentioned in the previous section: with axial tensile growth stresses in the outer layers balanced by compressive growth stresses in the core layers.

3.4 Stiffness: juvenile vs. mature wood

In addition to the strength characteristics, the stiffness characteristics of the tree are also important. As a general rule, a small sapling needs to be able to bend in the wind, and especially in windy gusts that may be considered as mild impact loading. On the other hand the mature tree needs to stand firm against the wind and against impact loading. Whilst the macroscopic scale level of the tree structure (the concentric layers of earlywood and latewood) has a very important effect on the stiffness of the tree, the main influences come at the cell-wall level. As described above, the main layer of the cell wall, the S2 layer, consists of a unidirectional-fibre laminate whose fibre direction spirals around the cell at an angle φ. Each cell wall is attached to the wall of a neighbouring cell, which spirals in the same rotation direction, thus giving at any point of the composite wall a pair of laminates attached to each other at angles of $+\varphi$ and $-\varphi$ respectively. The stiffness of this composite sheet for loads in the axial direction evidently depends on the spiral angle, φ: the larger the angle, the lower the stiffness. It is found that *juvenile wood*, meaning the wood of trees less than about ten years old, has a larger spiral angle than the new wood of older trees, although the change from juvenile wood to mature wood is gradual, taking place over a period of years. Since new wood is laid down at the circumference of the tree, this means that the juvenile wood in an older tree forms a central cylinder surmounted by a cone; and this part is surrounded by a cone of mature wood, as shown in Figure 9. This arrangement makes the parts of the tree trunk that are older than about ten years fairly rigid, while the thinner top and the branches are still flexible.

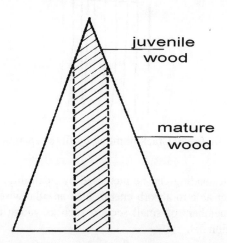

Figure 9: Juvenile wood and mature wood regions of tree trunk (taper exaggerated).

3.5 Reduction of stress concentration by change of grain direction

The geometrical configuration of the joint where a branch joins the main trunk inevitably results in a stress concentration, and such stress concentrations are always weak points in structures. However, the re-orientation of the grain direction in the region of knots and branch connections means that the effective elastic modulus in the axial direction is lowered in that region. For example it can be calculated that for a 20° change of grain angle the effective elastic modulus, E_θ, of spruce is lowered by about 30%. Since this knot region has adjacent regions in which the normal longitudinal modulus E_L applies, the assumption that the strains are uniform across the cross-section of the trunk means that the stresses are proportionately lower in the region of the joint. This has the effect of counteracting the geometrical stress concentration.

4 Tree geometry

The geometry of the tree, and the tree's mechanisms of adjustment of this geometry, have been extensively studied by Mattheck and his group of researchers, and published in a valuable book, *Design in Nature* [7]. Their research method consisted of using an iterative finite-elements procedure to optimise the geometrical forms of trees under various conditions from the points of view of structural safety. In particular they were looking for the *minimisation of stresses* in regions where failure was possible or likely, and especially in matching the stresses to the properties of the green wood. The idea behind these iterations was that during the increase in tree growth by annual increments, the tree adjusts the amount of new growth according to the need in any specific area. Although this adjustment is obviously slow, taking place year by year, as long as the tree is not exposed to the most extreme conditions during the adjustment, the necessary stiffening or strengthening can take place. Their finite-elements predictions were compared with visible growth adjustments in living trees, and found to be in good agreement.

4.1 Types of growth adjustments

The forms of growth adjustments for stiffness and strength include the production of reaction wood, the production of wider or narrower growth rings, changes in the ratio of cellulose to lignin (i.e. fibre to matrix), variations in grain directions including spiral growth, changes in the directions of the microfibrils in the cell wall, and variations in the growth-ring orientation. In addition, of course, there are the adjustments in growth above ground to maximise the access to light for photosynthesis and adjustments in the growth below ground to maximise the uptake of water and soil nutrients. These could include adjustments for changes in the environment such as increase or decrease in competition from neighbouring trees, and changes in the soil environment.

4.1.1 Tree shape to resist wind forces
The tree is able to control its trunk shape, by changing the rates of apical and diametrical growth, in reaction to its wind environment, among other factors. This wind environment is controlled by the exposure of the site in general, and the influence of surrounding trees or other wind-breaks. In an environment that is not exposed to high wind speeds, the tree tends to grow taller and thinner, especially if the tree is closely surrounded by other trees. The shape of the crown is also partly controlled by the wind environment. In some cases, for example when the trunk is supported by an artificial or natural brace part way up, there can be an anomalous change in diameter at that position: below the brace, where the trunk is not subject to wind loading, the trunk can have a considerably smaller diameter than above the brace. This can often be seen in parks, where young saplings are often supported by braces.

Mention has already been made in Section 2.4, above, of the increased flexibility of juvenile wood that results from the large angle of the structural microfibrils within the cell walls of the trunks of young trees. This allows the sapling to resist wind loads by bending rather than providing rigid resistance, as described in Section 3.

4.1.2 Reaction wood
The trunk of a tree standing vertically with little wind loading adds perimeter growth in a uniform manner. However, the tree will react to a permanent bending load by developing reaction wood, as mentioned in Section 2.4. In softwoods, where the thin-walled earlywood cells are especially sensitive to compression loading, the reaction wood occurs in the region of compressive stresses, and is known as 'compression wood'. The compression-wood cells have thicker walls and the S2-layer microfibrils are orientated at a larger angle to the axial direction, thus lowering the stiffness and therefore reducing the danger of localised stress concentrations. In the case of hardwoods, where the fibres are of smaller diameter and have relatively thick walls, there is less danger of compression failure, and the reaction wood occurs on the tensile side of the member, being known as 'tension wood'. Typical cases where reaction wood can be expected are the trunks of trees leaning at an angle, side branches, trees in which the prevailing wind causes nearly permanent bending loads in a certain direction, and disturbances in the growth caused by soil movements or obstacles to symmetrical root development.

Both types of reaction wood have large longitudinal shrinkages, which cause problems of warping of cut boards.

4.1.3 Width of growth rings
In a tree that suffers no abnormal stress situation, the annual rings are of a fairly uniform thickness. However, the type of stress that results in reaction wood also usually leads to a thickening of the annual ring in a region of higher stress. Mattheck [7] shows that this is an attempt by the tree to return the tree to a situation of uniform stress on its surface. Other cases of extra-wide growth rings may occur in the regions of thickening around the junction of side branches and forks, and at regions in which the tree is repairing external damage.

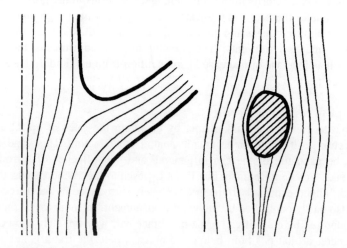

Figure 10: Fibre directions at branches from main trunk, as seen in transverse and plan sections.

4.1.4 Variations in grain directions

Under some special conditions, the growth takes place in such a way as to orientate the grain in a different direction from the normal axial direction. Such disturbances especially occur at stress concentrations such as joints with branches. Again Mattheck [7] has shown that the reorientation has the effect of lowering stress concentrations, and especially of reducing the danger of sufficiently high tensile stresses across the grain, or high shear stresses along the grain, to cause splitting. He generalises his result by stating that the actual grain orientations developed by the tree minimise the shear stresses between the tracheids or fibres. Figure 10 gives an example of the grain flow near a side branch.

In many cases a tree will develop 'spiral grain'. Its causes are not fully understood, although it has been noticed that this is sometimes associated with a site that is exposed to wind on one side of the tree only, causing torsional loading. Again, this can be visualised as a means of minimising the shear stresses between the fibres. Such spiral grain may be good for the tree but causes big problems of twisted timber after seasoning.

4.1.5 Healing of wounds

If the surface of a tree is damaged deeply enough to disrupt the cambium layer, that part of the trunk is unable to produce new wood to repair the damage. The result is that the cambium closest to the damage has to produce material to gradually close the wound, in order to discourage the ingress of decay organisms. This takes the form of a thickening of the growth rings in the region of the damage and a curvature as shown in the sketch of Figure 11. The effect is to close the wound from the sides rather than from the ends. Mattheck [7] has associated this with the compressive growth stresses in the circumferential direction.

Figure 11: Healing of damage (at top of picture), during successive years' growth.

4.1.6 Joints between trunk and branches

Any kind of joint in a structure is potentially a weak point. Besides the problem of actually making the joint, there is also the problem of the stress concentration, which could lead to static or dynamic stress failure. Mattheck [7] has shown that the tree overcomes these problems at branch junctions by control of both the external shape and the grain orientations. The shape of the branch at the junction is optimised both on the weaker compression side and the stronger tension side, in order to avoid stress concentrations and to maintain a uniform surface stress, giving the typical junction shape of Figure 12. Similarly the fibres or tracheids of the trunk are made to flow around the branch junction, whilst the branch fibres themselves flow into the trunk.

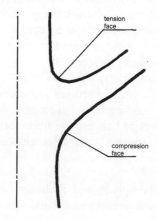

Figure 12: Junction of branch gives optimum shape to minimise stress concentrations.

At junctions of the upper branches, where the trunk and branch may be subject to similar loadings, the trunk and branch fibres tend to flow into the junction together. However, in the case of the lower branches, where the loading of the trunk is often primarily a weight loading, the grain curvature can be a source of weakness, a situation comparable to 'end loading of a curved bar'. To combat this, the trunk develops a thickening around the junctions of the lower branches.

Figure 13: Curved side branches can result in tensile stresses perpendicular to the grain.

4.1.7 Joint between trunk and roots

A well-known problem with curved beams is that when the bending load is such as to cause the beam to straighten, one result is to produce tensile stresses perpendicular to the beam axis. In the case of wood, such transverse tensile stresses can cause splitting or delamination. In trees it is important that such transverse tensile stresses are minimised. There are various positions where such a situation occurs. The two most important ones are the curved side branch that is loaded by its own weight, and the junction between the trunk and the spreading roots. The curved side branch is shown in Figure 13.

Figure 14: The tree in the foreground shows a typical tapered trunk shape. The one on the right shows a correction to its vertical standing, possibly for wind loading, with extra root growth on the left.

(a)   (b)

Figure 15: Formation of roots: (a) symmetrical cone formation; (b) buttress roots. In both cases they reduce the danger of tensile stresses perpendicular to the grain.

The tree's method of overcoming the problem is by an adjustment of the shape of the cross section to an ellipse. In the case of the junction between the trunk and the root, a common solution is to develop a suitable trunk profile, as shown in Figure 14 or to develop buttress roots as shown in Figures 15(a) and 15(b).

Figure 16: A common shape of root that optimises bending stresses about a horizontal axis.

One interesting aspect of root geometry is the cross section of some roots that are subject to large bending loads. The cross section is often of the form shown in Figure 16, which can be considered a specialised form of the I-beam. This form occurs when the bending loads are always in the same plane. Branches sometimes show this form, but less exaggerated, because the wind loadings of branches vary in direction more.

Figure 17: A 357-foot wood transmission tower at U.S. Naval Ordinance Test Station. U.S. Navy photo, courtesy of the Forest Products Research Society.

5 Engineering structures

It is of interest to consider some timber structures designed by engineers, which might be considered to be a kind of imitation of the tree. There are some large structures such as timber-framed buildings and timber bridges, which endure gravity and superimposed loadings, wind loadings, vibrational loadings, and so on. Some examples are given in Figures 17 and 18.

There are various reasons, both functional and ecological, for using timber for structures. As is well known, wood is a renewable resource and it is energy-efficient in production. Apart from cost considerations, there are technical advantages too. Its high strength-to-weight ratio and high stiffness-to-weight ratio compared with many alternative materials make it a useful material for long spans. For the same reasons it has been much used in transport structures: boats, road vehicles, aeroplanes. Although it burns easily in small sections, in large sections it retains its strength in a fire much longer than steel or concrete. It also has good toughness and impact resistance, having a work of fracture in the region of 10,000 J/m^2, comparable to man-made composites [8].

It is of interest to consider some of the methods that the engineer uses to try to overcome some of the problems that are also endured by a real tree.

Figure 18: Straight and curved glued laminated timber members at Local Examination Syndicate, Cambridge. By permission of Cowley Structural Timber Ltd, Lincoln.

5.1 Traditional methods

Traditional buildings used large sections cut from trees, often making use of the natural curvature of the tree itself. These suffered from deep cracks resulting from the drying shrinkage, but most of all there was the problem of joining the pieces together. Medieval craftsmen developed ingenious mechanical joints for transferring the loads from one piece to another, but the efficiency of such mechanical joints was inevitably low compared with that of the tree [9].

142 *Nature and Design*

5.2 Modern methods

The two problems mentioned above, namely drying cracks and efficient jointing, have been partly overcome by the use of modern methods, in particular those that make use of modern synthetic adhesives. Because the shrinkage is very different in the longitudinal, radial and tangential directions of the wood during drying, large stresses can build up in large sections, leading to shrinkage cracks. These high stresses do not occur in small sections, because the various shrinkages are unrestrained. Many modern structures are therefore constructed of laminated members, in which the log is cut into thin slices while still wet, the slices are dried, during which they can shrink freely, and then they are bonded together using modern adhesives that are generally resistant to moisture and weather. The laminated members can have various cross sections, and they may be straight or curved.

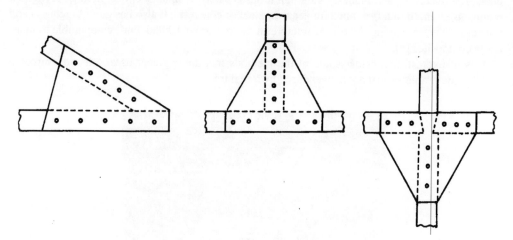

Figure 19: Some structural timber joints made using gusset plates.

Synthetic adhesives can also be used in the problem of joining one member to another; equivalent to the joint between the branch and trunk of a tree. The difficulty here is to try and transfer the stresses from the wood fibres or tracheids of one member to those of the next one. It is often not feasible to use adhesives directly, because they are not able to transfer stresses through the ends of the grain. For this reason, care has to be taken to design the joint so that the adhesive bonds to 'side-grain' surfaces. Examples with small members are dowelled or mortised-and-tenoned joints, and with larger load-carrying members are the use of 'gusset plates', as shown in Figure 19. In the latter, the adhesive provides most of the mechanical strength, rather than the nails.

Direct transfer of load between the end grain of two members can be done with 'finger joints', as shown in Figure 20. Joints can also be made by mechanical fasteners, without adhesives. Again the problem is to transfer the loads efficiently. Many of these joints use nail plates, in which the load is transferred by a metal plate from one member to the next: the load being transferred from the timber member to the metal plate by a large number of nails or spikes that are integral to the plate itself (Figure 21). Other joints use bolts, the load being transferred more efficiently by various load-spreading devices, an example of which is shown in Figure 22.

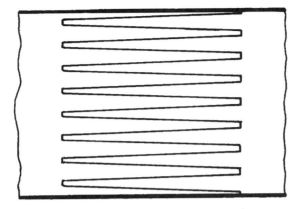

Figure 20: A finger joint for end-grain connections in timber.

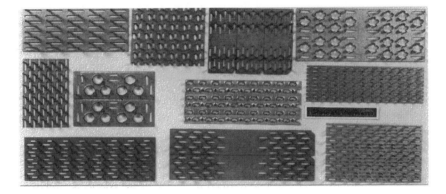

Figure 21: Examples of nail-plate connectors for dry joining of structural timber members. Reproduced by permission of the Building Research Establishment Ltd.

Figure 22: A toothed-ring connector for spreading the load in bolted timber joints.

However, in spite of the relatively efficient use of modern methods, man-made timber structures can never match the efficiency of the tree structure. The tree structure consists entirely of cantilevers, which in man-made structures are difficult to design efficiently. In the tree the cantilevered members allow a degree of flexibility in the wind that would be intolerable in a man-made structure. Such flexibility of the members and weakness of the joints is prevented in a man-made structure by the use of triangulation of members, as for example in the structure of Figure 17.

6 Conclusions

The main structural concerns of the part of a growing tree above ground are to grow in such a way as to maximise light for photosynthesis, but at the same time to prevent breakage from gravity and wind loadings. Since little or no adjustments can be made to the existing wood, any adjustments to its strength and stiffness, to allow for a changing environment, must be made in this year's new growth. In the new growth there are various possible adjustments, including growth rates, the production of internal growth stresses, reaction wood, adjustment of fibre or grain directions, changes in fibre/matrix ratio, and adjustment of cell-wall properties. In the case of traumatic changes to the environment, the tree may require several growing seasons. A record of the seasonal adjustments to its growth is contained within the structure of the tree, thus giving a year-by-year stored memory of its environment.

The tree is entirely made up of a system of cantilever beams, giving it flexibility compared with man-made structures that are triangulated. This flexibility is used by the tree to reduce the danger of breakage by dynamic wind loadings.

The growth mechanisms have been found to conform approximately to a principle of uniform surface stresses, and specifically to growing in such a way as to avoid all types of stress that are likely to lead to fracture. At the same time, the additional growth of excess or 'wasted' material is avoided.

References

[1] Mardrus, J. & Mathers, P., *The Thousand Nights and One Night*, Routledge & Kegan Paul: London, Boston & Henley, **3**, p. 305, 1986.
[2] Wilson, F., *The Growing Tree*, University of Massachusetts Press: Amherst, 1988.
[3] Alhasani, M., Thelandersson, S. & Runge, T., Initial, internal stresses in logs from Norway spruce. *Proc. 1st COST Action E8 Int. Conf. On Mechanical Performance of Wood and Wood Products*, ed. P. Hoffmeyer, EC DGXII: Copenhagen, pp. 281-294, 1997.
[4] Dinwoodie, J.M., *Timber: its nature and behaviour*, 2nd Ed., E. & F.N. Spon: London, 2000.
[5] Kollmann, F. & Côté Jr., A., *Principles of Wood Science and technology*, Springer-Verlag: Berlin, Heidelberg & New York, 1968.
[6] Armstrong, F.H., *The Strength Properties of Timber*, Forest Products Research Bulletin No. 45: Dept of Scientific & Industrial Research, HMSO, London, 1960.
[7] Mattheck, C., *Design in Nature*, Springer-Verlag: Berlin, Heidelberg & New York, 1998.
[8] Gordon, J.E. & Jeronimidis, G., Work of fracture of natural cellulose. *Nature*, **252**, p. 116, 1974.
[9] Hewett, C.A., *English Historic Carpentry*. Phillimore: London and Chichester, 1980.

CHAPTER 8

The homeostatic model as a tool for the design and analysis of shell structures

O.A. Andrés, N.F. Ortega & J.C. Paloto
Departamento de Ingeniería,
Universidad Nacional del Sur, Argentina.

Abstract

Physical models have played an important role in the development of structural shapes. During the last decades interesting advances have been made in conceiving and generating new shapes for shell structures by means of physical models.

Homeostasis is a biological principle according to which, when any living creature is attacked by an external agent, it reacts intelligently to recover its vital functions. We have demonstrated in a number of recent publications that homeostatic models may be generated by the simultaneous action of thermal and mechanical loads on thermoplastic materials. Such models, primarily conceived for designing new shapes, can be used also as tools for analysing their structural behaviour.

The analysis of shell structures can be performed by following two different approaches:

a) experimental, by means of electrical strain gauges applied on both sides of the model surface,

b) hybrid, by the combined action of a digital space arm with the computer and appropriated software.

In view of the fact that homeostatic models fulfil both the functions of design and analysis, they qualify to be regarded as "total" models. The proposed methodology has advantages in the varied fields of teaching, research and professional practice, and these advantages are highlighted and discussed.

This chapter gives an overview of various aspects of the application of the homeostatic model to problems in shell structures. The treatment is clarified with a good number of photographs and graphics illustrations.

1 Introduction

Throughout the history of construction, physical and numerical models have played important and complementary roles in all regarding development of structural engineering and architecture. However, the evolution of both model types shows well differentiated stages: while physical models, well known since ancient times, have experienced a relatively slow development, the more recently formulated numerical models have experienced a surprising and explosive development. In fact, the introduction of high-speed and high-capacity computers has greatly expanded the efficiency and the potential of structural analysis in numerical model algorithms. At the same time, this great potential of numerical models has opened new horizons to creativity in allowing and, indeed, demanding the conception and generation of new structural types and shapes. It is precisely within this field of conceptual design where physical models have an important role to play.

This chapter deals with a particular type of physical model, namely the *homeostatic models*. The first appearance of this model was as a design tool for the generation of non-conventional shell structure shapes. However, advances in technology have allowed the additional use of this model as a tool for structural analysis.

2 Physical models

In order to be clear and avoid confusion it is convenient to distinguish from the beginning the different physical model classes we are going to deal with:

- Scale models: these are used to make a three dimensional representation of a building under construction or already completed.

- Analysis models: these are used to study structural behaviour of a previously designed structure.

- Design models: these are used to create, generate and determine the geometry of a building to be built.

No doubt, the scale models were the first to be used by man. Ancient civilisations (the Egyptians among them) used models made to scale as a tool not only to represent their projects but also to study erection and construction procedures [1]. The use of models with the exclusive purpose of analysis corresponds to not so ancient stages in Construction History. An important landmark in the development of this technique was the funicular model constructed by Giovanni Poleni (1685–1761) in order to verify the stability of the dome of S. Peter's Cathedral in Rome [2, 3]. Even later, however, it has only been in the last century that the analysis model acquires real importance and becomes a valuable tool, especially due to the introduction of electrical strain measurement. Curiously enough another funicular model - A. Gaudí's stereo funicular (1852–1926) - became a new landmark in the development of design models proving its potentiality as a proper tool for the conception, generation and determination of new structural shapes. Later, other physical design models were proposed by using other materials instead of Gaudí's thread: soap solution films, elastic membranes, elastic thread nets and plastic materials, are all examples of these [4].

In the above paragraph the expression "funicular model" has been used. Simply this arises from the definition of *funicular* which is the equilibrium shape assumed by a hanging cable under the action of a set of loads. By analogy, we also employ the term *funicular polygon* or simply *funicular* which is the polygonal line used to find the resultant of a plane system of forces, when the graphical method is applied for the static resolution of that system of forces.

Hence, both the funicular and the funicular polygon are plane lines belonging, as they do, to the plane of the system of forces. Extending the consideration to multiple planes, we then have a set of multiple funiculars, which is termed a *stereo funicular*.

The first two concepts, namely *funicular* and *funicular polygon*, are well known in the French, Italian and Spanish technical literature. They also appear in the English literature, for example: *Shaping Structures: Statics* by Zalewshi and Allen [5]. They are applied indiscriminately in this chapter.

In what follows we will describe and explain a physical for multiple purposes: the homeostatic model. This was originally conceived as a design model, but can be used both as an analysis model and obviously as a scale model to make spatial representation of a structure.

3 The homeostatic model

The homeostatic model was conceived and materialized for the first time in 1988 [6], with the primary aim of generating new structural shapes, specifically shell roofs. Because subsequent presentations at IASS meetings and publications [7–14] exempt from the need to give further details, we are going to give here only a brief description of its principles and the technique for its application.

3.1 Principles

Homeostasis is a principle of Biology according to which, when any living creature is attacked by an external agent, it reacts intelligently in order to recover the balance of its vital functions. Something similar occurs with structures. When an external agent (such as increase in loading or debasing of the mechanical properties of its materials) attacks a structure, the structure "defends itself intelligently" in order to recover its capability to withstand such attacks. In the homeostatic model technique as specifically applied here, heat is used as an external agent to provoke the degradation of its material. The model thus attacked reacts intelligently, trying to find the most proper structural shape to continue resisting loads. For example, a small square thermoplastic plate (400 × 400 × 2 mm), suspended from its corners and submitted to a uniformly distributed load (approximately 0.02 N/cm^2), is heated to over 125°C. Under such conditions its compression Young's modulus decreases approximately 400 times, and therefore it loses its bending strength capabilities and experiences a metamorphosis: intelligently it changes its plate status, becoming a membrane and thus modifying its initial shape in order to resist the applied load. The model generates its own resisting shape and after being cooled, will recover its original stiffness, but keeping its new shape.

3.2 The homeostatic technique

In order to illustrate the specific technique used here to obtain a homeostatic model of a shell roof, the following procedure is applied:

- An acrylic material plate is cut in the shape of the perimeter to be covered (mother plate).

- This plate is suspended from the points or lines where links (boundary conditions) will be applied.

- This plate is suspended from the points or lines where links (boundary conditions) will be applied.

148 *Nature and Design*

- On the mother plate a loading carpet is applied Figure 1 shows the mother plate and loading carpet corresponding to models Nrs. 103, 104 and 105 shown on Figures 10, 11 and 12 respectively.
- Balancing forces are applied at the points (or lines) corresponding to the above links.
- The assembly (plate, load and balancing forces) is put into an oven to be heated to over 125°C.

After being cooled, the model thus obtained is either a prevailing tensile shape (when suspended) or a prevailing compression shape (when it is inverted and placed on the supports).

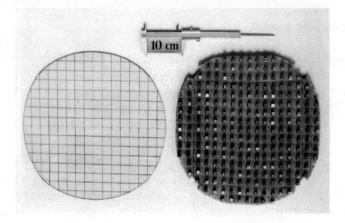

Figure 1: Mother plate and loading carpet.

4 Design model

The homeostatic model generated as explained above is a design tool. Just as the funicular model can be used to design the shape of a 2D tensile structure (or a compression one, if inverted), the homeostatic model can be applied to design the shape of a 3D continuous structure. In fact, it is possible to measure X, Y and Z coordinates of a network of points on the surface of the homeostatic model, that is, the data with which its geometry is fully determined. Then it is sufficient to expand these coordinates by means of the use of a scale factor in order to obtain the geometry of the structure to be built. This operation can be made rapidly and safely taking advantage of the facilities of a digitiser arm such as the SpaceArm, as shown in Figure 2.

All that is necessary to scan the model surface with the arm probe and to click at each point to obtain its three coordinates. SpaceArm ModSA48 has an accuracy of 0.4 mm, such that by applying a scale factor of 50 it is possible to obtain structure coordinates within an uncertainty of 20 mm, which in most cases is sufficiently good for the execution of the real work [15]. It is convenient to point out that:

- the model surface finish is perfectly smooth and continuous due to nature of the plastic material (acrylic), and
- the stability and stiffness of models are good enough as to admit probe scanning. Therefore, the determination of coordinates given by SpaceArm is reliable, precise and consistent.

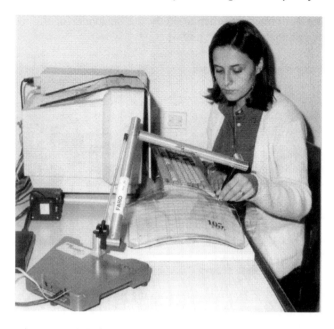

Figure 2: Digitalising a model by means of the SpaceArm.

5 Analysis model

It is well known that the aim structural analysis is to study the behaviour of a structure under loads with the purpose of evaluating its safety level. By using the homeostatic model, it is possible to reach this objective by two different ways: namely the experimental method and the hybrid method.

5.1 Experimental method

The homeostatic model allows for the application of the experimental technique of strain measurement by means of electrical strain gauges. In our case, the acrylic model Nr. 17 (Figure 3) was generated in a similar way to that described in Section 3.2 but starting from a square mother plate instead of a circular one. In general, for the application of this method, we follow the conventional guidelines for this technique [12, 16].

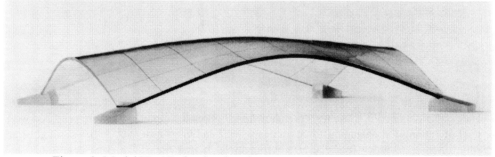

Figure 3: Model Nr. 17 after leaving the oven and supported on the corners.

Several tests were performed in order to study the behaviour of the model when it was subjected to different loading patterns (Figure 4).

Figure 4: Model Nr. 17 suspended from the corners to be analysed by means of electrical strain gauges.

These tests had the aim of quantifying the bending effects on the membrane behaviour of the model. To achieve this purpose the difference between the strains measured on the inner surface and those measured on the outer surface, both at the same point of the models, was obtained as:

$$SD = \delta_i - \delta_o \qquad (1)$$

where SD is the strain difference, δ_i and δ_o the inner and outer strains respectively.

These measurements were performed on three significant points, namely 1, 2, and 3 of model Nr. 17 shown on Figure 5.

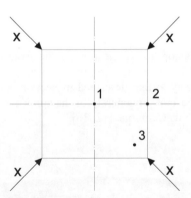

Figure 5: Points and balancing forces on model Nr. 17.

The results obtained in these tests are shown in Figures 6, 7 and 8, and were prepared with the goal of finding the balancing force values when bending is null (SD = 0). A straight-line equation of high accuracy (correlation coefficient R^2 higher than 95%) was determined for each of these cases.

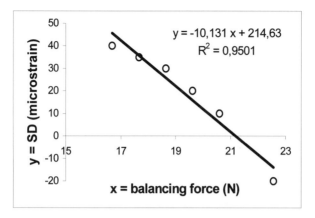

Figure 6: Strain differences at point 1 as a function of balancing force variations.

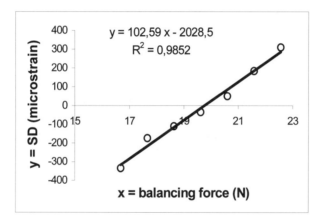

Figure 7: Strain differences at point 2 as a function of balancing force variations.

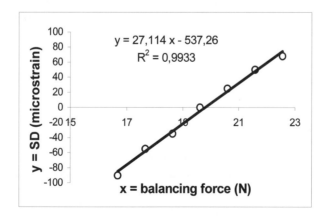

Figure 8: Strain differences at point 3 as a function of balancing force variations.

5.1.1 Information gathered from the tests

From the tests made by applying electrical extensometry to homeostatic model Nr. 17, the following information was gathered:

- It is convenient to take measurements right after constructing the model due to the fact that acrylic (PMMA) has an important creep magnitude which is evidenced by the reduction in model thickness. Another recommendation associated with this phenomenon is the convenience of unloading the model at the end of the test. These precautions did not completely eliminate the effects of the phenomenon, but tended to moderate it.

- More reliable results are obtained with the model subjected fundamentally to tensile stress. In other words, it is convenient to test the model in the same way that it was generated within the oven instead of doing it with the model placed in an inverted position. If the model is placed in the same position, under actual construction conditions (i.e. inverted), great difficulties are experienced in creating the correct support conditions. When determinations were made in this way a rearrangement of the supports was observed and this altered the measured values due to the important bending effect produced.

- In order to expedite the tests by reducing the time required for stabilisation of measurements at the Wheatstone bridge, and therefore to reduce the possibilities of reading errors in the results, several techniques are it is convenient:

   a) to use strain gauges with resistances between 120 to 350 ohms (the last one recommended by Perry [16]),

   b) to carry out load cycles, making an individual analysis at each one of the points at both model faces, instead of analysing all the points of only one face, and then to study the other face,

   c) to employ polyamide based strain gauges; these can be fixed directly to the acrylic surface by means of cyanoacrylate adhesive; the interposition of metallic sheet to dissipate heat not being necessary.

However, in spite of all these precautions, the galvanometer does not stabilise immediately, particularly during the first load cycle.

Another recommendation for obtaining reliable results is to work with loading patterns similar to those employed for the generation of the model. In addition to model Nr. 17, model Nr. 103 was also analysed. Both models showed good behaviour and reading repeatability with the same loading status. Results obtained by overlapping several loading patterns were not considered acceptable, especially when the model was placed in an inverted position, due to the fact that stresses were generated at the contact surface between loading carpet, and this interfered with the uniform distribution of load over all the surface.

In general, data gained by applying strain gauges on both faces (inner and outer) at different points of the models have shown a membrane (or quasi-membrane) behaviour in most part of the homeostatic model surface.

5.2 Hybrid method

This method involves the use of two different models in two consecutive steps:

- first, a physical model, namely the homeostatic model generated as explained and described in Section 3.2, and

- second, a numerical model whose geometry is generating by digitalising the homeostatic model by means of the SpaceArm. Files obtained in this operation and multiplied by the scale factor are input as data into the finite element software Algor [17]. These data, together with the mechanical constants of the material, edge and load conditions of the real structure are processed by the software. It achieves the generation of numerical and graphic results of displacements and stresses of the structure, which are then used to evaluate its conditions for safety (Figure 9).

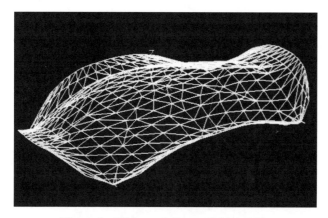

Figure 9: FEA mesh for model Nr. 103.

By applying both alternative methods (experimental and hybrid) on the same physical model, allows the possibility of checking them and therefore guarantees a better reliability of the results. In fact, we made such a check with a number of homeostatic models (among them Nr. 103, Figure 10), and in all cases obtained a good coincidence of results.

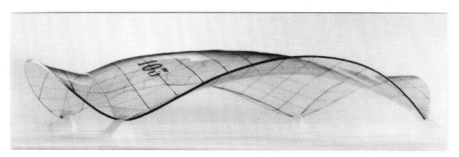

Figure 10: Free form shell roof supported on four lateral arches (model Nr. 103).

Figure 11: Free form shell roof supported on four points (model Nr. 104).

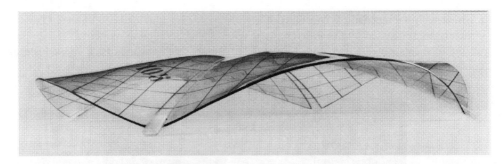

Figure 12: Free form shell roof supported on two diagonal arches (model Nr. 105).

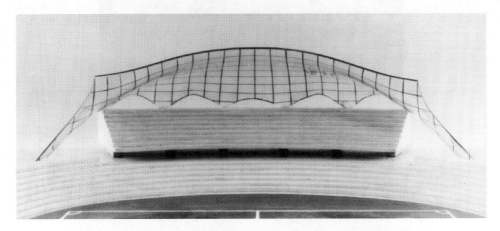

Figure 13: Shell roof model for a football stadium (model Nr. 45).

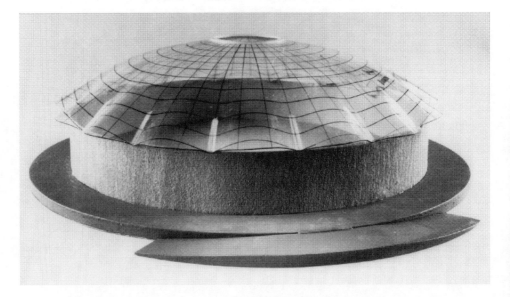

Figure 14: Shell roof model for a minor sport stadium (model Nr. 71).

6 Conclusions

The following conclusions may be drawn form our overall experience of widespread use of the homeostatic model.

a) From the viewpoint of structural design:

- It is a simple method for the generation of structural shape: to make a homeostatic model is a straightforward, fast and economical procedure.

- Generation of free forms with free edges is one of the most fertile field for the application of the method (Figures 10 to 14).

- The method is versatile in the production of structural shape: variations in value and direction of balancing forces, shape of mother plate perimeter, level of the suspension points, and other features, are immediately reflected in the geometry of the structure thus enabling the dialogue to take place between designer and structural shape. Figures 10, 11 and 12 show some of the generated shapes obtained from the same mother plate (Figure 1), with and without inner ribs and supported on four points. Completely different shapes are obtained when the perimeter shape of the mother plate is modified (Figures 5, 13 and 14).

- Interaction between shape and forces is direct and its visualisation is immediate, thus guaranteeing the permanent satisfaction of the requirement for force balance during the search for the best adaptation of shape-space-structure.

As any physical model, it presents obvious advantages from the viewpoint of promoting and exercising creativity [14] as well as from the viewpoint of its application to teaching and research [18].

b) From structural analysis viewpoint:

- Fast and safe generation of a numerical model from the homeostatic model.

- The possibility of a two approach (experimental and numerical) allows results to be checked, especially at points located at conflicting areas.

Finally, and from a general viewpoint, it is easy to see that the multiplicity of functions (scale model - Figures 13 and 14 - design model, analysis model) accepted by the homeostatic model allows us to consider this physical model as a total model.

References

[1] Ortega, N.F., Diseño y Análisis de Estructuras Laminares Mediante Modelos Físicos, Doctoral Thesis, Departamento de Ingeniería, Universidad Nacional del Sur, pp. 143–159, Bahía Blanca, 1998.

[2] Melaragno, M., *An Introduction to Shell Structures*, Ed. Van Nostrand Reinhold, New York, p. 243, 1991.

[3] Poleni, G., *Memorie Istoriche della Gran Cupola del Tempio Vaticano*, Padua, 1747, Ed. INTEMAC S.A., Madrid, 1982.

[4] Andrés, O.A., Experimental Design of Shell Structures, *Proc. IASS Symposium on Tension Structures and Space Frames*, Tokyo and Kyoto, Oct. 1971, Architectural Institute of Japan, Tokyo, Japan, pp. 845–854, 1972.

[5] Zalewski, W. & Allen, E., *Shaping Structures: Statics*, John Willey & Sons, Inc., New York, 1998.
[6] Andrés, O.A., Modelos Homeostáticos para Diseño de Cubiertas Laminares, *Memorias de las VIII Jornadas Argentinas de Ingeniería Estructural*, pp. 361–370, Asociación de Ingenieros Estructurales, Buenos Aires, 1988.
[7] Andrés, O.A., Homeostatic Models for Shell Roofs Design, *Proc. IASS Congress, CEDEX Laboratorio Central de Estructuras y Materiales*, Madrid, Vol. 1, 1989.
[8] Andrés, O.A. & Ortega, N.F., Experimental Design of Free Form Shell Roofs, *Proc. IASS Symposium, Kuntsakademiets Forlag Arkitetskolen*, Copenhagen,Vol. II, pp. 69–74, 1991.
[9] Andrés, O.A. & Ortega, N.F., Extensión de la Técnica Funicular de Gaudí a la Concepción y Génesis de Superficies Estructurales, *Informes de la Construcción*, Vol. 44, Nr. 424, pp. 9–34, Madrid, 1993.
[10] Andrés, O.A., Ortega, N.F. & Schiratti, C.A., Comparison of Two Different Models of a Shell Roof, *Proc. IASS - ASCE Congress*, Atlanta, pp. 320–329, 1994.
[11] Andres, O.A. & Ortega N.F., An Extension of Gaudi's Funicular Technique to the Conception and Generation of Structural Surfaces, *IASS Bulletin*, Vol. 35, Nr. 3, Dec. Nr. 116, 1994.
[12] Paloto, J.C. & Ortega, N.F., Experimental Analysis of Acrylic Models, *Strain, Journal of the British Society of Strain Measurement*, London, **34(3)**, pp. 95–98, 1998.
[13] Andrés, O.A., Ortega, N.F. & Paloto, J.C., The Homeostatic Model as a Total Model, *Congress I.A.S.S.*, Sydney, **2**, pp. 985–991, October 1998.
[14] Andrés, O.A., On the Aesthetics of Homeostatic Shell Shapes, *Proc. 40 th Anniversary Congress of the IASS*, CEDEX, Madrid, Vol. II, pp. F1–F9, 1999.
[15] *User's Guide SpaceArm*, Faro Tech. Inc., Lake Mary, Fl, USA, 1995.
[16] Perry, C.C., Strain gauge measurement on plastics and composites, *Strain, Journal of the British Society for Strain Measurement*, **23(4)**, pp. 155–156, 1987.
[17] *User's Guides and Reference Manuals*, Algor, Inc. Pittsburgh, PA, USA, 1992.
[18] Andrés, O.A. & Serralunga, R.E., Physical Models for Teaching and Researching Conceptual Design, *Proc. IASS International Symposium on Conceptual Design of Structures*, University of Stuttgart, Stuttgart, Vol. I, pp. 161–168, 1996.

CHAPTER 9

Adaptive growth

J. Platts
University of Cambridge, UK.

Abstract

Like wood, bone is created in response to stress fields. Unlike wood, bone is also removed when the stresses are removed. Throughout its life an animal's bone structure is constantly edited, adapting to circumstances to maintain peak efficiency as a structure. The algorithm which Nature uses to achieve this provides a design algorithm which can be used to create computer software which genuinely designs and optimises a structure for minimum mass. Arising from the combined observations of medical researchers, engineers and mathematicians, the development of this understanding has taken a century and a half and is deeply intertwined with the development of structural theory itself.

1 Introduction

Like wood, bone is an adaptive material. The difference between wood and bone is that bone is used to create dynamic structures. Animals are constantly on the move and have to carry their structures with them. Unnecessary mass uses unnecessary energy to move it. So a high performance to mass ratio is an important criteria for bone whereas it is not for wood. The cellular process which generates bone has thus developed around the need not only to create optimally designed structures but to maintain them throughout the animal's lifetime. As loading requirements change this involves the ability to edit the structure by removing material which is no longer necessary as well as adding material which is. It is this ability to remove material which distinguishes bone from wood.

The idea of designing minimum mass structures is central to good engineering design and so designers have always sought a way to do it. Unfortunately, science is about analysis, not design. Many analytical tools have been developed, but until recently design has remained a two step process of 'guess a shape and then analyse the stresses to see where and when it will break'. The design process should be 'define where and when you want it to break and then design a shape which ensures that it will do so'. Michell recognised this requirement and captured it in a descriptive theorem in 1904 but that did not help matters. For nearly a century Michell's minimum mass theorem has stood as a frustrating beacon for engineers, demonstrating great insight, yet at the same time offering a tantalisingly impractical structural ideal. With hindsight, an erroneous assumption about how to apply the theorem - a natural assumption to make at the time - generated the impracticality, which is not fundamental to the theorem itself. Correcting the error allows the basic resilience of the approach to re-emerge and yields a highly efficient design tool concerning structural optimisation. Correcting the error has come from the detailed understanding of the way bone structures are created. This is truly a design process and it produces minimum mass structures, and so following Nature's guidance allows a proper design process to be developed. This has required the development of an understanding of how bone does design itself and that is an illuminating story spanning nearly a century and a half which is intimately bound up with the development of engineering science itself.

Michell's theorem is one particular insight arising from the steady development of understanding about stress fields, which itself has an interesting history. Though of central importance to structural engineering, most of the progress in understanding has come from medical research into bone structure, and an interaction with mathematicians, over a century and a half, at the very slow rate of only about one interested person - and one contribution to the understanding - per generation. Michell, who was an Australian engineer, stands as one of the only two engineers in the sequence, and the one with the mathematical link, his elder brother being a Cambridge mathematician, and both of them being Fellows of the Royal Society, in whose *Philosophical Magazine* Michell published his theorem in 1904. The principle expressed in the theorem is said to have occurred to him some years earlier, while travelling on a train near Cambridge, while he was there with his brother, but the opportunity to completely develop the theorem and write the paper occurred when he was settled back in Australia, in Melbourne, where he ran a most innovative company designing and manufacturing hydraulic pumps.

2 Graphic statics

The first engineer in the sequence is Professor Culmann, in Zurich, whose development of Graphic Statics [1] is the beginning of the idea of analysing stress fields as a field. His approach was to picture in the mind the flow of tension and compression in a structure and then to draw the likely lines of flow of this tension and compression, by eye, on a drawing of the structure. Here we have all the essentials of an engineer's understanding already present in embryo. The mind's eye is engaged and the engineer is 'being' the structure and 'feeling' the stress. We already have the notion that stresses have direction, and are 'seeing' stress intensity, and we have the impression that the stress 'flows' or wants to flow - a dynamic impression conjured up at least in part by the actual dynamic of our own hand's movement in drawing the lines freehand, and also expressed in their name. Culmann called them *stress trajectories*. The basic idea of orthogonality is also present.

A key step came in 1867, when Professor Culmann was examining the bone samples which Herman von Meyer was presenting to the Natural Science Society in Zurich [2]. He observed that the cancellous architecture of the upper end of the human femur closely resembled the stress trajectories which he as an engineer would draw in a crane known as Fairbairn's crane, a crane of the same general external shape as the femur but of homogeneous structure (Figure 1). The trabecular lines cross approximately at right angles, suggesting that they are following lines of tension and compression that would be there in a homogeneous body of the same shape, and they were therefore readily described as *materialised trajectories*. Interestingly, the cause of the research which has given rise to this chapter was the same, the picture of Herman von Meyer's bone sample, seen in a medical textbook while a student over thirty years ago (Figure 2), still having enough power to demand attention after half a lifetime of other activity.

The key to the answer is also already there. It is in the biological detail of how those trajectories materialise. But that wasn't where the engineers looked.

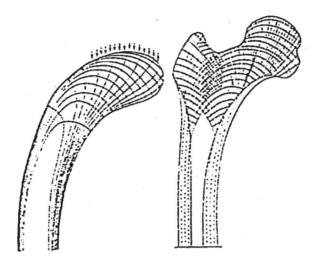

Figure 1: Culmann's comparison of the fibre geometry in a human femur with the stress trajectories in a Fairbairn crane.

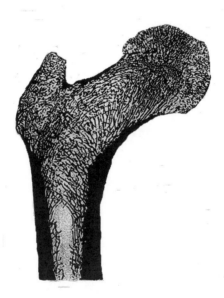

Figure 2: Section of the upper femur.

A key step came in 1867, when Professor Culmann was examining the bone samples which Herman von Meyer was presenting to the Natural Science Society in Zurich [2]. He observed that the cancellous architecture of the upper end of the human femur closely resembled the stress trajectories which he as an engineer would draw in a crane known as Fairbairn's crane, a crane of the same general external shape as the femur but of homogeneous structure (Figure 1). The trabecular lines cross approximately at right angles, suggesting that they are following lines of tension and compression that would be there in a homogeneous body of the same shape, and they were therefore readily described as *materialised trajectories*. Interestingly, the cause of the research which has given rise to this chapter was the same, the picture of Herman von Meyer's bone sample, seen in a medical textbook while a student over thirty years ago (Figure 2), still having enough power to demand attention after half a lifetime of other activity. The key to the answer is also already there. It is in the biological detail of how those trajectories materialise. But that wasn't where the engineers looked.

3 Medical insights

Following Culmann and Meyer, progress was slow, and the engineering and medical sides became separated. In 1895, in medical work rather similar to Meyer's work, Roux published an analysis of bone structure. His contribution is crucial in that he was concerned with the process of bone formation and he rightly saw bone not as a static product, but as something which evolves, which is the continuous product of a dynamic process and which is not only generated in the first place out of cartilage - effectively out of weak structural material - in response to a stress field, but is sustained, and indeed dynamically edited, continuously thereafter, in relation to the same repeating and evolving stress field.

He concluded that bones were constructed with the greatest economy of material because of this process, their shape and internal structure being adapted to the loading in this way. His theory of *maximum - minimum* (maximum strength and minimum material) [3], published in 1895, contained all the right concepts, including in particular the key focus on the dynamic nature of the problem. However, his work was not to be seen by Michell, who essentially produced a static solution to an idealised, and static, problem.

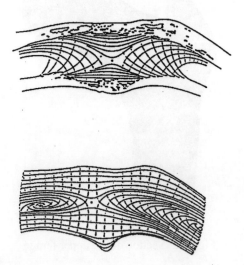

Figure 3: Cancellous architecture in an ankylosed knee and stress trajectories in a (solid) photoelastic model.

Roux' work remained untouched until in the 1950's Pauwells considered the problem of the ankylosed knee and added an important detail [3]. Pauwells tried to examine in more detail

Culmann's notion of bone fibres being materialised stress trajectories, by making a photo-elastic model of an ankylosed knee. Here he did indeed confirm the general principle (Figure 3) but with an important difference. He realised that while the fibre trajectories in the bone were *like* the stress trajectories in the photo-elastic resin they were not the same because the photo-elastic model consisted of a uniform sheet of constant thickness, whereas the real specimen was hollow, for one thing, and showed a varying density of fibres depending on the need. The photo-elastic model thus did not truly represent the bone structure it was supposed to model and naturally gave a somewhat different answer.

Pauwells did not phrase it this way, but what he identified is a key difference in principle which has to be understood in this sort of analysis, which is that in essence the photo-elastic resin and the bone were working to opposing criteria. The resin structure had a uniform distribution of material allowing a non-uniform distribution of stress in the material, whereas the bone developed a non-uniform distribution of material to achieve uniform stress within the material used. A second, more subtle observation is that the first result (the stress field in the resin) is produced simply by analysis. The second however (the fibrous shape of the bone) is the result of a creative process.

4 Achieving structural efficiency

In his continuing study Pauwells did much to add substance to Roux' perception of that dynamic process and looked at the way an existing structure (a strut, say) can be progressively edited to realign itself with a stress newly imposed on it, which loads it eccentrically (Figure 4).

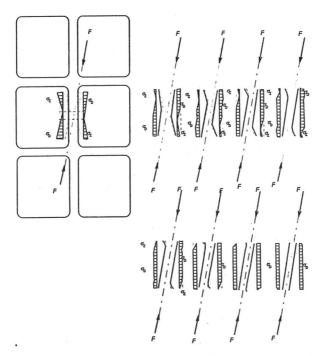

Figure 4: Bone's progressive geometric adaptation to load.

The eccentric load places unequal stresses on the two sides of the strut. Bone is both constructed and maintained by a pair of cells. One cell continuously feels the strains occurring locally and if the strain level is too high it will lay down calcium, providing additional stiffness to reduce that strain. Its colleague also continuously feels the strains occurring locally but does the reverse. If the strains are too low it removes calcium, allowing the strain to increase [4].

It is easy to see that these cells, working as a pair, sustain a dynamic equilibrium which will both grow bone structure out of nothing, and will eat it away back to nothing, in response to a varying stress field, and that they will dynamically sustain the bone structure at maximum structural efficiency. It is equally easy to see that the process demonstrated by these two cells is ideally suited for adaptation into any finite element analysis program, because finite element analysis programs model large structures as an assembly of small cells, each of which, in principle, can have different material properties appropriate to its local conditions.

5 Field theory

While the medical researchers were following this track the engineers were following a different line of thinking. Not only did Michell himself have considerable mathematical ability, he was moving in a circle of mathematicians who were between them steadily developing field theory in general. Stress fields are only one kind of field. Pressure and flow fields are of interest to all fluid dynamicists and magnetic fields to all physicists, as well as to electrical engineers. Michell's brother John was one of this group building on the earlier work of Airy, de Saint Venant, Maxwell and Ibbetson, as were J.J. Thomson, Love and Kirchoff, who developed Helmholtz' work on flow fields.

From this group emerged the mathematical, numerical representations of fields that we think of as field theory today. In particular, in relation to stress fields, all the work was related to stresses in uniformly distributed homogeneous materials. One understanding that emerged was about the three dimensional local stresses at a point: that there are three principal stress directions, which are mutually orthogonal, and which see only tension or compression and do not see shear stresses, and that the stress effects in all other directions, including shear stresses, are related to these principal stresses (familiarly represented by Mohr's circles). The other understanding that emerged was about the global field shape: that the stress lines would always form a curvilinear orthogonal mesh, with the three sets of stress lines always intersecting orthogonally.

This mathematical approach gave geometric precision to Culmann's original observations concerning stress trajectories and has enabled his understanding of stress concentration, etc., to become a familiar part of engineering vocabulary. Michell made the key step that Pauwells was also to make later, from thinking about lines of stress flow in a homogeneous material, to the geometry necessary for an open structure to carry a load efficiently. Michell's theorem thus retains the global geometry of a three-dimensional curvilinear orthogonal mesh, constructed of three sets of members set orthogonally to each other, and then he allows his members to become thicker or thinner as the local loading dictates, so that the material everywhere is uniformly stressed [5]. Here are all the correct observations to give mathematical substance to Pauwells' observations about bone structure. But that was not the context Michell was in. Orthogonal meshes and their possible transformations, not simply in three dimensions but also mathematically in higher dimensions, form a whole subject of their own which continues to excite mathematicians, and Michell therefore focused on the global geometry requirements and quickly established that in two dimensions only two classes of orthogonal curves exist that give global solutions:

- Systems of tangents and involutes derived from any evolute curves

- Orthogonal systems of equi-angular spirals, with systems of concentric circles and their radii, and rectangular networks of lines,

and these give rise to the standard illustrations of Michell structures (Figure 5), with which everybody is familiar.

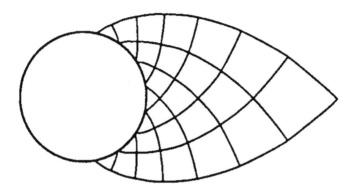

Figure 5: An analytically derived Michell truss cantilever.

6 Modelling growth

Many people have tried (for instance Prager in the 1950's) [6] but unfortunately Michell's use of a classical mathematical approach, and interest in the global field problem as the key conceptual issue, has proved to be a debilitating decoy.

Recent work on structural optimisation has started quite independently and has not benefited from all the observations that Culmann, Meyer, Roux, Pauwells, Michell and the mathematicians had collectively accumulated. Mattheck, working at Karlsruhe, has developed an approach akin to the natural process that bone uses, of adding and subtracting material [7], but with two critical differences. Firstly he used isotropic materials, so that until recently he did not take any directionality of material properties into account [8]. He has then taken wood, and tree growth, as his model.

The first decision has the unfortunate result of throwing away one of the key variables in the optimisation process, for the following reason. If a structure is being developed from nothing to match an imposed load field, at any given iteration stage, if a stress in a particular direction requires more material to be added at some location, it will be added. But once it is added, since it is isotropic, it will also contribute its stiffness in the other two orthogonal directions, even though these were not required. This lack of directional sensitivity and responsiveness significantly handicaps the optimisation process. A further handicap then comes from using tree growth as a model in that, whereas bone both adds and subtracts material, trees never subtract material. They are in that sense less evolved than bone and do not achieve the same level of optimisation.

In contrast to Mattheck, Kikuchi, in Michigan, has generated an approach which allows for directionality, but by a technique which is abstracted from any direct representation of material [9, 10]. He uses a specially developed finite element code that builds a structure out of a volume made up of cells (elements) with voids in (Figure 6). After each iterative stress analysis the void in each element is altered, changing its volume, proportions and orientation, to represent a change both in quantity and orientation of material. This provides a faster analysis than Mattheck in terms of the number of iterative steps required to generate a shape, but it does so at considerable cost in computational complexity, and it generates only a rough indication of shape, represented in an unrealistic way which has to be remodelled as real material before the final analysis can be done (Figure 7). Kikuchi notes with apparent surprise that his answers tend to confirm the sort of shapes Michell's theorem suggests. There is no reverence in Kikuchi's work for the lessons of Nature.

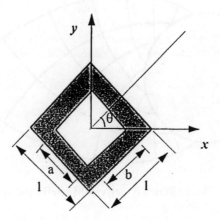

Figure 6: A volume element with a cavity whose proportions and direction can be varied.

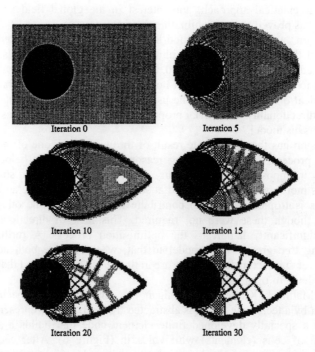

Figure 7: An indication of a cantilever shape developed using voided elements.

7 Honouring Nature

Cambridge has followed Nature closely, and in doing so has found great beauty [11, 12]. One of the things which the modern finite element approach to structural analysis offers which was not available to Michell is that complex geometries can be assembled from elements, each of which only needs to know about its local conditions and respond to them, yet we know that this assembled behaviour will generate the desired overall behaviour. It is the exact structural equivalent of Adam Smith's trusting of the invisible hand, in economics, that provided the rules of co-ordination are well founded, good co-ordination will result. Good overall results will emerge from appropriate local behaviour and global dictation is not necessary.

What this approach models is the *local* rather than the *global* principle captured in Michell's theorem. Locally in each element, analysis of the three-dimensional stresses suggests three principal load directions which can most efficiently be responded to by placing unidirectional fibres in the right quantities in these three directions. This generates a new structure, with new stiffness properties, which has to be analysed again, but after a few iterations a fibrous structure is achieved which is exactly a Michell structure. It has the right quantities of fibres, set in the right directions everywhere, to carry the load with all fibres fully stressed, producing a structural framework which is a Michell structure, with thick and thin members distributed appropriately in an orthogonal mesh.

But there is a difference. While Michell's theorem looked for smooth curvilinear mesh geometries which defined an overall structure in one go, these meshes have real, awkward corners, where some load or support point creates a particular need. Local orthogonality is observed - but it would not be appropriate to describe the overall shape as 'smooth' - 'flowing' perhaps, but not 'smooth'.

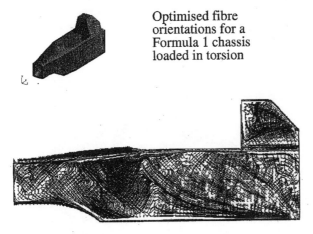

Optimised fibre orientations for a Formula 1 chassis loaded in torsion

Figure 8: Two layers of fibres optimised for torsion on a formula 1 Grand Prix car.

But then a bonus appears. The original 'elements' which described the general volume are now filled with material which is represented by the fibres forming this framework. These fibres are also uniformly stressed, and, in each element, they are mutually orthogonal to each other. If the finite element program is asked to plot stress flowlines, the computer draws the mesh which is at the same time both the structural framework and its own stress map (Figures 8 and 9). Michell's theorem is thus complete and usable - in one very simple, fast subroutine.

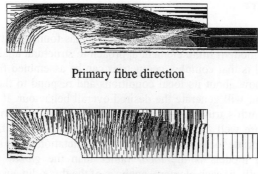

Figure 9: Two orthogonal layers of fibres for a lug, optimised for tension and compression.

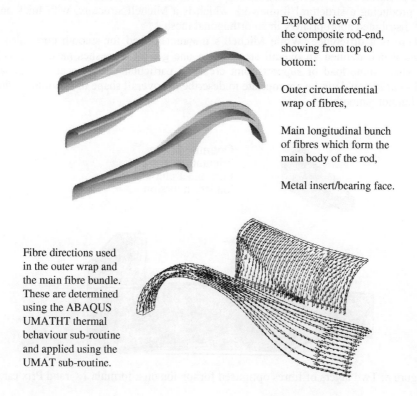

Exploded view of the composite rod-end, showing from top to bottom:

Outer circumferential wrap of fibres,

Main longitudinal bunch of fibres which form the main body of the rod,

Metal insert/bearing face.

Fibre directions used in the outer wrap and the main fibre bundle. These are determined using the ABAQUS UMATHT thermal behaviour sub-routine and applied using the UMAT sub-routine.

Figure 10: Composites eye at the end of a tubular strut/tie.

The focus of the research at the time it was done was trying to design optimum 3D fibre layouts for seriously three-dimensional components in composite materials, with carbon fibre aerospace components in mind (Figure 10). At the same time, the drive to want to do so came from the experience common to all civil engineers of genuinely thinking in three dimensions when designing a reinforcing cage, and dimensioning, orienting and spacing out the steel reinforcement appropriately to carry the tensile stress field, and sometimes part of the compressive stress field, in a reinforced concrete structure. Practicality intervenes, and sets limits, and commonly produces a certain rectilinearity, but civil engineers are the only ones who consistently remember, visualise and think as Culmann did, and consciously place material bar by bar to make Michell-like structures.

Always in the past that work has been from the pragmatic, reaching towards the ideal. What is suddenly in view is to reverse that process and generate a design route which generates the ideal minimum mass structure as its first step. Of course the real world will still intervene and aspects of that optimum will have to give way to pragmatism.... But this time the trade-off can be costed.

References

[1] Culmann, K., *Die Graphische Statik*, Zurik, 1866.
[2] Meyer, H., Die architectur der spongiosa. *Archiv f. Anat. u. Physi.*, 1867.
[3] Pauwells, F., *Biomechanics of the Locomotor Apparatus* (Chapters 9, 19). Springer-Verlag: Berlin, 1980.
[4] Wolff, J., *The Law of Bone Remodelling* (Chapter 2). Springer-Verlag: Berlin, 1986.
[5] Michell, A.G.M., The limits of economy of material in frame structures. *Phil. Mag.*, **8(6)**, pp. 589, 1904.
[6] Prager, W., Optimisation in structural design. *Mathematical Optimisation Techniques*, ed. Bellman, R., RAND Corp. Report P-396-PR, pp. 279–289, 1963.
[7] Mattheck, C., An intelligent CAD method based on biological growth. *Fatigue Fract. Engng Mater. Struct.*, **13(1)**, pp. 41–51, 1990.
[8] Reuschel, C. & Mattheck, C., Three dimensional fibre optimisation with computer aided internal optimisation (CAIO). *Multidisciplinary Design and Optimisation*, pp. 10.1–10.11, R. Ae. Soc.: London, 1998.
[9] Bendsøe, M.P. & Kikuchi, N., Generating optimal topologies in structural design using a homogenization method. *Computer Methods in Applied Mechanics and Engineering*, **71**, pp. 197–224, 1988.
[10] Bendsøe, M.P., Dfaz, A. & Kikuchi, N., Topology and generalised layout optimisation of elastic structures. *Topology Design of Structures*, eds. M.P. Bendsøe & C.A. Mota Soares, pp. 159–205, Kluwer: Netherlands, 1993.
[11] Makiyama, A.M. & Platts, M.J., Topology design for composite components of minimum weight. *Appl. Comp. Mater.*, **3(1)**, pp. 29–41, 1996.
[12] Jones, S.E. & Platts, M.J., Practical matching of principal stress field geometries in composite components. *Composites Part A*, **29A**, pp. 821–828, 1998.

CHAPTER 10

Optical reflectors and antireflectors in animals

A.R. Parker & N. Martini
Department of Zoology, University of Oxford, South Parks Road, Oxford UK.

Abstract

Structures that cause colour or provide antireflection have been found in both living and extinct animals in a diversity of forms, including mirror-reflective and diffractive devices. An overview of this diversity is presented here using invertebrates as examples. The behavioural and evolutionary implications of optical reflectors and antireflectors in animals are introduced.

1 Introduction

Animal pigments have long received scientific attention. Bioluminescence, or 'cold light', resulting from a chemical reaction, is also well understood. However, another major category of colour/light display in animals has more recently attracted the attention of biologists: structural colouration or reflectors. Structural colouration involves the selective reflectance of incident light by the physical nature of a structure. Although the colour effects often appear considerably brighter than those of pigments, structural colours often result from completely transparent materials. In addition, animal structures can be designed to provide the opposite effect to a light display: they can be antireflectors, causing 'all' of the incident light to penetrate a surface (like glass windows, smooth surfaces cause some degree of reflection).

Hooke [34] and Newton [51] correctly explained the structural colours of silverfish (Insecta) and peacock feathers respectively, and Goureau [25] discovered that the colours produced from the shells of certain molluscs and the thin, membranous wings of many insects, resulted from physical structures. Nevertheless, until the end of the nineteenth century pigments were generally regarded as the cause of animal colours. Accurate, detailed studies of the mechanisms of structural colours commenced with Anderson and Richards [1] following the introduction of the electron microscope.

Figure 1: Scanning electron micrograph of a cross section of the wall of a cylindrical spine of the sea mouse *Aphrodita* sp. (Polychaeta). The wall is composed of small cylinders with varying internal diameters (increasing with depth in the stack), arranged in an hexagonal array, that form a photonic crystal (Parker *et al.* [62]). Scale bar represents 8 μm.

Invertebrates possess the range of structural colours known in animals and will therefore be used to provide examples in this review. Within invertebrates, structural colours generally may be formed by one of three mechanisms: thin-film reflectors, diffraction gratings or structures causing scattering of light waves. Some structures, however, rather fall between the above categories, such as photonic crystals (Parker *et al.* [62]) (Figure 1). In some cases this has led to confusion in identification of the type of reflector. For example, the reflectors in some scarab beetles have been categorised by different authors as multilayer reflectors, three-dimensional diffraction gratings and liquid crystal displays. Perhaps all are correct! It is not always easy to predict how variations in the 'optical' dimensions or design of a structure will alter its effect on light waves. This is particularly the case when the dimensions of the structures are of the order of a few wavelengths of light. Since the above categories are academic, I place individual cases of structural colours in their most appropriate, not unequivocal, category.

The array of structural colours found in animals today results from millions of years of evolution. Structures that produce metallic colours have also been identified in extinct animals (e.g. Towe and Harper [70]; Parker [54]). Confirmation of this fact, from ultrastructural examination of exceptionally well-preserved fossils such as those from the Burgess Shale (Middle Cambrian, British Columbia), 515 million years old, permits the study of the role of light in ecosystems throughout geological time, and consequently its role in evolution.

This review introduces the types of structural colour found in invertebrates, and hence animals in general, without delving deeply into the theory behind them. Examples of their functional and evolutionary implications are introduced. However, this field is very much still in its infancy.

2 Mechanisms causing reflection/antireflection in animals

2.1 Multilayer reflectors (interference)

Light may be strongly reflected by constructive interference between reflections from the different interfaces of a stack of thin films (of actual thickness d) of alternately high and low refractive index (n). For this to occur reflections from successive interfaces must emerge with the same phase and this is achieved when the so-called 'Bragg Condition' is fulfilled. The optical path difference between the light reflected from successive interfaces is an integral number of wavelengths and is expressed in the equation:

$$2nd \cos \Theta = (m + \tfrac{1}{2}) \lambda$$

from which it can be seen that the effect varies with angle of incidence (Θ, measured to the surface normal), wavelength (λ) and the optical thickness of the layers (nd). There is a phase change of half a wavelength in waves reflected from every low to high refractive index interface only. The optimal narrowband reflection condition is therefore achieved where the optical thickness (nd) of every layer in the stack is a quarter of a wavelength. In a multilayer consisting of a large number of layers with a small variation of index the process is more selective than one with a smaller number of layers with a large difference of index. The former therefore gives rise to more saturated colours corresponding to a narrow spectral bandwidth and these colours therefore vary more with a change of angle of incidence. Both conditions can be found in animals – different coloured effects are appropriate for different functions under different conditions. For an oblique angle of incidence, the wavelength of light that interferes constructively will be shorter than that for light at normal incidence. Therefore, as the angle of the incident light changes, the observed colour also changes.

Single layer reflectors are found in nature, where light is reflected, and interferes, from the upper and lower boundaries (Figure 2). A difference in the thickness of the layer provides a change in the colour observed from unidirectional polychromatic light. The wings of some houseflies act as a single thin film and appear as different colours as a result of this phenomenon (Fox and Vevers [21]). A single quarter-wavelength film of guanine in cytoplasm, for example, reflects about 8% of the incident light (Land [40]). However, in a multilayer reflector with 10 or more high index layers, reflection efficiencies can reach 100% (Land [39]). Thus, animals possessing such reflectors may appear highly metallic.

The reflectance of the multilayer system increases very rapidly with increasing number of layers (Land [39]). If the dimensions of the system deviate from the quarter-wave condition (i.e. nd is not equal for all layers), then the reflector is known as 'non-ideal' (Land [39]) in a theoretical sense (may be 'ideal' for some natural situations). 'Non-ideal' reflectors have a reduced proportional reflectance (not always a significant reduction) for a given number of layers and this reflectance has a narrower bandwidth. A narrow bandwidth, less conspicuous reflection is sometimes selected for in animals, as will be discussed later in this paper. Multilayer reflectors polarize light incident at Brewster's angles. This is about 54° for a quarter-wave stack of guanine and cytoplasm. At very oblique angles, all wavelengths are strongly reflected.

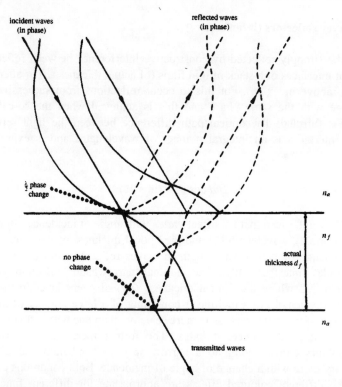

Figure 2: Schematic diagram of thin-film reflection. Direction of wave (straight line) and profile of electric (or magnetic) component are illustrated. Incident waves are indicated by a solid line, reflected waves by broken lines. Refraction occurs at each media interface. The refractive index of the film (n_f) is greater than the refractive index of the surrounding medium (n_a). Constructive interference of the reflected waves is occurring. As the angle of incidence changes, different wavelengths constructively interfere. At normal incidence constructive interference occurs where $n_f \times d_f = \lambda/4$.

Multilayer reflectors are common in animals. They are usually extra-cellular, produced by periodic secretion and deposition, but sometimes occur within cells. Guanine ($n = 1.83$) is a common component in invertebrate reflectors because it is one of very few biological materials with a high refractive index and is readily available to most invertebrates as a nitrogenous metabolite (Herring [30]). However, arthropods, including insects, crustaceans and spiders, have largely ignored guanine in favour of pteridines (Herring [30]). Also surprising is that the reflector material of closely related species, e.g. the molluscs *Pecten* (scallop) and *Cardium* (cockle), may differ (Herring [30]).

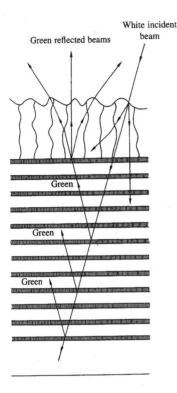

Figure 3: Generalized diagram of a multilayer reflector in the cuticle of the green beetle *Calloodes grayanus* (high refractive index material is shown shaded). Outer layer causes scattering (Parker *et al.* [59]).

The crustaceans *Limnadia* (Conchostraca), *Tanais tennicornis* (Tanaidacea), *Ovalipes molleri* (Decapoda) and the males of *Sapphirina* (Copepoda) all bear multilayer reflectors in their cuticles, in different forms. In contrast to the usual continuous thin layers, male sapphirinids have ten to fourteen layers of interconnecting hexagonal platelets within the epidermal cells of the dorsal integument (Chae and Nishida [10]). The reflector of *O. molleri* comprises layers that are corrugated and also slightly out of phase (Figure 4). At close to normal incidence this structure reflects red light, but at an angle of about 45° blue light is reflected. The corrugation, however, functions to broaden the reflectance band, at the expense of reducing the intensity of reflection (Parker *et al.* [60]).

A broadband wavelength independent reflectance, appearing silver or mirror-like to the human eye, can be achieved in a multilayer stack in at least three ways in invertebrates (Figure 5) (see Parker *et al.* [59] and Parker [56]). These are (A) a composite of regular multilayer stacks each tuned to a specific wavelength, (B) a stack with systematically changing optical thicknesses with depth in the structure, termed a 'chirped' stack, and (C) a disordered arrangement of layer thicknesses about a mean value, termed a 'chaotic' stack (Figure 5).

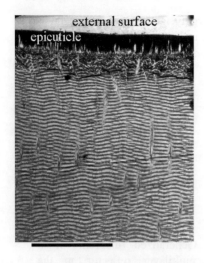

Figure 4: Transmission electron micro-graph of a multilayer reflector in the cuticle of a swimming paddle of the crab *Ovalipes molleri* (Crustacea: Decapoda). Layers of reflector are slightly sinuous and out of phase; note the unusual side branches of each high refractive index (dark) layer, which provide support for these solid layers within a liquid matrix (Parker *et al.* [60]). Scale bar represents 5 μm.

Figure 5: Three ways of achieving a broadband, wavelength-independent reflector in a multilayer stack (high refractive index material is shown shaded) (Parker *et al.* [59]). **A**, Three quarter-wave stacks, each tuned to a different wavelength. **B**, A 'chirped' stack. **C**, A 'chaotic' stack.

The nauplius eye of the copepod *Macrocyclops* (Crustacea) has regularly arranged platelets about 100 nm thick in stacks of 20–60 (Fahrenbach [18]), achieving the first condition. Silver beetles and the silver and gold chrysalis' of butterflies in the genera *Euoplea* and *Amauris* owe their reflection to the second condition (Neville [49];

Steinbrecht and Pulker [68]). The mirror-like reflectors in the scallop *Pecten* eye comprise alternating layers of cytoplasm ($n = 1.34$) and guanine crystals ($n = 1.83$) and approximate an 'ideal' quarter-wave system in the same manner as within fish skin (Land [40]) using the third mechanism. The ommatidia of the superposition compound eyes of *Astacus* (Crustacea) are lined with a multilayer of isoxanthopterin (a pteridine) crystals (Zyznar and Nicol [77]), which again fall into the third category. Multilayer reflectors can also be found in the eyes of certain spiders, butterflies and possibly flies, where they assist vision, as discussed further in this review.

Figure 6: Transverse section through an iridophore from the skin of *Sepia officinalis* (cuttlefish). Reflecting platelets are viewed from their edges, except one viewed from the side (hatched); nucleus is stippled. Platelets form a sinuous but regular stack (Land [39]). (After Schafer [65]).

Squid and cuttlefish, for example, possess mirror-like reflectors in photophores (light organs) and iridophores (Land [39]). Iridophores are cells that in this case contain groups (iridosomes) of flexible layers of thin lamellae with cytoplasm between, forming a quarter-wave stack (Parker [63]) (e.g. Figure 6). The platelets of both squids and octopods develop from the rough endoplasmic reticulum and are separated by extra cellular space (Arnold [2]). Euphausiid crustaceans possess photophores with very elaborate mirror-like reflectors (Herring [30]). Up to 60 dense layers, about 70 nm thick and 75–125 nm apart, are formed from the aggregation of granules (probably a type of chitin) and surround the main photogenic mass (Harvey [27]). An intricate ring consisting of very flattened cells, forming the dense layers of a multilayer reflector (about 175 nm thick, separated by 90 nm) surrounds the lens of the photophore and reflects blue light at acute angles of incidence (Herring [30]).

Dead invertebrates may not display their original colours. Following death, one (or both) of the layers in a multilayer reflector may become gradually reduced. For example, water may be lost from the system. This occurs in beetles of the genus *Coptocycla*; their brassy yellow colour quickly changes through green, blue and violet until the brown of melanin is finally observed. The colour progression may subsequently be reversed by water uptake (Mason [44]). This is an important consideration when examining fossils for multilayer reflectors (see below).

2.2 Diffraction gratings

When light interacts with a periodic surface consisting, for example, of a series of parallel grooves, it may be deviated from the direction of simple transmission or reflection. For this to happen, light which is scattered or diffracted from successive grooves should be out of phase by integral values of 2π. This occurs when for a given direction of propagation the optical path difference via successive grooves is $M\lambda$ where M is an integer known as the circle number. This may be expressed by the grating equation

$$2d (\sin \alpha - \sin \beta) = M\lambda$$

where α and β are angles of incidence and diffraction, and d is the period.

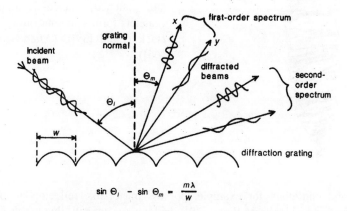

Figure 7: Reflection-type diffraction grating (w = periodicity, m = order of reflected beam). For white light of fixed angle of incidence (Θ_i), the colour observed is dependent on the point of observation (e.g. violet light can be seen at point x, red at point y, within the first order spectrum).

As with multilayers the effect gives rise to colouration because different wavelengths are diffracted into different directions. Although the effect changes with angle of incidence it is less critical than it is with thin films and the visual appearance is different. For a parallel beam of white light incident upon a multilayer, one wavelength will be reflected as determined by the Bragg Condition. The same beam incident upon a grating will be dispersed into spectra (e.g. Figure 7). The complete spectrum reflected nearest to the perpendicular (grating normal) is the first order. The first order spectrum is reflected over a smaller angle than the second order spectrum, and the colours are more saturated and appear brighter within the former. Diffraction gratings have polarizing properties, but this is strongly dependent on the grating profile.

Diffraction gratings were believed to be extremely rare in nature (Fox and Vevers [21]; Fox [20]; Nassau [47]), but have recently been revealed to be common among invertebrates. They are particularly common on the setae or setules (hairs) of Crustacea, such as on certain first antennal setules of male Myodocopina ostracods or 'seed shrimps' (Crustacea) (Parker [52]). Here, the grating is formed by the external surface of juxtaposed rings with walls circular in cross-section (Figure 8) (Parker [55]).

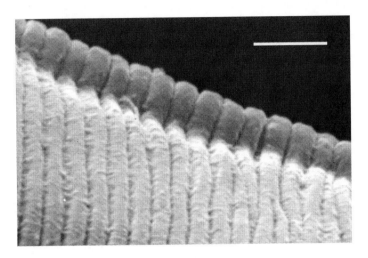

Figure 8: Scanning electron micrograph of diffraction grating on a halophore (setule) of *Azygocypridina lowryi* (Ostracoda) (Parker [52]). Scale bar represents 2 μm.

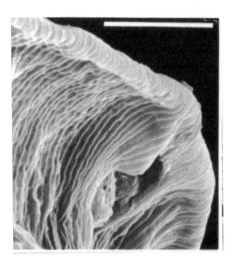

Figure 9: Scanning electron micrograph of diffraction grating on the selvage (of carapace) of *Euphilomedes carcharodonta* (Ostracoda). Scale bar represents 5 μm.

The width of the rings, and consequently the periodicity of the grating, is about 700 nm in *Azygocypridina lowryi*. Different colours are viewed with varying angles of observation under a fixed light source. The ostracod *Euphilomedes carcharodonta*, for example, additionally houses a diffraction grating on the rostrum, a continuous flattened area of the carapace that is corrugated to form periodic ridges (Figure 9). The dark brown beetle *Serica sericea* bears gratings on its elytra with 800 nm periodicity, which causes a brilliant iridescence in sunlight (Anderson and Richards [1]).

Figure 10: Scanning electron micrograph of diffraction gratings on a seta of *Lobochesis longiseta* (Polychaeta). Scale bar represents 10 μm.

Many polychaetes possess gratings on their setae (hairs). For example, the opheliid *Lobochesis longiseta* bear gratings with periodicities in the order of 500 nm (Figure 10), appearing iridescent. The wings of the neurochaetid fly *Neurotexis primula* bear diffraction gratings only on their dorsal surfaces, and the iridescent effect remains after the insect is gold coated for electron microscopy. These gratings cause iridescence with a higher reflectance than the iridescence of the membranous wings of other insects, which reflect light by interference. Iridescence caused by interference disappears after gold coating because transmission of light through the outer surface is prevented.

Very closely spaced, fine setules may also form the ridges of a diffraction grating. Cylindroleberidid ostracods (seed shrimps) possess a comb on their maxilla bearing numerous setules on each seta, collectively forming a grating with a periodicity of about 500 nm (Figure 11).

The 'helicoidal' arrangement of the microfibrils comprising the outer 5–20 μm of the cuticle (exocuticle) of certain scarabeid beetles, such as *Plusiotis resplendens*, also gives rise to metallic colours (Neville and Caveney [50]). Here, the fibrils are arranged in layers, with the fibril axis in each layer arranged at a small angle to the one above, so that after a number of layers the fibrillar axis comes to lie parallel to the first layer. Thus, going vertically down through the cuticle, two corresponding grating layers will be encountered with every 360° rotation of the fibrils - the 'pitch' of the system. Polarized light encounters an optically reinforcing plane every half turn of the helix. The system can be treated as a three-dimensional diffraction grating (Nassau [47]), with a peak reflectance at λ = 2*nd*, where *d* is the separation of analogous planes, or half the pitch of the helix. The diffracted light resembles that from a linear grating except for the polarization; the three-dimensional grating reflects light that is circularly or elliptically polarized. It should be noted that the diffracted colour does not depend on the total film thickness as it does in interference, but on the layer repeat distance within the film as in the diffraction grating (Nassau [47]) (analogous to 'liquid crystals').

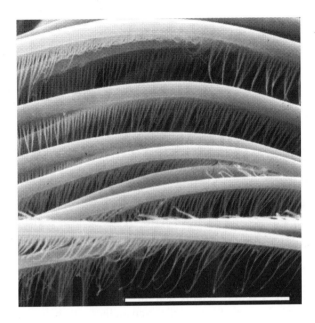

Figure 11: Scanning electron micrograph of *Tetraleberis brevis* (Ostracoda), setae of comb of maxilla; setules (orientated vertically) form the ridges of a grating. Scale bar represents 2 μm.

When each groove of a grating is so formed that it independently, by means of geometrical optics, redirects the light in the direction of a chosen diffracted order, it is known as a blazed grating. In a blazed reflection grating, each groove consists of a small mirror (or prism) inclined at an appropriate angle (i.e. the grating has a 'sawtooth' profile). Blazed gratings have been identified on the wing scales of the moth *Plusia argentifera* (Plusinae).

When the periodicity of a grating reduces much below the wavelength of light, it becomes a zero-order grating and its effect on light waves changes (see Hutley [35]). This difference in optical effect occurs because when the periodicity of the grating is below the wavelength of light the freely propagating diffracted orders are suppressed and only the zero-order is reflected when the illumination is normal to the plane of the grating. To describe accurately the optical properties of a zero-order grating, rigorous electromagnetic theory is required. In contrast to gratings with freely propagating orders, zero-order structures can generate saturated colours even in diffuse illumination (Gale [23]). Such structures occur on the setae of the first antenna of some isopod crustaceans, such as the giant species of *Bathynomus*. Here, there are diffracted orders and the spectral content of the light within the grating is controlled by the groove profile. In an optical system that only accepts the zero-order, what is seen is white light minus that diffracted into the ±1 orders.

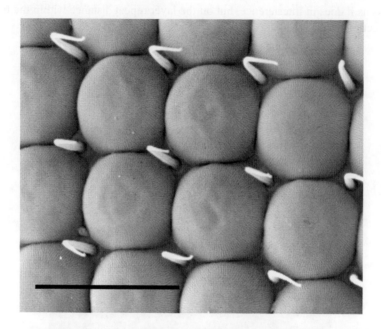

Figure 12: Scanning electron micrograph of the compound eye of *Zalea minor* (Diptera). Lenses of six whole ommatidia (with hairs between), showing antireflection gratings on the corneal surface (Parker *et al.* [61]). Scale bar represents 20 μm.

Figure 13: Tri-grating (antireflector) on the corneal surface of a butterfly eye. Scale bar represents 2 μm.

A zero-order grating can cause total transmission (i.e. there is no reflection). Such antireflective structures are found on the corneal surfaces of each ommatidium (visual unit) in the eye of *Zalea minor* (Diptera) (Figure 12). The periodicity of the corneal gratings of this fly is 242 nm (Parker *et al* [61]. Another form of antireflection grating is formed on the transparent wings of the hawkmoth *Cephonodes hylas* (Yoshida *et al.* [75]), on the corneal surface of each visual unit (ommatidium) of the eyes of moths (Miller *et al.* [46]) and butterflies (e.g. Figure 13). Here, optical-impedance matching is achieved by means of an hexagonal array of tapered cylindrical protuberances, each of about 250 nm diameter (Miller *et al.* [46]), thus forming a 'tri-grating' with grooves transecting at 120°. The protuberances provide a graded transition of refractive index between the air and the cornea/wing. Hence the refractive index at any depth is the average of that of air and the corneal/wing material.

The grooves of a grating may also create parallel rows in two directions, forming a bi-grating. Bi-gratings can be found in some crustaceans and flies. In the amphipod crustacean *Waldeckia australiensis*, two effectively superimposed gratings subtend angles of about 60/120° (Figure 14).

Figure 14: Bi-grating on the callynophores (setae) of the amphipod crustacean *Waldeckia australiensis*. Scale bar represents 2 μm.

2.3 Scattering

Simple, equal scattering of all spectral wavelengths results in the observation of a diffuse white effect. This commonly arises from the effects of a non-periodic arrangement of colloidally dispersed matter where the different materials involved have different refractive indices (Figure 15), or from solid colourless materials in relatively concentrated, thick layers (Fox [20]). In the colloidal system, the particles are larger than the wavelength of light and can be thought of as mirrors oriented in all directions. The reflection is polarized unless the incident light is at normal incidence on the system and, in the colloidal system, spherical or randomly arranged particles are involved.

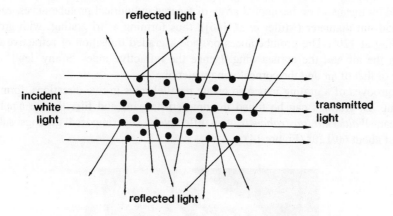

Figure 15: Scattering of white light by small particles (represented by black circles). The reflected light is randomly directed at different angles.

The colloidal system involves either a gas-in-solid, gas-in-liquid, liquid-in-liquid (emulsions) or solid-in-liquid (Fox [20]). For example, the gas-in-liquid system is partly responsible for the white body and/or tentacles of certain anemones (Fox [20]). Light is reflected and refracted at the surfaces of the particles of matter or spaces (with dimensions >1 μm), regardless of the colour of the materials involved (except opaque brown and black compounds, such as melanin) (Mason [44]). In insects, the materials involved typically have very low transparencies (Mason [44]).

From some scales of butterfly wings, light is scattered uniformly and completely in all directions, due to the chaotic disposition of the surfaces. Matt or pearly whites may be observed depending on the complexity or the arrangement of the structures, which affects the relative degree of scattering (Mason [44]). The structures may be so small that the molecular topography of the surface has an effect. The chromatic effects of the butterfly scales are greatly intensified if a dark, absorbing pigment screen lies beneath (Fox [20]). This screen prevents reflection of white or coloured light from the background that would dilute or alter the colour. Additionally, if a dark pigment such as melanin is interspersed with the scattering elements, the reflection will appear a shade of grey or brown. The cells of the reflector in the photophore of a beetle ('fire-fly') are packed with sphaerocrystals of urate that cause a diffuse reflection (Lund [42]).

Reflection and refraction that occur at the interfaces of strata with different refractive indices may result in the display of white light. The degree of whiteness depends upon the difference in refractive indices (Fox [20]). This mechanism is evident in the shells of many lamellibranch molluscs (Verne [71]). Between the outer, often pigmented layer and the mantle is a thick middle layer of crystalline calcium carbonate.

The inner surface of this (nacreous) layer is lined with multiple laminations of the same salt. In most species these laminations are sufficiently thick (>1 μm) to render the inner lining white, although in some species they become thin so as to form a multilayer reflector. Calcium carbonate similarly produces whiteness in Foraminifera and in calcareous sponges, corals, echinoderms and crustacean cuticles. Also in the class of white solids is silica in diatom tests and skeletons of hexactinellid sponges (Fox [20]).

An unordered (as opposed to periodic) group of closely spaced setae, such as those in patches on the fly *Amenia* sp., may form a white reflection via random scattering or reflection (Figure 16). However, if the arrangement becomes periodic to some degree, a diffraction grating may be formed, such as the grating of *Tetraleberis brevis* (Ostracoda) (Figure 11).

Figure 16: Scanning electron micrograph of a section of a white reflecting patch on the abdomen of the fly *Amenia* sp., showing closely packed, randomly arranged setae that scatter incident light in all directions (cf. Figure. 11). Scale bar represents 20μm.

Other forms of scattering also exist and result in a blue coloured effect (red when the system is viewed in transmission). Tyndall or Mie scattering occurs in a colloidal system where the particle size approximates to the wavelength of light. Here, diffraction is important. Rayleigh scattering occurs in molecules in a two-photon process by which a photon is absorbed and raises the molecule to an excited electronic state from which it re-radiates a photon when it returns to the ground state. Diffraction is not involved here.

Tyndall scattered light is polarized under obliquely incident light. The intensity of the resultant blue is increased when it is viewed against a dark background, such as melanin. The relative sizes of particles determine the shade of blue. If the particles responsible for the scattering coalesce to form particles with a diameter greater than about 1 μm, then white light is observed (see above). A gradation from blue to white scattering ('small' to 'large' particles) occurs on the wings of the dragonfly *Libellula pulchella* (Mason [44]).

Scattered blues can also be found in other dragonflies. In the aeschnids and agrionids, the epidermal cells contain minute colourless granules and a dark base. The males of libellulids and agrionids produce a waxy secretion that scatters light similarly over their dark cuticle. The green of the female *Aeschna cyanea* is the combined result of Tyndall scattering and a yellow pigment, both within the epidermal cells (degradation of the yellow pigment turns the dead dragonfly blue) (Fox and Vevers [21]).

Scattered blues are also observed from the skin of the cephalopod *Octopus bimaculatus* (Fox [20]), where a pair of ocelli are surrounded by a broad blue ring. Blue light is scattered from this region as a result of fine granules of purine material within cells positioned above melanophore cells. The colour and conspicuousness of the ring are controlled by the regulation of the melanophores, by varying the distribution of melanin and consequently the density of the absorbing screen. The squid *Onychia caribbaea* can produce rapidly changing blue colours similarly (Herring [30]). The bright blue patterns produced by some nudibranch molluscs result from reflecting cells containing small vesicular bodies, each composed of particles about 10 nm in diameter and therefore appropriate for Rayleigh scattering (Kawaguti and Kamishima [37]).

3 Functions of animal reflectors in behavioural recognition and camouflage

Structural colours may provide either conspicuousness or camouflage under this category. The metallic coloured effect of invertebrate structural colours is often very distinct in environments where light is present, during daylight hours. Therefore in such situations, metallic colours may be functional if they are displayed externally in the host animals' natural environment, where other animals capable of detecting light coexist. This statement is made because needlessly attracting attention to oneself carries obvious disadvantages (Verrell [72]). Thus, a structure producing incidental metallic colour may become modified, by the action of selective pressures, to prevent the external display of metallic colour. For example, the shells of many molluscs and the exoskeleton of certain crabs have an opaque outer layer. This opaque layer prevents ambient light becoming incident on the internal structural layers, which contain the materials and dimensions of a multilayer reflector. Iridescence is displayed from the internal surfaces of these shells and exoskeletons in the presence of an incident light source, but this is biologically insignificant because it is not visible in the external environment. These internal structural layers, which provide a function other than iridescence (i.e. contribute structural strength), comprise an incidental multilayer reflector ('non-ideal'; Land [39]). However, conspicuous colour may not necessarily contribute to an increased mortality rate via predation. For example, blackbirds with a bright red wing patch have a lower rate of predation compared to those without a red patch (Götmark [24]).

Another point to consider regarding the function of structural colours is the efficiency of reflection. Simply because the resulting reflectance is theoretically sub-optimal ('weak') does not imply that the reflected light is not functional. In fact this condition is sometimes appropriate. The cryptic beetle *Calloodes grayanus* appears a weak structural green colour from its complete dorsal hemisphere. This matches exactly the background radiation of leaves; since leaves reflect light omnidirectionally, to appear camouflaged and deceive predators the beetle must match this reflection. Such an effect cannot be achieved from the theoretically optimal quarter-wave stack. However, if one rule can be made on the functions of structural colour it is that each case should be considered independently and comparisons with other cases minimised since no two species live under exactly the same conditions.

3.1 Terrestrial invertebrates

The metallic colours of butterflies and beetles are known to provide warning colouration (e.g. the eye spots of butterfly wings), deceptively changing images in the eyes of predators (Hinton [32]), and/or an attractant to conspecifics. Some beetles (Hinton [33]) and butterflies (Nekrutenko [48]; Brunton and Majerus [8]) also reflect ultraviolet light to further enhance a pattern conspicuous in the human visual range.

Some flies such as *Amenia* sp., have areas that scatter white and ultraviolet light in all directions. This light probably assists conspecific recognition. The predacious fly *Austrosclapus connexus* has multilayer reflectors in its exoskeleton, including the cornea of its eyes (Bernard and Miller [6]), which reflect only green light in certain directions and blue in others. These colours may be used partly to provide camouflage from predators or prey. The metallic colours from the membranous wings of some insects may also be functional. For example, the wasp *Campsoscolia siliata* is a pollinator of *Ophrys vernixia*, a bee orchid. The flower of the orchid mimics the colour (and to some extent shape) of the female wasp. The male wasp is deceived and attempts to mate with the flower (subsequently transporting pollen). However, in addition to mimicking the body colouration, the flower also mimics the wings of the wasp as a blue central region with a red outline. This is how the metallic coloured wings appear from certain directions. Therefore, the metallic blue colour of the wings of the female *C. siliata* appears to be an important character during mating (Paulus and Gack [64]).

Some beetles appear green as a result of structural colours. True green pigments are generally rare in insects, particularly in beetles, and a structural green may be the most easily achieved substitute (Crowson [72]). Additionally, infrared radiation is reflected, possibly as part of a thermoregulatory mechanism.

3.2 Aquatic invertebrates

Water absorbs light, but selectively for different colours. For example, red is the first colour to disappear with increasing oceanic depth; blue the last. In fact some deep-sea animals can detect blue light down to about 1000 m (Denton [85]). From a given point in the water column, light travelling directly downwards will be least attenuated with depth because it will travel the least distance through the water (Denton [14]). However, a downwardly directed light would be the least visible from far away because it would be viewed against the brightest background. Many deep-sea animals make themselves less visible by shining lights downwards to diminish the shadow that they cast below themselves (Denton [14]). Therefore, selective pressures may determine the position and orientation of the structures causing structural reflectances, and a modification of behaviour, so that light is reflected in a direction that causes the maximum effect. This explains the distribution of iridescence over the body of the crab *Ovalipes molleri* (Decapoda). This crab, which usually orientates its carapace (shell) at an angle of about 45° when on the sea floor, houses multilayer reflectors in its dorsal exoskeleton. These multilayers reflect blue light at angles of about 45°. Therefore blue light, the main light present at the depths where these crabs live (Denton [15]), is reflected approximately laterally when the crab is on the sea floor. Additionally, due to the corrugation of layers in its multilayer stack (Figure 4) the reflectance of the crab is over a wide angle and therefore permits some degree of directional flexibility. Hence, an individual of *O. molleri* on the sea floor could attract a conspecific also on the sea floor and at the same time remain invisible to predators above. However, other species of *Ovalipes* that live in shallow waters do not possess reflectors, or contain them in very restricted areas of the total body surface. This is because in shallow waters longer wavelengths of light (e.g. red) are also present, and these would be reflected almost vertically upwards from a crab on the sea floor as its carapace approaches the horizontal position (such as during burial). Therefore the crab would be conspicuous to predators, which hunt with more emphasis on visual cues in shallower waters. However, the fact that some species of *Ovalipes* that inhabit shallow waters display at least some degree of structural colour, despite the dangers, indicates that the reflectors are probably functional (Parker *et al.* [60]).

The maximum reflectance (theoretically optimal) situation may not always be practical; a structurally coloured animal does not always know the position of the recipient animal prior to signalling (e.g. a predator could approach from any angle, in which case structural colour used to provide warning colours or camouflage must have a broad angular field). Alternatively, an advantage of reflecting light maximally towards the surface (i.e. when the recipient animal is above the animal displaying the structural colour in the water column) is that the background is usually darker than the ocean surface, and therefore the contrast of the structural colour against its environment is greater. The whole system is very much a compromise.

Male copepods (Crustacea) in the family Sapphirinidae display different, species specific, colours as a result of multilayer reflectors in their dorsal integument (Chae and Nishida [10]). The daytime depths at which each species of Sapphirinidae lives depends on the light conditions of the ocean; species reflecting all spectral colours live in near surface waters where all colours are present in the incident light, and species reflecting only blue live in deeper waters where the incident light is mainly blue. It is believed that male structural colour, the well-developed eye and the daytime shoaling of Sapphirinidae are closely related and constitute a mate-finding mechanism (Chae and Nishida [11]).

Ostracoda (Crustacea) is a good example of a group of animals where their bioluminescent light display has attracted much attention since the seventeenth century. Nevertheless, the metallic colour of ostracods, often so bright it appears like a neon light, has until recently (Parker [52]) remained unnoticed. This may be due to the unusual orientation of the ostracod required for observation of the metallic colour. Most importantly, this metallic colour is known to be functional. In at least one species of *Skogsbergia,* when a male ostracod approaches a female its 'iridescent fan' (collection of metallic coloured setules) is displayed, which is otherwise held within the carapace which encloses the body. The female then becomes sexually receptive to the light displayed and mating follows (Parker [52]; Parker [53]).

4 Mirror and antireflection functions of animal structures

In addition to providing a direct light display function for behavioural recognition and camouflage, structural reflectors may act to focus light (usually all incident wavelengths, i.e. in a mirror-like manner) to increase the efficiency of a light system, or provide protection of the host from harmful intensities of light produced by a light system. Reflectors may also act as filters in a system, screening out unwanted wavelengths from incident light. In these cases, the system may be a photophore (light organ), an eye, or a light director. Under this category the structures are usually the multilayer type. Sometimes light is required to be optimally absorbed into tissue, and antireflection structures may be employed.

4.1 Mirrors in photophores

Photophores are light emitting organs where the chemical reaction that produces bioluminescence occurs within the organ itself. They are present in many invertebrate taxa (see Harvey [28]). Photophores often consist of a layer of luminous cells with an underlying concave reflector, and sometimes an overlying lens or lens system (in euphausiid and decapod crustaceans, and squid; Herring [30]). The luminous cells of invertebrate photophores may belong to the host (photocytes), such as in the beetle *Pyrophorus* (Dahlgren [13]) and the shrimp *Sergestes prehensilis* (Terao [69]) (Figure 17), or may be symbiotic luminous bacteria, such as in the squid *Sepiola intermedia* (Skowron [67]). The mirror-like reflector beneath the main photogenic mass serves to: (i) direct the bioluminescent light in a precise direction, and/or (ii) protect the body tissue beneath from the harmful effects of intense light. In some cases there are dark pigments (e.g. melanin), sometimes contained within chromatophores (e.g. melanophores), behind the reflector to enhance the reflective effect. The ring which surrounds the lens in euphausiid photophores reflects blue light, i.e. bioluminescence, at oblique angles of incidence and therefore functions to collimate light that escapes round the edge of the lens (Herring [30]). Similarly, in some enoploteuthine squid photophores, collagen fibres forming a multilayer reflector encircle the photogenic crystals and provide light guides, enabling total internal reflection (Herring [30]). In other enoploteuthine photophores, the collagen fibres are lost from the ring (or torus), leaving membranous lamellae to form this reflective tissue (Young [76]). Lamellae formed of endoplasmic reticulum with a periodicity of about 25nm occur in the squid *Selenoteuthis* and probably act as additional reflective diffusers (Herring *et al* [31]). The distal iridosomes (platelet groups) of many squid photophores form multilayer reflectors when aligned parallel to the skin surface. These probably function as interference filters limiting the spectral emission of the photophores (Herring [30]). The axial stack of iridosomes in the squid *Abralia* and *Enoploteuthis* probably act similarly (Young [76]). In some of the large subocular photophores in squid, e.g. in *Bathothauma*, the distal iridosomes function to spread the bioluminescent light over a large surface area (Dilly and Herring [16]).

The mirror-like reflector layer in the photophore of a 'fire-fly' beetle (e.g. Figure 17A) achieves scattering rather than directional reflection, and therefore its function is presumed to be protection (Lund [42]). However, when reflectors are present in photophores, they are more generally multilayer types (e.g. Figure 17B).

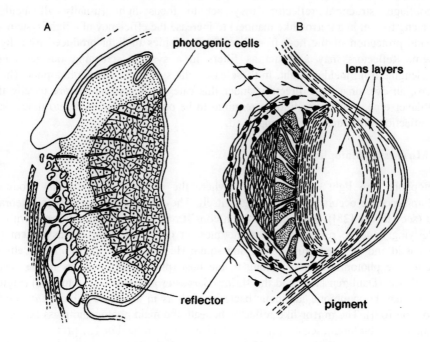

Figure 17: Sections of photophores, external surface of animals on the right. **A** The 'fire-fly' beetle *Pyrophorus* sp., abdominal photophore of male (after Dahlgren [13]). **B** The shrimp *Sergestes prehensilis*, showing lens layers and absorptive pigment (after Terao [69]).

4.2 Mirrors in eyes

Broadband multilayer reflectors are found in the eyes of many invertebrates to focus incident light onto the retina (e.g. Figure 18). Reflectors perform the function of light path doubling in the tapeta of lycosid spiders (Baccetti and Bedini [3]) and butterfly ommatidia (Miller and Bernard [45]), and image-forming in some molluscs and crustaceans (Land [40]).

The eyes of the scallop *Pecten* (Mollusca) and the ostracod *Gigantocypris* (Crustacea) contain a concave reflector with a retina positioned at the focal point(s). The reflector is almost spherical in *Pectin*, but more complex in *Gigantocypris* (Figure 18A) (Land [40]). These simple mirror eyes, especially those of *Gigantocypris*, are exceptionally good at gathering available photons and, therefore, detecting extremely low levels of light (Land [40]). *Gigantocypris* is found, for example, at 1000 m depths, where such a property is critical in achieving vision. However, the rarity of the *Pecten* and *Gigantocypris* type eyes in nature (a similar design may also be found only in some copepod and a *Notodromas* ostracod (Crustacea) eyes (Land [40])) suggests a disadvantage in this design; they are probably poor at resolving low-contrast patterns (Land [40]).

Reflecting superposition compound eyes of macruran crustaceans contain an array of reflector-lined ommatidia (Figure 18B). The reflectors are aligned so that they always focus incident light at a single point on the retina (Land [40, 41]). These eyes greatly intensify the image, and so are again useful in low light regimes, but also provide good resolution (Land [40]).

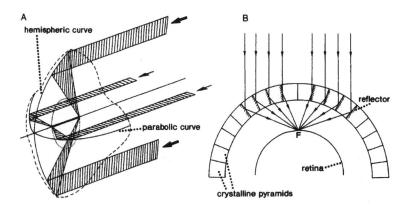

Figure 18: Schematic diagrams of eyes with mirror optics. A *Gigantocypris* (Crustacea: Ostracoda) reflector (modified 'dish', dashed line indicates edge) with a part spherical and part parabolic profile. This reflector produces a line image within the retina (dotted). B Reflecting superposition compound eye, with reflectors coating the crystalline 'pyramids' which focus incident light onto the same point on the retina (F) (Land [41] Incident light beams are arrowed. Reproduced (modified) from [41] by permission of M. Land.

Reflecting pigment cells occur in the compound eyes of many crustaceans and often undergo substantial movements during light and dark adaptation (Douglass and Forward 17]). The reflective elements are usually aggregated or withdrawn below the basement membrane during light adaptation, and are dispersed around the rhabdom and crystalline cones during dark-adaptation (Ball *et al.* [4]). The tapetum of some pelagic crustaceans is often not uniformly distributed around the eye; the ventral tapetum is most developed because of the comparatively low light intensities below the host (Shelton *et al.* [66]). This distribution of reflectors may also provide a camouflage function by making the opaque eye less visible (Douglass and Forward [17]).

Multilayer reflectors composed of alternating layers of high and low density chitin occur in the cornea of certain Diptera, especially horseflies (Tabanidae) and long-legged flies (Dolichopodidae), and produce colour patterns from the eye surface (Bernard and Miller [7]). These reflectors may serve to (A) reduce glare caused by sources outside of the ommatidial visual field, (B) optically enhance contrast for coloured objects in a background of dissimilar colour, or (C) provide colour vision by filtering different colours in different regions (Bernard [5]).

190 *Nature and Design*

Anti-reflection, zero-order gratings may be present on the corneal surface of the eyes of certain flies. These permit almost total transmission of incident light, thus maximising the number of photons entering the eye and are potentially incident on the retina (Parker *et al.* [61]). Similar anti-reflective structures (an hexagonal array of cylindrical protuberances, effectively forming a gradual change in refractive index) are found on the cornea of the night moth eye (Miller *et al.* [46]), although here the function is also believed to be an addition to the moths' stealth system (Gale [23]).

4.3 Mirrors in other body parts

A new species of amphipod related to the genus *Danaella* (Crustacea: Lysianassidae) has a source of bioluminescence in its head, but also a quite separate reflector in the form of an expanded joint of the second antenna (Figure 19). This joint is shaped like a shallow cup or shield, and the concave side displays a high, broadband reflection. The concave side points downwards in the relaxed state, and may be used to precisely direct the bioluminescence, in a narrow beam, from its source in the head (Parker [57]).

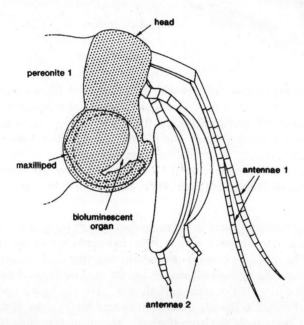

Figure 19: Diagram of the head region of the amphipod crustacean Gen. Nov., aff. *Danaella*, sp. C1, juvenile, lateral view. Shaded areas appear black; expanded, dish-like joints (about 1.5mm long) of the second antenna appear silver on their concave surfaces (facing the head), resulting from a 'chirped' multilayer reflector (Parker [57]).

Giant clams in the genus *Tridacna* are usually positioned with their valves held wide apart (Mansour [43]). In this position, the outer reflecting edge of the shell directs light onto the mantle edge. At the mantle edge, the light intensity becomes high enough to support the physiology of zooxanthellae (algae), which are 'farmed' for consumption by the clam (Mansour [43]; Wilkens [74]). Iridophores in *Tridacna* may also function as reflectors of harmful light waves in tropical sunlight (Kawaguti [36]).

The chrysalis of the butterfly *Euploea core* has a mirrored surface to provide camouflage. The surrounding environment is reflected from this surface so that the chrysalis cannot be seen (Parker [56]). However, this means of camouflage can only be achieved in an environment with diffuse light to prevent a strong, direct reflection from the sun. *Euploea core* indeed lives in forests with diffuse light conditions. Many fishes take advantage of such conditions, to achieve the same effect, in the sea (Denton [15]). Similarly, iridophores (broadband reflecting cells) camouflage the parts of squids and cuttlefishes that cannot, by their nature, be made transparent, such as eyes and ink sacs (Land [39]). Iridophores in some echinoderms and cephalopod and bivalve molluscs may also appear as sand grains to an observer (Kawaguti and Kamishima [38]; Land [39]). Here, light is re-directed to prevent illumination of the host animal and support a stealth system.

5 Evolution of animal reflectors

Some invertebrate taxa may have evolved with light as the major stimulus. In this situation, the evolution of structural colours may correlate with the evolution of species.

The least derived extant (living) Myodocopida (Crustacea: Ostracoda) appears to be *Azygocypridina* (about 350 million years old), within the family Cypridinidae (Parker [52]). The evolution of Cypridinidae following *Azygocypridina* shows a consistent improvement in the physics of the diffraction gratings that produce the metallic colour. One group of Cypridinidae continued this trend to the point where the most derived species have very dense 'iridescent fans' (collection of metallic coloured hairs) with theoretically near-perfect reflectors in males and live in very shallow water to obtain maximum incident light. The females of these derived species have very sparse iridescent fans, appearing similar to those of less derived male and female species of Cypridinidae. Another group of Cypridinidae, following the evolution of an eye maximally attuned to blue light, produce bioluminescence, which is also blue. Although this bioluminescent group probably made metallic colour functionally redundant, they continue the light adaptation story. In fact the whole of the Cypridinidae appear to have evolved with light as the major stimulus. Using non-bioluminescent cypridinids, a cladogram (a 'tree' diagram inferring evolutionary relationships) made using many morphological characters, not linked to light, reveals exactly the same sequence as a cladogram made using only characters linked to light adaptation (Parker [52]). Divergence in sexual light displays may have generated sufficient sexual isolation among populations to lead to speciation (see Verrell [72]). Therefore, cypridinid metallic colour is probably a precursor of cypridinid bioluminescence (Parker [52]).

The molecular processes underlying the generation of structural colour patterns are also under investigation, beginning with the eyespots of butterfly wings (Carroll *et al.* [9]). Additionally, we are now considering evolution as a process for producing optimal designs for a light reflector; this may be supported by the case of the beetle and the fern. The intricate 'multilayer' reflectors of certain scarabeid beetles, such as *Plusiotis resplendens* (Neville and Caveney [50]), are identical to those of some ferns (Graham *et al.* [26]). A blue reflection, therefore, has been achieved independently by evolution of the same design, which may be inferred as highly efficient. An alternative hypothesis, however, is that the reflector is just efficient enough, but simple so that it is easily evolved.

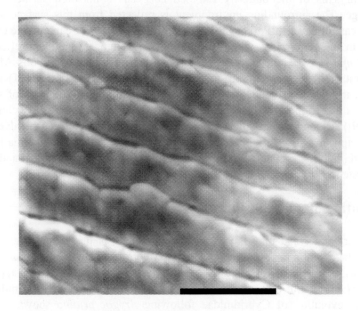

Figure 20: Transmission electron micrograph of a multilayer reflector in the shell of an 80 million year old ammonite. Scale bar represents 250 nm.

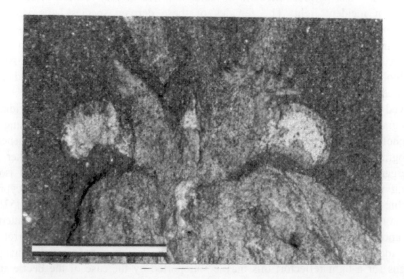

Figure 21: Light micrograph of the lateral eyes of the arthropod *Waptia fieldensis* (Middle Cambrian, 515 million years old, British Columbia). Scale bar represents 2 mm.

Diffraction gratings are responsible for the nacreous lustre of pholidostrophiid brachiopods, such as those from the Devonian, around 360 million years old (Towe and Harper [70]). Here, tabular aragonite platelets averaging 600 nm in thickness, each comprising a linear diffraction grating, form layers (Towe and Harper [70]) and consequently a three-dimensional diffraction grating. Multilayer reflectors occur in the shells of some ammonites, such as in a specimen known from South Dakota, 80 million years old (Figure 20).

Antireflective, zero-order gratings have been identified on the eye of an Eocene fly, 45 million years old, preserved in Baltic amber (Parker *et al.* [61]). Linear diffraction gratings causing colour have been discovered on the sclerites of *Wiwaxia corrugata* from the Burgess Shale (Middle Cambrian, 515 million years old, British Columbia) (Parker [54]). This polychaete lived (Fritz [22]) where ambient light levels may have been sufficient for the gratings to be effective in reflecting colours. Animals with eyes are also known from the Burgess Shale (e.g. Figure 21), such as *Anomalocaris canadensis* and *Opabinia regalis*, which probably include predators of *Wiwaxia*. Therefore, any light reflected from *Wiwaxia* may have served as warning colouration. Diffraction gratings have been identified on the defensive parts ('spines' and 'shields') of other fossils from the Burgess Shale, such as *Canadia spinosa*, a polychaete (Figure 22), and *Marrella splendens*, a relative of trilobites (Parker [54]). These diffraction gratings have not survived in their entirety, rather as mosaics (e.g. Figure 22). Therefore, to observe the original colours, the surface must be reconstructed in a photoresist. Then accurate reconstructions of these animals can be made in colour.

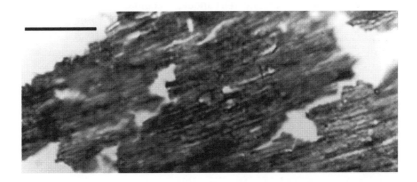

Figure 22: Light micrograph of the surface of a seta of *Canadia spinosa* (Middle Cambrian, 515 million years old, British Columbia), isolated by acid maceration of the rock matrix, showing gratings running longitudinally. Scale bar represents 10 μm.

Prior to the Cambrian period, any incidental iridescence would have been neutrally selective because predators with eyes did not exist. However, during (or just prior to) the Cambrian, predators and eyes (capable of producing visual images) (e.g. Fordyce and Cronin [19]) began to evolve. The sudden evolution of predators and vision would have effectively 'turned on the light' for the Cambrian animals. Metazoan (multicellular) animals were suddenly visually exposed to predators for the first time. The abrupt addition of this new, yet most powerful, stimulus to metazoan behaviour would have caused extreme disorder in the system, and may have been the major cause of the explosion in evolution which occurred in the Cambrian (Parker [54, 58]). Light displayed from the defensive parts of Cambrian animals was probably a response to predators with eyes, i.e. to advertise their armour. Light has probably been a major selection pressure in the subsequent evolution of metazoan animals, and has driven the evolution of the diversity of optical reflectors found in animals today.

Acknowledgements

This work was funded by The Royal Society (University Research Fellowship), The Ernest Cook Trust (Somerville College, Oxford, Fellowship) and The Australian Research Council (Large Research Grant).

References

[1] Anderson, T.F. & Richards, A.G., An electron microscope study of some structural colours of insects. *Journal of Applied Physics*, **13**, pp. 748–758, 1942.

[2] Arnold, J.M., Organellogenesis of the cephalopod iridophore: cytomembranes in development. *Journal of Ultrastructure Research*, **20**, pp. 410–421, 1967.

[3] Baccetti, B. & Bedini, C., Research on the structure and physiology of the eyes of a lycosid spider. *Archives Italiennes de Biologie*, **102**, pp. 97–122, 1964.

[4] Ball, E.E., Rao, L.C., Stone, R.C. & Land, M.F., Eye structure and optics in the pelagic shrimp *Acetes sibogae* (Decapoda, Natantia, Sergestidae) in relation to light-dark adaptation and natural history. *Philosophical Transactions of the Royal Society of London, Biological Sciences*, **313**, pp. 251–270, 1986.

[5] Bernard, G.D., Evidence for visual function of corneal interference filters. *Journal of Insect Physiology*, **17**, pp. 2287–2300, 1971.

[6] Bernard, G.D. & Miller, W.H., Interference filters in the corneas of Diptera. *The Journal of Physiology*, **191**, pp. 416–434, 1967.

[7] Bernard, G.D. & Miller, W.H., Interference filters in the corneas of Diptera. *Investigative Ophthalmology*, **7**, pp. 416–434, 1968.

[8] Brunton, C.F.A. & Majerus, M.E.N., Ultraviolet colours in butterflies: intra- or inter-specific communication? *Proceedings of the Royal Society of London, Biological Sciences*, **260**, pp. 199–204, 1995.

[9] Carroll, S.B., Gates, J., Keys, D.N., Paddock, S.W., Panganiban, G.E.F., Selegue, J.E. & Williams, J.A., Pattern formation and eyespot determination in butterfly wings. *Science*, **265**, pp. 109–114, 1994.

[10] Chae, J. & Nishida, S., Integumental ultrastructure and color patterns in the iridescent copepods of the family Sapphirinidae (Copepoda: Poecilostomatoida). *Marine Biology*, **119**, pp. 205–210, 1994.

[11] Chae, J. & Nishida, S., Vertical distribution and diel migration in the iridescent copepods of the family Sapphirinidae: a unique example of reverse migration? *Marine Ecology Progress Series*, **119**, pp. 111–124, 1995.

[12] Crowson, R.A., *The Biology of the Coleoptera*, Academic Press, Glasgow, 1981.

[13] Dahlgren, U., The production of light by animals. *Journal of the Franklin Institute*, **183**, pp. 79–754, 1917.

[14] Denton, E.J., On the organization of reflecting surfaces in some marine animals. *Philosophical Transactions of the Royal Society of London: Biological Sciences*, **258**, pp. 285–313, 1970.

[15] Denton, E.J., Light and vision at depths greater than 200 metres. *Light and Life in the Sea*, eds. P.J. Herring, A.K. Cambell, M. Whitfield & L. Maddock, Cambridge University Press: Cambridge, pp. 127–148, 1990.

[16] Dilly, P.N. & Herring, P.J., Ultrastructural features of the light organs of *Histioteuthis macrohista* (Mollusca: Cephalopoda). *Journal of Zoology*, **195**, pp. 255–266, 1981.

[17] Douglass, J.K. & Forward, R.B., The ontogeny of facultative superposition optics in a shrimp eye: hatching through metamorphosis. *Cell and Tissue Research*, **258**, pp. 289–300, 1989.

[18] Fahrenbach, W.H., The fine structure of a nauplius eye. *Zeitschrift für Zellforschung und Mikroskopische Anatomie*, **62**, pp. 182–197, 1964.

[19] Fordyce, D. & Cronin, T.W., Comparison of fossilized schizochroal compound eyes of phacopid trilobites with eyes of modern marine crustaceans and other arthropods. *Journal of Crustacean Biology*, **9**, pp. 554–569, 1989.

[20] Fox, D.L., *Animal Biochromes and Structural Colours*, University of California Press: Berkeley, 1976.

[21] Fox, H.M. & Vevers, G., *The Nature of Animal Colours*, Sidgwick and Jackson Ltd.: London, 1960.

[22] Fritz, W.H., Geological setting of the Burgess Shale. *Proceedings of the North American Paleontological Convention, September 1969*, **1**, pp. 1155–1170, 1971.

[23] Gale, M., Diffraction, beauty and commerce. *Physics World*, **2**, pp. 24–28, 1989.

[24] Götmark, F., Does a novel bright colour patch increase or decrease predation? Red wings reduce predation risks in European blackbirds. *Proceedings of the Royal Society of London: Biological Sciences*, **256**, pp. 83–87, 1994.

[25] Goureau, M., Sur l'irisation des ailes des insects. *Annuaire de la Société Entomologique de France*, **12**, pp. 201–215, 1842.

[26] Graham, R.M., Lee, D.W. & Norstog, K., Physical and ultrastructural basis of blue leaf iridescence in two neotropical ferns. *American Journal of Botany*, **80**, pp. 198–203, 1993.

[27] Harvey, B.J., Circulation and dioptric apparatus in the photophores of *Euphausia pacifica*: some ultrastructural observations. *Canadian Journal of Zoology*, **55**, pp. 884–889, 1977.

[28] Harvey, E.N., *Bioluminescence*, Academic Press Inc.: New York, 1952.

[29] Herring, P.J., Studies on bioluminescent marine amphipods. *Journal of the Marine Biological Association of the United Kingdom*, **61**, pp. 161–176, 1981.

[30] Herring, P.J., Reflective systems in aquatic animals. *Comparative Biochemistry and Physiology*, **109A**, pp. 513–546, 1994.

[31] Herring, P.J., Dilly, P.N. & Cope, C., The photophore morphology of *Selenoteuthis scintillans* Voss and other lycoteuthids (Cephalopoda: Lycoteuthidae). *Journal of Zoology*, **206**, pp. 567–589, 1985.

[32] Hinton, H.E., Some little-known surface structures, *Insect ultrastructure*, ed. A.C. Neville, Royal Entomological Society, London, 1970.

[33] Hinton, H.E., Plastron respiration in bugs and beetles. *Journal of Insect Physiology*, **22**, pp. 1529–1550, 1976.

[34] Hooke, R., *Micrographia*, Martyn and Allestry: London, 1665.
[35] Hutley, M.C., *Diffraction Gratings*, Academic Press: London, 1982.
[36] Kawaguti, S., Electron microscopy of the giant clam with special reference to zooxanthellae and iridophores. *Biological Journal of Okayama University*, **12**, pp. 81–92, 1966.
[37] Kawaguti, S. & Kamishima, Y., Electron microscopic study on the iridophores of opisthobranchiate molluscs. *Biological Journal of Okayama University*, **10**, pp. 83–91, 1964a.
[38] Kawaguti, S. & Kamishima, Y., Electron microscopic structure of iridophores of an echinoid. *Diadema setosum*, *Biological Journal of Okayama University*, **10**, pp. 13–22, 1964b.
[39] Land, M.F., The physics and biology of animal reflectors. *Progress in Biophysics and Molecular Biology*, **24**, pp. 75–106, 1972.
[40] Land, M.F., Animal eyes with mirror optics. *Scientific American*, **239**, pp. 126–134, 1978.
[41] Land, M.F., Crustacea. *Photoreception and Vision in Invertebrates*, ed. M.A. Ali, Plenum Publishing Corporation: New York, pp. 401–438, 1984.
[42] Lund, E.J., On the structure, physiology and use of photogenic organs, with special reference to the Lampyridae. *Journal of Experimental Zoology*, **11**, pp. 415–467, 1911.
[43] Mansour, K., The zooxanthellae, morphological peculiarities and food and feeding habits of the Tridacnidae with reference to other lamellibranches. *Proceedings of the Egyptian Academy of Sciences*, **1**, pp. 1–11, 1945.
[44] Mason, C.W., Structural colours in insects, II and III. *Journal of Physical Chemistry*, **31**, pp. 321–354, 1856–1872, 1927.
[45] Miller, W.H. & Bernard, G.D., Butterfly glow. *Journal of Ultrastructure Research*, **24**, pp. 286–294, 1968.
[46] Miller, W.H., Moller, A.R. & Bernhard, C.G., The corneal nipple array. *The Functional Organisation of the Compound Eye* (C.G. Bernhard, ed.), 21–33, Pergamon Press: Oxford, 1966.
[47] Nassau, K., *The Physics and Chemistry of Colour*, John Wiley and Sons: New York, 1983.
[48] Nekrutenko, Y.P., Gynandromorphic effect and the optical nature of hidden wing-pattern in *Gonepteryx rhamni* L. (Lepidoptera: Pieridae). *Nature*, **201**, pp. 417–418, 1965.
[49] Neville, A.C., Metallic gold and silver colours in some insect cuticles. *Journal of Insect Physiology*, **23**, pp. 1267–1274, 1977.
[50] Neville, A.C. & Caveney, S., Scarabeid beetle exocuticle as an optical analogue of cholesteric liquid crystals. *Biological Review*, **44**, pp. 531–562, 1969.
[51] Newton, I., *Opticks*, reprinted from the 4th edition by Dover Publications Inc.: New York, 1730.
[52] Parker, A.R., Discovery of functional iridescence and its coevolution with eyes in the phylogeny of Ostracoda (Crustacea). *Proceedings of the Royal Society of London: Biological Sciences*, **262**, pp. 349–355, 1995.
[53] Parker, A.R., Mating behaviour in Myodocopina ostracods (Crustacea): results from video recordings of a highly iridescent cypridinid. *Journal of the Marine Biological Association of the UK*, **77**, pp. 1223–1226, 1997.

[54] Parker, A.R., Colour in Burgess Shale animals and the effect of light on evolution in the Cambrian. *Proceedings of the Royal Society of London: Biological Sciences*, **265**, pp. 967–972, 1998a.

[55] Parker, A.R., Exoskeleton, distribution and movement of the flexible setules on the myodocopine (Ostracoda: Myodocopa) first antenna. *Journal of Crustacean Biology*, **18**, pp. 95–110, 1998b.

[56] Parker, A.R., Light-reflection strategies. *American Scientist*, **87**, pp. 248–255, 1999a.

[57] Parker, A.R., An unusually isolated reflector for host bioluminescence on the second antenna of a lysianassoid (Amphipoda: Gammaridea), *Crustaceans and the Biodiversity Crisis (Crustaceana)* (Brill, Leiden), eds. F.R. Schram & J.C. von Vaupel Klein, pp. 879–887, 1999b.

[58] Parker, A.R., The Cambrian light switch. *Biologist,* **46**, pp. 26–30, 1999c.

[59] Parker, A.R., McKenzie, D.R. & Large, C.J., Multilayer reflectors in animals using green and gold beetles as contrasting examples. *Journal of Experimental Biology*, **201**, pp. 1307–1313, 1998a.

[60] Parker, A.R., McKenzie, D.R. & Ahyong, S.T., A unique form of light reflector and the evolution of signalling in *Ovalipes* (Crustacea: Decapoda: Portunidae). *Proceedings of the Royal Society of London: Biological Sciences*, **265**, pp. 861–867, 1998b.

[61] Parker, A.R., Hegedus, Z. & Watts, R.A., Solar-absorber type antireflector on the eye of an Eocene fly (45Ma). *Proceedings of the Royal Society of London: Biological Sciences*, **265**, pp. 811–815, 1998c.

[62] Parker, A.R., McPhedran, R.C., McKenzie, D.R., Botten, L.C. & Nicorovici, N.-A.C., Aphrodite's iridescence. *Nature*, 409, pp. 36–37, 2001.

[63] Parker, G.H., *Animal Colour Changes and their Neurohumours*, Cambridge University Press: Cambridge, 1948.

[64] Paulus, H.F. & Gack, C., Pollinators as pre-pollinating isolation factors: evolution and speciation in *Ophrys* (Orchidacea). *Israel Journal of Botany*, **39**, pp. 43–79, 1990.

[65] Schafer, W., Bau, entwicklung und farbenenstehung bei den flitterzellen von *Sepia officinalis*. *Zeitschrift für Zellforschung und Microscopische Anatomie*, **27**, pp. 222–245, 1937.

[66] Shelton, P.M.J., Gaten, E. & Herring, P.J., Adaptations of tapeta in the eyes of mesopelagic decapod shrimps to match the oceanic irradiance distribution. *Journal of the Marine Biological Association of the United Kingdom*, **72**, pp. 77–88, 1992.

[67] Skowron, S., On the luminescence of some cephalopods (*Sepiola* and *Heteroteuthis*). *Rivista di Biologia*, **8**, pp. 236–240, 1926.

[68] Steinbrecht, R.A. & Pulker, H.K., Interference reflectors in golden butterfly cuticle – a system to test for shrinking or swelling in EM preparation. *Electron Microscopy*, **2**, pp. 754–755, 1980.

[69] Terao, A., Notes on photophores of *Sergestes prehensilis*, Bate, *Annotationes Zoologicae Japonenses*, **9**, pp. 299–316, 1917.

[70] Towe, K.M. & Harper Jr., C.W., Pholidostrophiid brachiopods: origin of the nacreous luster. *Science*, **154**, pp. 153–155, 1966.

[71] Verne, J., *Couleurs et pigments des êtres vivants*, Armand Colin: Paris, 1930.

[72] Verrell, P.A., Illegitimate exploitation of sexual signalling systems and the origin of species. *Ethology, Ecology and Evolution*, **3**, pp. 273–283, 1991.
[73] Vukusic, P., Sambles, J.R. & Lawrence, C.R., Colour mixing in wing scales of a butterfly. *Nature*, **404**, p. 457, 2000.
[74] Wilkens, L.A., The visual system of the giant clam *Tridacna*: behavioral adaptations. *Biological Bulletin*, **170**, pp. 393–408, 1986.
[75] Yoshida, A., Motoyama, M., Kosaku, A. & Miyamoto, K., Nanoprotuberance array in the transparent wing of a hawkmoth. *Cephonodes hylas*, *Zoological Science*, **13**, pp. 525–526, 1996.
[76] Young, R.E., Oceanic bioluminescence: an overview of general functions. *Bulletin of Marine Science*, **33**, pp. 829–845, 1983.
[77] Zyznar, E.J. & Nicol, J.A.C., Ocular reflecting pigments of some malacostraca. *Journal of Experimental Marine Biology and Ecology*, **6**, pp. 235–248, 1971.

CHAPTER 11

A medical engineering project in the field of cardiac assistance: a lumped-parameter model of the Guldner muscle-powered pump trainer and its use with a ventricular assist device

C.D. Bertram & J.P. Armitstead
Graduate School of Biomedical Engineering,
University of New South Wales, Sydney, Australia.

Abstract

Guldner *et al.* [1] have developed the Frogger, a saline-filled rubber sac device for use during the transformation of skeletal muscle (*latissimus dorsi*) wrapped into a barrel configuration for cardiac assist use. The main advantage of the device is that it offers a compliant rather than a resistive load against which the stimulated muscle contracts, which is reported to result in less muscle mass loss and superior final contractile characteristics. We are using numerical modelling in the design of our device and its connection as a possible power source to an Australian VAD (ventricular assist device), the Spiral Vortex. In this chapter, we firstly review the concept of transforming human skeletal muscle into fatigue-resistant mode, able to act as a power source for an auxiliary ventricle. The alternatives of SMV (Skeletal Muscle Ventricle) and MPP (skeletal-Muscle Powered Pump) are explained. Then, after describing our current design in detail, we explain the Hill-based method of using muscle force-length-velocity data, as a basis for predicting pump behaviour in engineering terms. For overall consideration of pressure-volume-power relationships, a quasi-static lumped parameter model involving simultaneous solution of equations describing the passive material stresses, the wrapped muscle force-length-velocity relationship and the geometric constraints, at a series of muscle activation levels, has been found useful.

1 Introduction

1.1 Heart transplantation and the need for 'assist' devices

Heart disease is a major cause of morbidity and mortality in the western world. According to the Australian Institute of Health and Welfare it was responsible for 41% of all deaths in Australia in 1999. Ironically, improvements in the treatment of cardiovascular and other diseases and the subsequent increase in lifespan have seen an increase of the end-stage form of heart disease known as heart failure. As the population ages, the inability of the heart to pump adequately will become a limiting factor for both quality of life and longevity.

The cost to society also is likely to increase, both in terms of treatment and the number of hospital beds required. Figure 1 shows the enormous increase in the hospitalisation rate for those over 60; the current prevalence of heart failure in this group has been reported by to be as high as 13% [3]. The cost of treating heart failure in Australia exceeds that of all cancers combined.

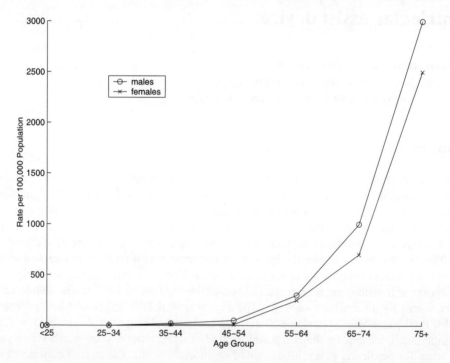

Figure 1: Hospital use in Australia due to heart failure (Data from the Australian Institute of Health and Welfare morbidity database [2].)

The prognosis for heart failure is poor: the USA five-year mortality rate is approximately 50%, the rate and prevalence both increasing with age [4]. Between one third and one half of heart-failure patients will die of sudden cardiac death. Despite rapid improvements in pharmacological therapy, the National Heart Lung and Blood Institute (NHLBI) in the USA has estimated that around 100,000 people in the USA currently have 'refractory heart failure', that is, heart failure that does not respond to current optimal medical therapy [5]. Those patients who then have an unacceptable prognosis for one-year survival or unacceptable quality of life because of cardiac symptoms are normally considered for cardiac transplantation. The paucity of donor hearts (2500 transplants annually in the USA, around 100 in Australia) means that the selection criteria for admission to a transplant program are stringent. Generally a patient will be accepted only if: aged less than 55, free of diseases such as diabetes and peripheral vasculopathy, no other major organ dysfunction is present and commitment to the program can be shown. Clearly this means that many patients die who might otherwise be transplanted.

Cardiac transplantation offers good outcomes for end-stage heart failure with survival rates of 85% immediately, 65% at five years and 40% at ten years. Cardiac transplantation has become commonplace thanks to modern immunosuppressive therapies, but therapeutic limitations and complications cannot be ignored. These include [6] the poor specificity and significant toxicity of the necessary drugs, the increased risk of infections and malignancy, and the possibility of coronary artery disease in the graft (the major factor affecting long-term survival). A host of new immunosuppressive drugs that may benefit heart recipients and improve graft longevity are in various phases of clinical trials. However, they will do nothing to improve the discrepancy between patients who would benefit from transplantation and those patients who actually receive one, which is growing at an alarming rate.

The prospect of an endless supply of organs, either from another animal (xenotransplantation) or from a tissue-engineered source, while promising, is beset by scientific and ethical dilemmas in each case. This paper deals with another technique for assisting or replacing the pumping capabilities of the heart: using the body's own substantial skeletal muscle reserves as a hydrodynamic power source.

1.2 The skeletal muscle graft: historical background

DeJesus [7] was perhaps the first to describe the use of skeletal muscle as a graft to repair damaged tissue (myoplasty) in his 'Brief considerations on a case of penetrating wound of the heart'. He used a flap of pectoral muscle to repair the heart wall as did Leriche & Fontaine [8] who, after testing the method in dogs, proposed the technique (cardiomyoplasty) as a routine, if heroic, cardiac intervention. Two decades later, Kantrowitz and McKinnon [9] saw the potential of 'stimulated' skeletal muscle to augment the cardiac function of heart-failure patients (dynamic cardiomyoplasty). In two series of groundbreaking (acute) experiments in healthy dogs they used the diaphragm muscle, stimulated via the phrenic nerve, as an active muscle-wrap around the heart and as an aortic counter-pulsator when wrapped around the thoracic aorta. A significant increase in diastolic and mean blood pressure was achieved by counterpulsation, indicating assistance to the heart, but the preparation fatigued rapidly. Such rapid muscle-fatigue was seen as an insuperable obstacle at the time by Kantrowitz and other workers in the field [10].

It had long been known that there are three types of striated muscle: cardiac, slow skeletal muscle and fast skeletal muscle. Cardiac muscle fibres are arranged in an interconnecting network, or syncytium, and contract with the familiar all-or-nothing twitch as the electrical stimulus passes in a wave through the myocardium. While skeletal muscle resembles cardiac muscle at the micro-structure level, it is arranged not as a syncytium, but rather in individually innervated motor units. Motor units can be recruited over both space and time to elicit the desired strength and velocity of contraction. Fast skeletal muscle is voluntary muscle which is usually called upon for short bursts of activity. Slow skeletal muscles are involved in posture, and may contract continuously at a low level over many hours. Cardiac muscle is involuntary and is called upon to contract approximately once every second for the duration of one's life. This leads to differences in the chemical makeup of cardiac and skeletal-muscle cells—cardiac muscle cells have many more mitochondria, for example. Also, a significant preload (initial stretching) is required in order to generate useful power with skeletal muscle. Geddes [11] called this the 'paradox in using electrically stimulated skeletal muscle to pump blood', alluding to the (comparatively) low preload with which the heart operates under normal physiologic conditions.

Early experiments on the differentiation of muscle fibres by Buller *et al.* [12] determined that in general all fibres are slow at birth and that mature slow fibres are prevented from conversion to fast fibres by constant neural stimulation. Salmons & Vrbová [13] subsequently showed in landmark work that chronic, artificial low-frequency (5–10 Hz) stimulation could also convert mature fast-twitch muscles to slow-twitch types. Their battery-powered stimulator, implanted in rabbits, operated for two minutes out of every three and over a period of weeks elicited muscle transformation. It was recognised that a burst of pulses would be necessary to develop the required (summation of) contractile force from skeletal muscle (tetany). The first stimulator that could produce a frequency burst synchronised to the cardiac cycle was produced by Chiu and coworkers [14].

Three sub-cellular systems are involved in the process of muscle transformation: up-regulation of the metabolic system from a primarily anaerobic (glycolytic) type to an aerobic (oxidative) system, down-regulation of the calcium system with expression of slow Ca^{++} ATPase, and shifting of the contractile myosin (muscle protein) isoforms from type 2 or fast fibres to type 1 or slow fibres. These last are virtually indistinguishable from the β-myosin isoforms of cardiac muscle. The resulting 'transformed' skeletal muscle is fatigue-resistant, but also weaker and (obviously) slower to contract, equating to less power per contraction. The commercial development of burst stimulators opened the way for the experimental and clinical application of transformed skeletal muscle to cardiac assist.

1.3 Skeletal muscle: application alternatives and the SMV

Carpentier & Chachques were the first to apply dynamic cardiomyoplasty (CMP) clinically in 1985 [15]. They applied a wrap of *latissimus dorsi* (LD) muscle to the heart of a patient who had had a resection of an invading fibrous tumour. Stimulation of the muscle commenced after five days on every other cardiac cycle, continued as such intermittently for ten days and, after a further 15 days continuous stimulation, was changed to a continuous one-for-one cycle. The authors reported an improved left ventricular ejection fraction with the stimulator on. The left LD muscle is most commonly used for dynamic CMP, although the right LD is occasionally used and other muscles have been tested experimentally in animals. While patients who undergo CMP continue to show quality-of-life improvement, the mechanisms underlying any functional improvement remain uncertain. However, it is now accepted that CMP can halt the remodelling caused by elevated wall stress in the failing heart through some combination of elastic 'girdling' and dynamic assist [16].

Other approaches to providing mechanical assistance to the failing heart by using skeletal muscle have included dynamic aortomyoplasty [17, 18], skeletal muscle ventricles (SMVs) [19–21], and skeletal-muscle-powered pumps (MPPs) [22]. Aortomyoplasty, as previously tested by Kantrowitz, uses transformed muscle to compress the aorta during diastole (an artificial aneurysm may be introduced), thus increasing mean blood pressure and coronary circulation and unloading the heart (although not to the same extent as a balloon pump which actively inflates and deflates). SMVs comprise a pumping chamber, usually manufactured from an elastic polymer, around which a transformed wrap of muscle contracts to expel the blood within. Typically, the device is anastomosed in parallel with the descending aorta (or some other vessel) via Dacron grafts with integral valves. Thrombus formation is minimised by the use of treated polymer surfaces internally or by endothelial seeding to form a pseudo-intima.

1.4 The muscle-powered pump

The MPP concept aims to isolate the skeletal-muscle power-source from the blood-contacting pump. This has several advantages. (1) An optimised device such as a ventricular assist device (VAD) can be used for blood pumping rather than the muscle-wrap chamber which may have thrombogenic flow patterns, i.e. some stasis. (2) An energy converter may be used such that the muscle performs hydraulic work and the pump is driven electrically, for example. (3) The blood pump can be driven by an external source while muscle training is in progress. Patients who require this type of therapy are frequently sick enough to require immediate cardiac assist. An example of an energy converter is the linear-pull MPP of Farrar's group which couples a muscle-driven cylinder hydraulically to a modified Thoratec VAD and has shown promising potential [23]. An obvious disadvantage is energy loss in conversion or transmission, so optimisation of any such technique is vital.

If a muscle-powered pump is formed surgically as a pumping chamber (as opposed to a linear configuration), the method is akin to the skeletal muscle ventricle except insofar as the fluid within the chamber is not blood. Similar problems thus arise in terms of obtaining adequate pressure development, pressure-flow work output and non-fatiguability [24]. The most-often-used muscle for this purpose, the *latissimus dorsi*, is like most voluntary muscles adapted for occasional maximal contractions that are both fast and forceful. Before it can be used for an onerous repetitive contractile task such as cardiac assist, it must be transformed biochemically [25, 26] through a training regimen of several weeks duration. The problems identified above of slow and weak contractions, and considerable loss of muscle mass with consequent loss of work capability are thus again problematic. Whalen *et al.* [27] have developed a MPP similar in spirit to our own, where a muscle powered hydraulic source is used to externally pressurise another chamber which, on collapsing, expels blood via a valved conduit. An advantage of this device is that, while the muscle is being transformed, the collapsible chamber can be driven by a second temporary extracorporeal power source.

1.5 The 'Frogger' device

For use in either SMV or MPP mode, Guldner and colleagues [28, 29] have proposed and developed a method of obtaining a transformed barrel of muscle with superior contractile properties and reduced muscle wastage. The central feature of the method is the use of a compliant training device called the Frogger, consisting of a barrel capped at each end by a spherical cheek, all three communicating parts being saline-filled silicone-rubber. In conjunction with a muscle-stimulation regimen during transformation that emphasises avoidance of excessive load on the muscle too early, Guldner *et al.* report that this compliant load promotes development of a transformed wrapped muscle with undiminished muscle mass, able to produce greater pressure-flow power than a muscle trained in conjunction with an inertial or resistive load. In the device as originally developed by Guldner *et al.* themselves, the Frogger is eventually extracted from the muscle wrap. The wrap is then incorporated in a circulatory loop either in parallel with the left heart (atrial or left ventricular apical input, aortic output) or in series (aortic input, aortic output) (see Figure 2). Guldner *et al.* [30] have recently developed an alternative procedure for use of the device which involves only one surgical operation.

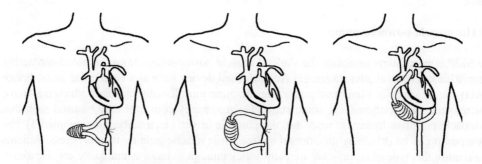

Figure 2: Sketches of the heart being assisted by an SMP in aorto-aortic mode (left and centre) and apex-to-aorta mode.

We are conducting experiments on a version of the Frogger where the compliant cheeks can be removed after the completion of the training period, leaving the barrel in place beneath the muscle wrap [31]. The barrel and wrap are then coupled to a pulsatile ventricular assist device (VAD), the Spiral Vortex, developed in Australia [32] which is normally pneumatically powered. Via a redesigned connector to the VAD, the muscle wrap is stimulated to pump saline to and fro from the barrel to the underside of the VAD diaphragm, thereby replacing the pneumatic pressure/vacuum cycle and inducing circulatory assist. In parallel with the production of the redesigned Frogger and the conduct of the animal experiments, we have simulated the performance of the muscle-wrapped Frogger barrel applied to this VAD-powering role.

1.6 Modelling using computational fluid dynamics (CFD) and lumped-parameter approaches

For certain specific purposes we have modelled the motions of the fluid in the Frogger in detail, using a commercial finite-volume CFD package [33]. Briefly, CFD involves the numerical solution of the partial differential equations describing the fluid flow, in this case the unsteady Navier-Stokes equations. The equations and appropriate boundary conditions are solved using a discretisation of the fluid domain into a number of cells. These cells are made small enough so that the error associated with the approximate representation of the fluid within the cells is small compared to the overall result. Because the equations are nonlinear an iterative approach is required to solve the system and, because the system is dynamic, many iterations must be performed at each time step through which the system is marched. This whole procedure is computationally intensive, requiring significant computer resources and time to perform each run. The results go to a level of fluid-dynamic detail which is not necessarily supportable here in terms of measurable data on the Frogger-VAD system available for comparison.

The problem posed here is not simply a fluid-dynamic one; it involves consideration of the coupled fluid motions and structural deformations. There are multiple structures involved: the Frogger itself, the VAD (in particular the flexing diaphragm which separates the blood from the driving fluid), and the properties of the system of blood vessels into which the VAD ejects. Each of these parts poses a separate fluid-structure-interaction problem. The combined solution of all of these coupled problems simultaneously is beyond the current abilities of even software which can handle some fluid-structure interactions. Thus it becomes necessary to model the global system at a lower level of detail.

A suitable alternative to CFD for the global system is to construct a lumped-parameter model. This approach has proved most useful to us in showing the nature of the design trade-offs we face. The lumped-parameter method may be described as analysing the system discretised in time, but not in space. This approach may be readily accommodated within the standard scientific and engineering software package Matlab (The MathWorks Inc., Natick, MA USA). Matlab (Matrix-Laboratory) is a high-level computation, visualisation and programming environment. Because the basic element of Matlab is an array and because dozens of built-in routines, such as equation integrators, are readily available to the developer, Matlab allows the quick and efficient exploration of a mathematical model. The purpose of this paper is to describe this lumped-parameter model, its realisation in Matlab, and the results obtained from it.

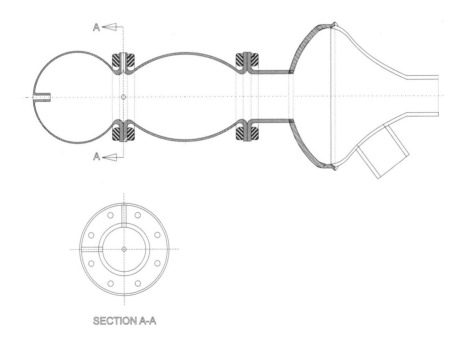

Figure 3: A drawing of the Sydney version of the Frogger, coupled to the Spiral Vortex VAD. At the other end of the barrel from the VAD connector is shown one of the cheeks that are attached to either end of the barrel during transformation and subsequently detached for VAD pumping

2 Methods

The Frogger barrel was assumed coupled to the VAD as shown in Figure 3, where the Frogger axis is also the axis of revolution of the VAD chamber (excluding the VAD inflow port). The problem was set up in Matlab as that of finding a solution to the equation

$$f(dV_{brl}) = 0, \qquad (1)$$

where dV_{brl} is the change of volume of the barrel. The three terms of the function were assembled in three subroutines, each starting from the current value of dV_{brl}, and each dealing with a separate component of the modelled system.

2.1 The Frogger barrel

The first subroutine resulted in a value of the pressure of the fluid in the Frogger, $P_{brl,int}$. From the extent of the barrel volume change from a rest state was calculated the barrel transmural pressure, $P_{tm,b}$, according to the function shown in Figure 4. This was a curve fitted to experimental pressure-volume data from the silicone-rubber Frogger tested on the laboratory bench. In the region of negative transmural pressure, the barrel of the real Frogger usually collapsed with a single asymmetrical dimple.

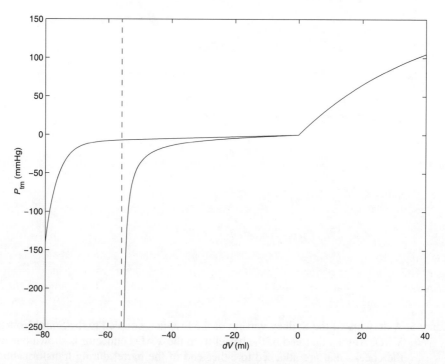

Figure 4: The pressure-volume relation for the Frogger barrel. A curve fitted to experimental measurements was constrained to volumes greater than those which would have produced opposite wall contact if the collapse was axi-symmetric.

The function used in the model was subjected to a constraint such that the volume could not go below that value which, had the collapse occurred axi-symmetrically, would have resulted in opposite-wall contact. This constraint was not strictly necessary in the lumped-parameter model, but was included for compatibility with the axi-symmetric finite-volume model.

The current and previous values of dV_{brl} were also used to calculate the length and velocity of the wrapped muscle, whence was found a value of the circumferential tensile force from the muscle according to a Hill-based force-velocity-length surface sub-model. This sub-model will be described in more detail below. Since an originally sheet-like muscle surgically formed into a barrel-like wrap is highly unlikely to produce its optimal contractile performance, an arbitrary coefficient η with value between zero and one was included to account for this loss of active force.

The application of the circumferential force from the muscle to the outside of the barrel was dealt with by defining a hydrostatic pressure, $P_{brl,ext}$, as that existing in an imagined thin layer of serous fluid lubricating contact between the underside of the muscle wrap and the outside surface of the barrel. The thickness of the wrap in surgical reality is considerable; not only is the original muscle thick, but the procedure defined by Guldner involves folding the outsides of the muscle back over the middle so as to form a double-thickness half-width wrap. No account of this was taken in the model, where a simple Laplace relation converted circumferential force to barrel external pressure. Similarly crudely, in this relation a single barrel radius was used to characterise the whole wrap, irrespective of the reality that the barrel (and wrap) radius decreased from the centre to a minimum at each end.

With the quantities $P_{brl,ext}$ and $P_{tm,b}$ thus evaluated, the required value for $P_{brl,int}$ followed immediately.

2.2 The VAD diaphragm

The second subroutine resulted in a value of the diaphragm transmural pressure, P_{elast}. Since the space inside the Frogger and below the VAD diaphragm was supposed completely filled with saline, any change in Frogger (barrel) volume is immediately reflected in an equal and opposite change in the volume below the diaphragm in the rigid VAD chamber. The VAD diaphragm in reality is a thin polyurethane membrane moulded in approximately the shape of a saucer, which moves between domed-up and domed-down configurations via buckled states which may be asymmetrical. Such buckled states cause stress concentrations in the diaphragm material which can be foci for calcification and degenerative breakdown; a design aim is to achieve diaphragm dimensions which minimise such repetitive folding as the diaphragm moves through the central position. In the model, axi-symmetric deformation keeping the original (spherical) shape of the curvature and varying only the radius of curvature was assumed. Thus a trigonometrical relation was obtained between the change in VAD diaphragm volume from its own rest state and the point where the diaphragm cut the extended axis of the Frogger. This point, x_{vnd}, defined the diaphragm position. The latter was then used to calculate the diaphragm transmural pressure, P_{elast}, according to the cubic algebraic relation shown in Figure 5. This was defined to represent the reality that the diaphragm offered almost no elastic restoring force except when transmural pressure inflated it to one or other domed extreme of its deflection range.

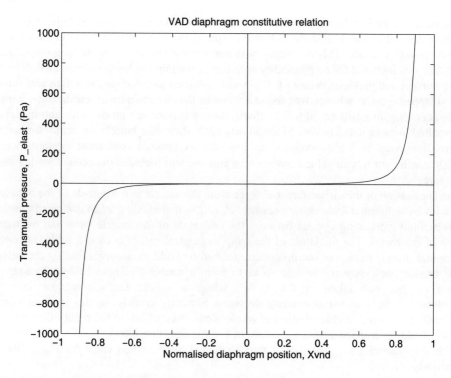

Figure 5: The assumed relation between VAD diaphragm centre axial position and diaphragm transmural pressure.

2.3 The blood in the VAD and beyond

The third subroutine resulted in a value of the pressure of the blood in the VAD above the diaphragm, P_{blood}. As per its intended use, the VAD was assumed filled with blood above the diaphragm and connected via inflow and outflow ports to a real or mock circulatory load. Changes in dV_{brl} (and accordingly volume below the VAD diaphragm) then result in a volume flow-rate of blood into or out of the VAD. By keeping track of the current, the last previous and the next previous values of dV_{brl}, both flow-rate Q and rate of change dQ/dt could be calculated. Depending on whether this flow-rate represented blood entering the VAD or leaving it, the circulatory mean pressure P_{circ} was defined as either a low pressure representative of left atrial pressure, P_{atrium}, or a high pressure representative of the systemic arterial load, P_{eject}. This amounts to assuming ideal operation of the non-return valves incorporated in the inflow and outflow tracts of the VAD. Then the pressure of the blood in the VAD above the diaphragm was calculated as

$$P_{blood} = P_{circ} + R_{blood} \cdot Q + I_{blood} \cdot dQ/dt , \qquad (2)$$

where R_{blood} was based on the expected resistance to outflow through the VAD outflow tract with the valve open, and I_{blood} was an estimate of the inertia of the blood being accelerated, based on the volume of blood in the VAD. This equation amounts to assuming that the task of moving the VAD diaphragm is dominated by the resistance and inertia offered by the blood in or leaving the VAD, and that the circulatory load manifests itself only through the agency of the mean pressure, P_{circ}. For the VAD itself, which has rigid walls and rigid inflow and outflow conduits, such an assumption makes sense. However, the circulation beyond is normally thought of as presenting primarily a resistive and compliant (capacitive) rather than inertial load, owing to the elasticity of the large arteries. The use of eqn. (2) above thus also rests on the assumption that the compliance of the circulatory load is sufficient that little (in the model, no) pressure increase results from the systolic ejection of blood into the systemic circulation. The compliance of the circulation is therefore essential to the model, although excluded from explicit algebraic description. Eqn. (2) also implies that either one or the other valve is always open; this is not strictly true when the pressure in the VAD is between P_{atrium} and P_{eject}. In this case, corresponding to isovolumic contraction and relaxation in ventricular parlance, $Q = 0$ until enough pressure has developed in the VAD to open the outflow valve, or has decayed to allow the inflow valve to open, and another variable would be needed at these times to describe the pressure in the VAD; however the omission makes no practical difference to the results presented here.

2.4 Muscle performance

The muscle was modelled as having varying degrees of active tensile force depending on length, velocity and level of activation, superimposed on a passive tensile force that depended only on length. Classically, active force peaks at the optimum sarcomere length L_0 and falls away on either side, while passive force increases nonlinearly but monotonically with length. In myocardium, passive stretch is sufficient already at L_0 that the total force (sum of active and passive) increases monotonically with length, whereas in voluntary muscles, passive stretch is small enough at L_0 that total force is sigmoidal with length [34]. No data appear to be available for the shape of this curve in transformed *latissimus dorsi*. The active and passive length-tension curves were approximated by exponential functions, giving the relation between total force and length shown in Figure 6a.

An important parameter here is the tightness with which the muscle is wrapped around the Frogger barrel. This is in theory under the surgeon's control, but is subject to later muscle fibre length remodelling. On the basis of advice from muscle physiologists, the relaxed length of the muscle was taken to reflect a majority of sarcomeres at L_0. A single dimensionless constant, the wrap looseness parameter, defined as

$$k_w = L_{max}/(2\pi.a_u) \tag{3}$$

where L_{max} is the whole-muscle length corresponding to sarcomeres at L_0, and a_u is the unstressed barrel radius, then determined where on the muscle-fibre length axis the wrapped muscle lay.

Hill's offset-hyperbola equation was employed for the relation between active force and shortening velocity. The model also required specification of a force-velocity relation for the lengthening or eccentric regime. Following McMahon [35], the Hill relation was initially adapted for this purpose by steepening (by a factor of 4) the slope dF/dV where shortening velocity V was negative, and constraining the force in muscle extension not to exceed $1.8P_0$, where P_0 is the active force at L_0. This gave the curve shown in Figure 6b. Further adaptation was necessary to overcome the singularity in the Hill equation at $V/V_{max} = -a/P_0$. An alternative curve lacking the singularity but closely corresponding to the Hill-based curve in the region $-a/P_0 < V < 0$ and still respecting the maximum force asymptote was substituted, allowing the force-velocity relation to be extended arbitrarily in the muscle-lengthening direction.

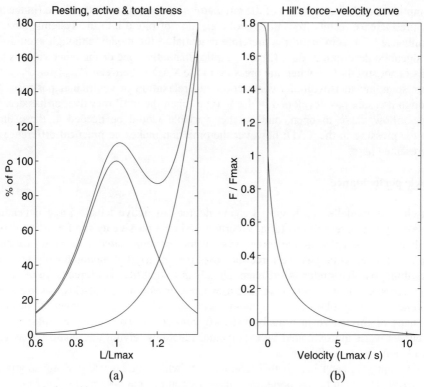

Figure 6: Length-tension and force-velocity relations for the muscle wrap.

Combining this force-velocity relation with the length-tension relation allowed specification of a force-length-velocity surface. Following Krylow & Sandercock [36], one further modification was made, in requiring V_{max} to scale with muscle length on either side of L_0 in the same way as active isometric tension. The overall active force at maximum activation then depended on velocity and length as shown in Figure 7. Activation was modelled as a dimensionless factor on active force between zero and one which followed a smoothly rising and falling prescribed waveform vs. time as shown opposite. No series elastic element was included in the muscle model.

2.5 Overall model

As described in the sub-sections above, from the current value of dV_{brl} was evaluated the three pressure terms in the function $f(dV_{brl})$, namely $P_{brl,int}$, P_{elast} and P_{blood}. The residual after root-finding by the van Wijngaarden-Dekker-Brent algorithm [37] was $(P_{blood} - P_{elast}) - P_{brl,int}$, where in the full (finite volume) problem, $P_{brl,int}$ would have varied spatially.

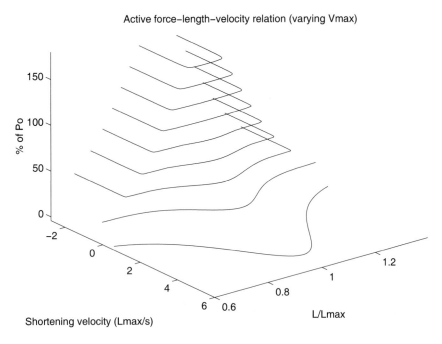

Figure 7: Active force (relative to P_0) is shown as a function of muscle length (relative to L_0) and shortening velocity (relative to V_{max}), in terms of equi-force contours.

The outcome of the running the model was a function of the activation waveform chosen and the values of several adjustable constants. The most important of these were the Hill muscle parameters P_0, V_{max} and a/P_0, along with a_u the unstressed barrel radius (actually replaced by R_a the original barrel-apex radius), h_m the thickness of the wrapped muscle, k_w the wrap looseness parameter, η the efficiency of wrapped muscle, and the pressures P_{atrium} and P_{eject}. The usual values taken for these variables were $P_0 = 0.235$ MPa, $V_{max} = 1.5\ L_0/s$, $a/P_0 = 0.25$, $R_a = 21.2$ mm, $h_m = 16.2$ mm, $k_w = 1$, $\eta = 0.5$, $P_{atrium} = 5$ mmHg (0.67 kPa) and $P_{eject} = 100$ mmHg (13.34 kPa). A large number of other constants controlled such fixed quantities as the shape of the curve in Figure 4; it would be tedious to detail all of these. SI values calculated for other constants defined above were $R_{blood} = 6.288 \times 10^5$, $I_{blood} = 2.333 \times 10^6$, and $dV_{min} = -55.69 \times 10^{-6}$ (the original barrel volume was taken as 59.98 ml).

3 Results

Figure 8 shows time-traces computed over 300 steps of 12.5 ms with all parameters as specified above. Note that the symbols o and × are trace identifiers applied to every twentieth time-step only; the underlying curve was computed at more steps than are so marked. The muscle activation waveform is shown in the top panel; it defines four equal episodes of muscle stimulation, each one being a flat-topped pulse reached and departed from via a cosine ramp.

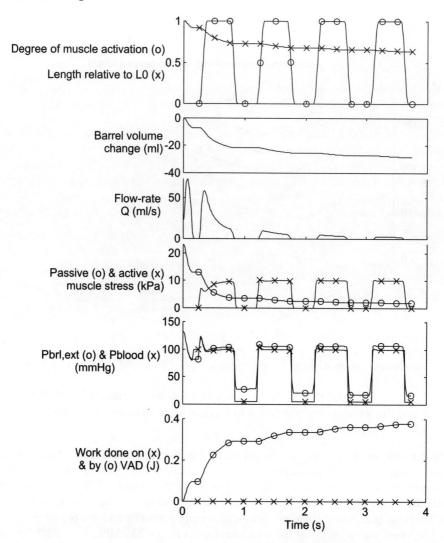

Figure 8: Time-traces of (numbering the graph panels from the top) (i) the degree of muscle activation, k_m (o), and normalised muscle length, L/L_0 (×), (ii) the change in barrel volume from the unstressed, as-moulded state, dV_{brl}, (iii) the volume flow-rate out of the barrel, Q, (iv) the passive (o) and active (×) tensile stress in the muscle, (v) the pressure on the outside of the barrel resulting from muscle stress, $P_{brl,ext}$ (o), and the pressure in the blood in the VAD, P_{blood} (×), (vi) the integral of the product of P_{blood} and Q, presented separately as positive and negative components, giving the flow work done on the VAD and systemic load by the barrel/muscle (o) and work done on the barrel/muscle by the VAD and atrial reservoir (×).

Because the passive elastic tension in the muscle is already considerable at L_0, the system reacts to the initial conditions ($L = L_0$) by moving to a smaller barrel volume and ejecting some fluid from the Frogger to the VAD. A small additional contribution to this process is made by the passive elasticity of the VAD diaphragm, taken as starting at normalised position $x_{vnd} = -0.513$, which creates a transmembrane pressure difference of 7.64 Pa (see Figure 5). The process is completed before the first activation starts. The first active contraction is effective; the VAD diaphragm is displaced by 14 ml against the 100 mmHg ejection pressure, and does 0.19 J of useful work (defined as the time-integral of the product of Q and P_{blood} when the product is positive).

However, the barrel fails to refill; the assumed atrial pressure is entirely insufficient to overcome the residual elastic tension in the now shortened muscle. The passive muscle stress is simply a nonlinear function of L/L_0 (shown in the top panel); its value in kPa is shown in the fourth panel, and it determines $P_{brl,ext}$ during muscle relaxation as shown in the fifth panel. An alternative and instructive view of the situation is provided by plotting muscle stress against L/L_0 as shown in Figure 9.

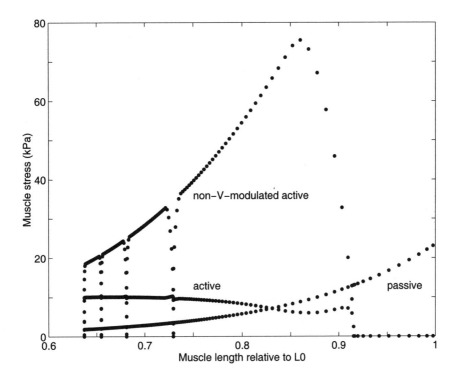

Figure 9: The passive and active components of the tensile stress in the muscle, plotted at every time-step against normalised muscle length, L/L_0, for the simulation run of Figure 8. Also shown is the active muscle stress that would have occurred if muscle tension were not reduced by shortening velocity.

In addition to the passive and active muscle stresses is also plotted the active stress that would have occurred isometrically, i.e. if the active tension were not reduced by shortening at a finite rate, according to the Hill force-velocity curve. The curves for non-velocity-modulated and actual active stress would, of course, converge if the contraction were allowed to proceed to completion, i.e. if the activation were maintained until shortening ceased. It does not pay to wait this long, but observing the form of these curves allows one to assess the extent to which more work can be obtained from a contraction by maintaining activation longer. Although the later contractions occur further down the shoulder of the length-tension curve in Figure 6a, the maximum active tension developed is actually greater, because the velocity of shortening is much reduced (relative velocity can be deduced from the distance between data points on Figure 9). In the absence of refilling, each succeeding contraction only drives the muscle further down the length-tension curve shoulder and is less and less effective in creating outflow, despite the creation of more active tension, as shown also in the fourth panel of Figure 8.

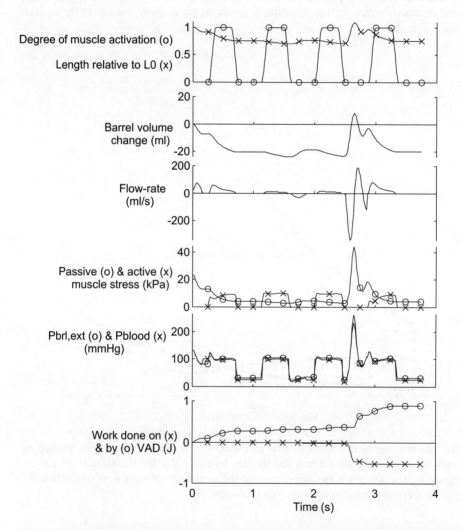

Figure 10: Time-traces as in Figure 8, but with P_{atrium} increased to 25 mmHg, and with P_{atrium} augmented to $P_{atrium} + P_{eject}$ (i.e. 125 mmHg, higher than the circulatory pressure faced by the VAD during ejection) during most of the interval between the third and fourth muscle activations (see text for exact timing).

The difficulty of refilling is emphasised in Figure 9 by the fact that the passive stress at $L = L_0$ is greater than the peak active tension reached in any of the contractions; some outside agency must overcome this much resistance to muscle lengthening if the muscle is to be allowed to develop maximal active tension in subsequent contractions. Figure 10 shows the outcome of a run similar to that shown in Figures 8 and 9, but in which two changes have been made. Firstly, the filling pressure has been increased from 5 to 25 mmHg (3.34 kPa). This is already pathologically high, yet the effect is clearly insufficient; flow-rate out of the barrel during the second and third muscle contractions is minimal (where previously it was zero), and the corresponding values of dV_{brl} and useful work done are also unsatisfactory.

Secondly, an outside agency is employed to refill the Frogger before the fourth contraction (the first three contractions establish steady-state conditions prior to this intervention). The intervention consists of increasing the pressure P_{circ} from P_{atrium} to $P_{atrium} + P_{eject}$ during the time interval $2.525 < t < 2.85$s (this occupies most but not all of the period of zero activation between beats 3 and 4, which occupies the interval $2.5 < t < 2.875$s). This is supposed crudely equivalent to manually squeezing a valved bladder inserted in the atrial inflow system. The effect is to induce an under-damped transient of flow surging into, back out of, then again into the Frogger barrel before the next muscle activation. At the start of the activation, the barrel volume is almost back where it started at $t = 0$, and the muscle is at a length only slightly less than L_0. The contraction is accordingly effective in doing work in expelling fluid from the Frogger against P_{eject}; work done on the VAD after 2.875s rises by 0.223 J. However, rather more work than this has been done on the Frogger by the hand which squeezed the bladder. The two barrel-inflow transients amount to 0.513 J of work done on the Frogger. The outflow transient in between gives back 0.287 J, but one must assume that this is energy lost rather than regenerated.

Clearly some other strategy is needed to assure refilling. The experimental version of the barrel in the Sydney Frogger was deliberately created to have as much compliance as possible commensurate with expectations of the necessary wear resistance and fatigue life. Compliance was achieved by using a silicone rubber with low stiffness modulus and by minimising the wall thickness. However, in the computer model, the stiffness of the Frogger in the direction of collapse was now deliberately increased (by attenuating any negative transmural pressure by a factor $i_{Ptm,b}$). Figure 11 shows the outcome of a model run with $i_{Ptm,b} = 35$. Somewhat surprisingly, the added stiffness is well tolerated by the muscle, which must now shorten against an artificially increased load. The stratagem achieves the desired object of ensuring refilling after each contraction; indeed refilling occurs rather more rapidly than ejection as soon as the activation decays and the muscle relaxes. Useful work stabilises at around 0.193 J per beat. Close inspection of the bottom panel of Figure 11 reveals that this is made up of two components: one (much the larger of the two) the product of positive P_{blood} and positive Q during the contraction, and the other the positive product of negative P_{blood} and Q during the refilling. This second component corresponds to the passive spring-like action of the stiffened Frogger barrel in drawing blood into the VAD from the atrial reservoir by withdrawing saline from the VAD diaphragm underside. It is legitimately regarded as a component of useful work, even though it is not part of the ejection, because the energy so used is stored in the barrel elasticity during the muscular contraction. The muscle now works in the range $0.79 < L/L_0 < 0.94$; although this is still shorter than the ideal range disposed around $L = L_0$, it is sufficiently close to the top of the length-tension curve to allow development of large active tensions. The stroke volume is 12.1 ml, which is again less than ideal; a stroke volume of 60 ml would be needed to operate the adult-sized VAD effectively. Any lesser stroke volume runs the strong risk of incurring further penalties in practice by failing to ensure that the VAD valves open and close fully and quickly, thereby leading to increased pumping losses through regurgitation. This is outside the scope of the present model. However, this rather small stroke volume is probably a realistic result of using as a contractile source the *latissimus dorsi* muscle. The dimensions assumed (in particular the thickness) are taken from an initial surgical exploration of this muscle in a sheep cadaver. The discrepancy in size between the sheep muscle and the adult human VAD is a drawback of the experiment modelled rather than the model itself.

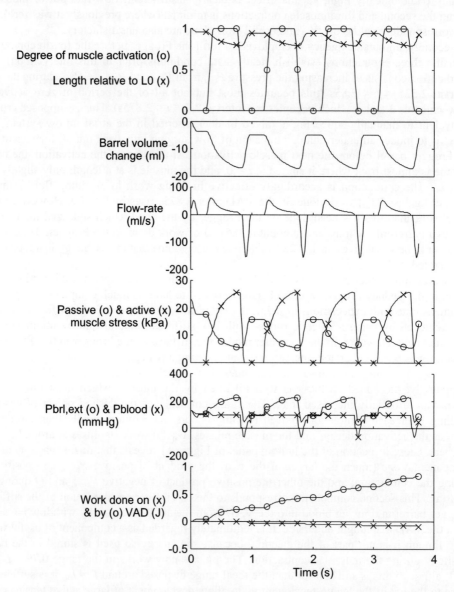

Figure 11: Time-traces as in Figure 8, but with the relation between $P_{tm,b}$ and dV_{brl} modified for $dV_{brl} < 0$ such that the apparent transmural pressure applied to the barrel is only $P_{tm,b}/i_{Ptm,b}$, where $i_{Ptm,b} = 35$. This simulates the reinforcement of the barrel with a spring-like structure to resist collapse.

Examination of the fifth panel of Figure 11 reveals a potential disadvantage of increasing the barrel stiffness in collapse. The pressure $P_{brl,ext}$ now briefly drops to 47 mmHg at the time when the barrel is drawing in fluid, but thereafter rebounds to 96 mmHg during the rest of the relaxation phase, before rising to higher pressures still during the next contraction. This is the pressure on the underside of the muscle; its value during relaxation determines whether the inner layers of the muscle wrap receive adequate perfusion. (Previously, $P_{brl,ext}$ was 32 mmHg in the relaxation phase when P_{atrium} was 25 mmHg, and less than 20 mmHg in the later stages of the run shown in Figure 8, with $P_{atrium} = 25$ mmHg.) Such elevated values suggest that in the absence of some additional safeguard the muscle would in practice rapidly become ischaemic [11].

Such a safeguard could be provided by a construction whereby the muscle was re-lengthened by the barrel 'spring' as far as a point beyond which the spring no longer exerted any outward force on the muscle. For instance the barrel might be surrounded by a netting sac of inextensible fibres.

4 Discussion

The model described herein is in many ways extremely crude. Most notable is the expression of each compartment pressure as a single time-variable, as indicated in the title of this paper. Pressure gradients associated with the driving of fluid between the Frogger and the VAD were excluded altogether in this model. A lumped-parameter approximation of these could have been included, but our parallel finite-volume investigation of these pressure distributions (to be reported elsewhere) indicated that the pressure drop from the blind end of the Frogger barrel to the VAD diaphragm was always small enough relative to the pressure differences involved elsewhere in the model that this addition would not have changed the results significantly.

Secondly, the muscle was treated as a medium which generated tensile stress and thereby force in proportion to its thickness, but the thickness of the muscle was not considered further in applying that force to the underlying barrel. This would certainly have given an over-optimistic view of how much the barrel can be squeezed by the muscle. In any thick-walled near-cylindrical contractile tube, progressively less shortening is possible in the outer layers. The effect is well-known in thick-walled models of the left ventricle [38–40], although there it is modified by other geometric factors. Since not all of a thick contractile wall can shorten equally, the effect must reduce the work done by muscular contraction, relative to a situation where all of the muscle is assumed located at the inner radius of the wrap. In the case of our Frogger design, a second source of interference with effective shortening is the rigid collars at each end of the barrel which allow removal of the compliant cheeks used during training and their replacement with a blanking plate at one end and the VAD connection at the other. The muscle wrap extends laterally as far as the collar, and is secured in position by sutures joining the sides of the wrap to the velour muff which protects the muscle from the collar edges. Thus at each end of the barrel in reality, the muscle is partially prevented from shortening, another effect which is omitted from consideration. Together with the unknown extent to which the muscle successfully remodels to accommodate the changed geometrical situation it finds itself in, these effects may collectively mean that a much smaller value of η than 0.5 should have been used.

Another issue which finds no explicit expression in the model is the geometry of collapse of the barrel, and the effect this has on the surrounding muscle wrap. Our tests *in vitro* [31] indicate that when acted upon by slowly increasing external pressure the barrel collapses in the lowest possible mode, i.e. with one large concavity. (At higher rates of pressure increase, a higher mode, with two or three dimples, is apparently preferred, but we find empirically that this has only minor effects on the pressure-volume relation.) Whether collapse occurs *in vivo* when the external compression is provided instead by a muscle wrap has yet to be established for either the original Frogger or our version, but it seems unlikely that a cavity arises between the muscle and the barrel. The effect of this constraint may be to make the barrel stiffer than our pressure-volume data suggest.

Along with the uncertainty attached to the parameters assumed for the muscle itself, the above factors mean that the results must be interpreted with great caution, and almost certainly are cumulatively responsible for gross quantitative error. However, the difficulty in ascribing accurate values to even the limited number of parameters that delineates this model is itself what justifies the use of a lumped-parameter model. If so much uncertainty attached to even the limited data needed for a lumped-parameter model, there is little to be gained, except for specific defined purposes, in going to a more detailed distributed model. An example of such a specific purpose, which has been addressed in our finite-volume model, is the calculation of the importance of viscous energy dissipation in the fluid in the Frogger. As indicated already above, in that model, the pressure drop from the blind end of the Frogger to the VAD diaphragm was found to be always relatively small, and the viscous losses calculated directly from the fluid shears were assessed also to be relatively insignificant. This is important to know, and cannot be found from the present model; however, once found, the information allows one to proceed with the present model by neglecting these secondary energy sinks. The utility of the present model lies in its ability to indicate rapidly the critical trade-offs between the various desirable and undesirable outcomes, and to signal the importance of parameters which might otherwise escape attention until an experiment is far advanced and expensive design decisions have already been made. The present model also had significant tutorial capacity for us. In an early manifestation, before the equations for the muscle behaviour had been included, the force exerted on the outside of the Frogger barrel was specified as a pressure time-course. Muscle of course reacts to both length and shortening velocity, and in consequence the optimal strategy for gaining useful work is not necessarily that which might be anticipated without considering these properties. An example of this is the extent to which useful work was increased by lengthening the activation time-course in each modelled ejection. The importance of considering force-velocity relations is further emphasised by the extent to which muscle conditioning by repetitive stimulation produces fatigue resistance but reduces shortening velocity [41]. A further obvious factor central to the outcomes of this paper is the extent of passive elasticity in muscle, giving rise to the difficulty in achieving adequate muscle chamber refilling which has been noted in other attempts to make use of skeletal muscle for cardiac assist [42, 43, 11].

5 Conclusion

In this chapter we have demonstrated how transformed skeletal muscle, coupled with a mechanical ventricular assist device, may be analysed as an engineering pumping system. The system in question is designed to be inserted into the human body for cardiac assistance, in a number of alternative positions/functions. In our model, the skeletal muscle itself, as surgically rearranged for SMV use, was modelled using Hill's approach, enabling it to be regarded as a power source with special characteristics. The treatment of muscle as a contractile system providing force which varied with shortening velocity was vital; quite different behaviour was predicted compared to that which would have been found had the wrap been treated simply as a time-varying pressure. The results illuminated many of the problems which are encountered in experimental use of skeletal muscle power in a convenient learning environment, and thereby enhanced our interpretation of the biological experiments in which the system was physically realized. Future work should involve more refined models, possibly involving a finite-element method representation of the thick muscle wrap utilising discrete Hill-type units, a more comprehensive set of physiological parameters taking into account muscle training, drug therapy and the effects of scar tissue, and a more refined model of the fluid circuit.

References

[1] Guldner, N.W., Eichstaedt, H.C., Klapproth, P., Tilmans, M.H.J., Thuaudet, S., Umbrain, V., Ruck, K., Wyffels, E., Bruyland, M., Sigmund, M., Messmer, B.J. & Bardos, P., Dynamic training of skeletal muscle ventricles. A method to increase muscular power for cardiac assistance. *Circulation*, **89**, pp. 1032–1040, 1994.

[2] Australian Institute of Health & Welfare & The Heart Foundation of Australia, *Heart, Stroke and Vascular Diseases: Australian Facts.*, Canberra, 1999.

[3] O'Rourke, M.F., Cardiovascular disease at the turn of the century. *Medical Journal of Australia*, **171(6)**, p. 279, 1999.

[4] Schocken, D.D., Arrieta, M.I., Leaverton, P.E. & Ross, E.A., Prevalence and mortality rate of congestive heart failure in the United States. *Journal of the American College of Cardiology*, **20(2)**, pp. 301–306, 1992.

[5] Willman, V.L., Anderson, J.M., Klesges, R., Kottke-Marchant, K., Lorell, B., Pennington, D.G., Platt, J., Schulman, K.A., Stevenson, L.W., Watson, J.T., Berson, A., Altieri, F., Nickens, P.D., Kraft, S. & Agnew, S., Expert panel review of the NHLBI total artificial heart (TAH) program. NHLBI, Washington, 1999.

[6] Costanzo-Nordin, M.R., Cooper, D.K., Jessup, M., Renlund, D.G., Robinson, J.A., & Rose, E.A., 24th Bethesda conference: cardiac transplantation task force 6: future developments. *Journal of the American College of Cardiology*, **22(1)**, pp. 1–64, 1993.

[7] DeJesus, F.R., Breves consideraciones sobres un caso de herida penetrante del corazon. *Boletin de la Asociacion medica de Puerto Rico*, **23**, pp. 380–382, 1931.

[8] Leriche, R. & Fontaine, R., Essai expérimental de traitement de certains infarctus du myocarde et de l'anérvrisme du cœur par une greffe de muscle strié. *Bulletin Societe Nationale de Churigie*, **9**, pp. 229–232, 1933.

[9] Kantrowitz, A. & McKinnon, W., The experimental use of the diaphragm as an auxiliary myocardium. *Surgical Forum*, **9**, pp. 266–268, 1959.

[10] Hume, W.I., Construction of a functioning accessory myocardium. *Transactions of the Southern Surgical Association*, **79**, pp. 200–202, 1968.

[11] Geddes, L.A., The paradox in using electrically stimulated skeletal muscle to pump blood. *Pacing & Clinical Electrophysiology*, **20**, pp. 1993–2002, 1997.

[12] Buller, A.J., Eccles, J.C. & Eccles, R.M., Differentiation of fast and slow muscles in the cat hind limb. *Journal of Physiology*, **150**, pp. 399–416, 1960.

[13] Salmons, S. & Vrbová, G., The influence of activity on some contractile characteristics of mammalian fast and slow muscles. *Journal of Physiology*, **201**, pp. 535–549, 1969.

[14] Neilson, I.R., Brister, S.J., Khalafalla, A.S. & Chiu R.C., Left ventricular assistance in dogs using a skeletal muscle powered device for diastolic augmentation. *Journal of Heart Transplantation*, **4(3)**, pp. 343–347, 1985.

[15] Carpentier, A. & Chachques, J.C., Myocardial substitution with a stimulated skeletal muscle: first successful clinical case. *Lancet*, **1(8440)**, p. 1267, 1985.

[16] Acker, M.A., Dynamic cardiomyoplasty: at the crossroads. *Annals of Thoracic Surgery*, **68(2)**, pp. 750–755, 1999.
[17] Chachques, J.C., Grandjean, P.A., Cabrera Fischer, E.I., Latremouille, C., Jebara, V.A., Bourgeois, I. & Carpentier, A., Dynamic aortomyoplasty to assist left ventricular failure. *Annals of Thoracic Surgery*, **49**, pp. 225–230, 1990.
[18] Pattison, C.W., Cumming, D.V., Williamson, A., Clayton-Jones, D.G., Dunn, M.J., Goldspink, G. & Yacoub, M., Aortic counterpulsation for up to 28 days with autologous latissimus dorsi in sheep. *Journal of Thoracic & Cardiovascular Surgery*, **102**, pp. 766–773, 1991.
[19] Stevens, L., Badylak, S.F., Janas, W., Gray, M., Geddes, L.A. & de Voorhees, W.D., A skeletal muscle ventricle made from rectus abdominis muscle in the dog. *Journal of Surgical Research*, **46**, pp. 84–89, 1989.
[20] Badylak, S.F., Wessale, J.E., Geddes, L.A., Tacker, W.A. & Janas, W., The effect of skeletal muscle ventricle pouch pressure on muscle blood flow. *ASAIO Journal*, **38**, pp. 66–71, 1992.
[21] Thomas, G.A., Isoda, S., Hammond, R.L., Lu, H., Nakajima, H., Nakajima, H.O., Greer, K., Gilroy, S.J., Salmons, S. & Stephenson, L.W., Pericardium-lined skeletal muscle ventricles: up to two years' in-circulation experience. *Annals of Thoracic Surgery*, **62**, pp. 1698–1707, 1996.
[22] Mizuhara, H., Oda, T., Koshiji, T., Ikeda, T., Nishimura, K., Nomoto, S., Matsuda, K., Tsutsui, N., Kanda, K. & Ban, T., A compressive type skeletal muscle pump as a biomechanical energy source. *ASAIO Journal*, **42**, pp. M637–M641, 1996.
[23] Reichenbach, S.H., Gustafson, K.J., Egrie, G.D., Weidman, J.R., Farrar, D.J. & Hill, J.D., Evaluation of a skeletal muscle energy converter in a chronic animal model. *ASAIO Journal*, **46**, pp. 482–485, 2000.
[24] Salmons, S. & Jarvis, J.C., Cardiac assistance from skeletal muscle: a critical appraisal of the various approaches. *British Heart Journal*, **68**, pp. 333–338, 1992.
[25] Salmons, S. & Henriksson, J., The adaptive response of skeletal muscle to increased use. *Muscle & Nerve*, **4**, pp. 94–105, 1981.
[26] Odim, J.N.K., Li, C., Desrosiers, C., Chiu, R.C.-J., O'Brien, P.J., Hamilton, N. & Ianuzzo, C.D., The remodelling of skeletal muscle for indefatigable hemodynamic work. *Canadian. Journal of Physiology & Pharmacology*, **69**, pp. 230–237, 1990.
[27] Whalen, R.L., Richards, C.L., Lim, G.W., Sherman, C.W., Norman, J.C., Bearnson, G.B., Burns, G.L., & Olsen, D.B., A ventricular assist device powered by conditioned skeletal muscle. *Annals of Thoracic Surgery*, **68**, pp. 780–784, 1999.
[28] Guldner, N.W., Tilmans, M.H.J., De Haan, H., Ruck, K., Bressers, H. & Messmer, B.J., Development and training of skeletal muscle ventricles with low preload. *Journal of Cardiac Surgery*, **61**, pp. 175–183, 1991.
[29] Klapproth, P., Guldner, N.W. & Sievers, H.H., Stroke volume validation and energy evaluation for the dynamic training of skeletal muscle ventricles. *International Journal of Artificial Organs*, **20**, pp. 580–588, 1997.
[30] Guldner, N.W., Klapproth, P., Grossherr, M., Brügge, A., Sheikhzadeh, A., Tölg, R., Rumpel, E., Noel, R. & Sievers, H.-H., Biomechanical hearts: muscular blood pumps, performed in a 1-step operation, and trained under support of clenbuterol. *Circulation*, **104**, pp. 717–722, 2001.
[31] Armitstead, J.P., Bertram, C.D. & Schindhelm, K., Design and testing of a muscle-powered cardiac assist device. *Abstr. 3rd World Congr. Biomech.*, eds. Y. Matsuzaki *et al.*, Sapporo, Japan, p. 361, 1998.

[32] Umezu M., Ye C.-X., Nugent A.H. & Chang V.P., Preliminary study-optimisation of the Spiral Vortex blood pump. *Artificial Heart 3*, eds. Akutsu T., Koyanagi H., Springer-Verlag, Tokyo, pp. 107–116, 1991.

[33] Bertram, C.D. & Armitstead, J.P., A finite-volume model of the Guldner 'Frogger', a training device for skeletal muscle in cardiac assist use, both in training mode and coupled to a ventricular assist device. *Advances in Computational Bioengineering*, Vol. 6, eds. M.W. Collins & M.A. Atherton., WIT Press: Southampton, UK, 2002.

[34] Jewell, B.R., Introduction: the changing face of the length-tension relation. *The Physiological Basis of Starling's Law of the Heart*, Ch. 1. Ciba Foundation Symposium 24 (new series). Amsterdam: Associated Scientific Publishers, 1974.

[35] McMahon, T.A., *Muscles, Reflexes, and Locomotion*, Princeton University, 1984.

[36] Krylow, A.M. & Sandercock, T.G., Dynamic force responses of muscle involving eccentric contraction. *Journal of Biomechanics*, **30**, 27–33, 1997.

[37] Press, W.H., Flannery, B.P., Teukolsky, S.A. & Vetterling, W.T., *Numerical Recipes: The Art of Scientific Computing*, Cambridge University Press, 1986.

[38] Pelle, G., Ohayon, J., Oddou, C. & Brun, P., Theoretical models in mechanics of the left ventricle. *Biorheology*, **21**, pp. 709–722, 1984.

[39] Taber, L.A., On a nonlinear theory for muscle shells: part II—application to the beating left ventricle. *ASME Journal of Biomechanical Engineering*, **113**, pp. 63–71, 1991.

[40] Pelle, G., Ohayon, J. & Oddou, C., Trends in cardiac dynamics: toward coupled models of intracavity fluid dynamics and deformable wall mechanics. *Journal of Physiology*, **111**, pp. 1121–1127, 1994.

[41] Jarvis, J.C., Kwende, M.M., Shortland, A., Eloakley, R.M., Gilroy, S.J., Black, R.A. & Salmons, S., Relation between muscle contraction speed and hydraulic performance in skeletal muscle ventricles. *Circulation*, **96**, pp. 2368–2375, 1997.

[42] Niinami, H., Hooper, T.L., Hammond, R.L., Ruggiero, R., Pochettino, A., Colson, M. & Stephenson, L.W., A new configuration for right ventricular assist with skeletal muscle ventricle. Short-term studies. *Circulation*, **84**, pp. 2470–2475, 1991.

[43] Niinami, H., Hooper, T.L., Hammond, R.L., Ruggiero, R., Lu, H., Spanta, A.D., Pochettino, A., Colson, M. & Stephenson, L.W., Skeletal muscle ventricles in the pulmonary circulation: up to 16 weeks' experience. *Annals of Thoracic Surgery*, **53**, pp. 750–756, 1992.

CHAPTER 12

Leonardo da Vinci

W. Grassi[1] & M. Collins[2]
[1]*Department of Energetics, University of Pisa, Pisa, Italy.*
[2]*Faculty of Engineering, London South Bank University, London, UK.*

Abstract

Leonardo da Vinci, besides being a household name half a millennium after his death, is the embodiment of the ethos of this volume and series. He was the epitomy of Renaissance man, having a completely holistic view of the almost unbelievable range of his interests. After giving an overview of his life and activities, we have sought to stress how this holism developed, and then how it worked out. We have reviewed the academic context of Leonardo's lifetime, and briefly attempted to compare him with his Renaissance peers. While no assessment of this length can do justice to our subject, we hope that our joint Italian/British engineering-biased appreciation will prove of interest to our readers.

1 Introduction

Trying to synthesise Leonardo's life is like attempting to trace the correct path through a complex labyrinth in a short time. You try to keep the direction as much as possible (a hopeless task), but you have so many paths around, whose relevance you are not able to evaluate unless you have gone through them. And once you have so evaluated, you may be either back to the starting point or very far away.

Leonardo's life is like this. He was interested in everything, be it painting, sculpture, music, anatomy, mechanics, optics, architecture, geology, hydraulics, and so on, without attributing a real priority to any of these subjects. For example: at one moment you leave him working at the "Last Supper" fully concentrated on painting Christ's face and, a few minutes later, you may meet him in the streets of Milan, hurrying up in order to fix some tubing for Ludovico il Moro. In addition, following Nardini [1], *"Leonardo's whole life is based on 'it seems' and 'they say'…. It is thus impossible to jump from one fact to another…without the doubt, or rather the certainty, that between these two episodes of which we have proof, there are many others of which we know nothing…"*.

Such a situation has its roots, at least partly, in the history of the material he left to posterity. On his death (23 April 1519) Leonardo left around 13,000 pages of annotations (White [2]), to his friend Francesco Melzi. We must add to this certain wooden and metal models together with his paintings. Notwithstanding his firm intention to catalogue the whole material, Francesco was only able to accomplish this by way of the volume on painting, now known as the "Treatise on Painting". After Francesco's death (1570), his son Orazio, who had no interest in Leonardo's work, dispersed this material by selling it, piece by piece, to various treasure-hunters, who took profit from his ignorance and readiness to sell. As a consequence, of the initial 13,000 pages, only a little more than half is known at present.

One more feature astonishing the newcomer is Leonardo's contradictory attitudes. Either you are deeply touched watching him beautifully assisting a dying man and, just a few seconds after the death, you become disappointed seeing him calmly dissecting the still warm body. You can hear him saying he does not eat meat so as not to kill animals, but, while saying this, he is designing powerful weapons or war machines that *"strongly will scare the enemy, with his great damage and confusion"*. Some historians interpret this "dual personality" to a matter of daily arguing between two of his disciples: Giovanni Beltraffio (who committed suicide, while still young) and Cesare da Sesto.

Among many other things, Leonardo's life seems to tell us the endless story of the duality (or conflict) between the detached nature of the scientist and the humanity of the common man. To the duality must be added the mind-set of the technologist so fond of his machines as to completely disregard their social impact. His complete indifference for human vicissitudes is stressed by several authors, and is probably attested to by the ease with which he changed his protectors: from Lorenzo de' Medici to Ludovico il Moro, from Pier Soderini to Cesare Borgia and then Giuliano de' Medici and Louis XII and Francis I. As a further example Nardini [1] quotes a note Leonardo made at the catastrophic time when the whole of Italy was a battlefield (between 1511 and 1513), his friend Della Torre had recently died of the plague and his protector Giuliano de' Medici was severely ill: *"Ask Biagio Crivelli's wife how the capon hatches the hen's egg while drunk"*. So it seems the point has been made: Leonardo was quite insensitive and we can conclude with Nardini that it was a case either of supreme indifference or childish irresponsibility. But what about his profound love for his mother (he supported her until her death), his Uncle Francesco (he helped financially throughout his life) and his pupils? And, moreover, what about the fact that he refused to divulge his studies about "submarines" and "divers", when in Venice, *"because of the evil nature of men, who might use them for murder on the sea-bed"*?

A final issue, continually raised, is whether Leonardo was a scientist or a technologist? Or could he have been both? This is a difficult question and many words have been written on this point. The question itself can be inserted into a qualification: is such an assessment of Leonardo meaningful, or is it simply an anachronistic effort on our part to describe him on our terms?

So then, we have raised undoubted problems, on one side suggesting a fractured personality, and on the other, querying the true nature of his ability. Yet in seeking to provide a general insight into Leonardo's character within the framework of "common law and design in the natural and engineered worlds" (the purpose of this book), we consistently find a deeper level is suggested. In the conclusion of this chapter we will address these issues again and, in fact, we will pose the *converse* question, namely to what extent there is an underlying unity to Leonardo's person and work.

2 Life and works

Leonardo da Vinci was born in Vinci, a small village in the Tuscan countryside a few kilometres north of the town of Empoli. Empoli is located midway between Florence and Pisa and can be easily reached, nowadays, both by rail and motorway. We know the exact date and time of Leonardo's birth, 10.30 p.m. on Saturday 15 April 1452, thanks to his grandfather, Ser Antonio da Vinci, who carefully wrote this note in his (Antonio's) father's notary book. This habit of detailed recording of events, which Leonardo also adopted, was probably related to the long tradition of notaries in his father's family. According to what we know this tradition, which dated from the mid-14th century, was interrupted by Ser Antonio and restarted with Leonardo's father, Ser Piero. It was common for Leonardo to make detailed notes on daily home expenses (which, in reality, was of little interest to him) besides the observations he made about nature, techniques of painting, machines and so on. It is thanks to this habit that we have such detailed knowledge about Leonardo, as he did not properly publish anything during his lifetime.

Leonardo was illegitimate as he was born from a short liaison of Ser Piero with a peasant girl Caterina, daughter of Boscaiolo Guidi. She was then married by Ser Antonio to a certain Accattabriga (nickname) di Piero del Vacca, with (or better thanks to) a dowry of thirty florins, provided by Antonio himself. In addition ser Piero was sent back to Florence, where he already worked as a notary, and compelled to marry Albiera di ser Giovanni Amadori, from a noble family. There has been some debate about the influence of such illegitimacy on Leonardo's personality. Sigmund Freud even coined the term "Leonardo complex".

As a favour to his son, Antonio agreed to keep the grandson with him and he granted the custody of Leonardo to his wife, the good "granny" Lucia di Piero da Bacaretto. So the childhood of Leonardo in Vinci and Anchiano began. There is a lack of agreement among historians on how long he lived in Vinci before his move to Florence, and how much time he spent with his Uncle Francesco, who was seventeen years older than Leonardo. To give an idea of the uncertainty on this point alone the suggested Leonardo's age at the time of his move is given by different authors as: sixteen or seventeen (C. Pedretti in Antoccia *et al.* [3]), thirteen (Mereskovskij, [4]), thirteen or fourteen (White [2]). Despite this age difference, Leonardo was strongly influenced by Francesco, a philosophic idler, who introduced the child to the careful observation of nature, thus following Aristotle's precepts on concrete learning from what exists in reality, and to a love for the whole of "Creation".

Leonardo spent all the time he could wandering in the countryside, observing variously birds and their flight, flowers and their structure, the lives of insects and their anatomy, and the flow of moving water. Some specific episodes might have exerted an especial influence on his future life. In 1466 the River Arno flooded the entire region and this probably had a great impact upon Leonardo, perhaps being the trigger for his lifelong, almost obsessive, interest in water. Of course this interest was also influenced by the "Sienese engineers" such as Taccola and Franceso di Giorgio, whom Leonardo personally met in Milan around 1490. Whatever the causes Leonardo's unusual conviction was of water as the most powerful element in nature, that could be either a catastrophic destroyer or a powerful means to be controlled and used for man's purposes.

A completely different aspect of his future activity might be due to the occasional visits he paid to the villa of Pandolfo Rucellai near Vinci, when it was still under construction. He usually spent his time observing the bricklayers align the large stone blocks with a square and lifting them by pulleys. There he came to know the architect Biagio da Ravenna, who had been a pupil of the famous Leon Battista Alberti. At first almost as a joke, but later on, in

Only at the end does the "advertisement" stress that, in peacetime, Leonardo could stand comparison with anybody else with regard to architecture, painting, water management as well as marble or bronze sculpture.

In Milan Leonardo lived with the De Predis brothers, who were painters. The power of his new painting style was revolutionary compared with the crude, static style current in Lombardy, and his "Virgin of the Rocks" (1483) became immediately famous. The following words have been used in order to describe effectively this work: (Antoccia et al. [3]): *"...Enigmatic is the same iconography of the group of the Virgin with the baby, the little Saint John (the Baptist) and the angel within a magic half-light of a peerless scenery of rocks, water and vegetation..."*; and Nardini [1]: *"...This emotional fusion is heightened still further by the position of the woman and the two children, while all of the personages are contained within an ideal circumference. In the background is an unreal, fantastic vision of rocks; above, an interwoven pattern of branches; far away, in the rocky landscape, flows a stream..."* It was so innovative and fascinating that the same Ambrogio De Predis, already well known in Milan for his portraits, became a pupil of Leonardo and was closely followed by several young painters.

Among other works of Leonardo of the same period we should mention the "Musician", the "Lady with the Ermine" and "La Belle Ferroniere". *"The 'Lady with Ermine' is a typical example of the so called 'ritratto di spalla' ('bust length portrait'), an approach developed by Leonardo from the dynamics of the human body, like the woman's bust sketched in eighteen different attitudes found in a sheet of the Royal Collection at Windsor, England"* (Antoccia et al. [3]). The lady portrayed is the beautiful and well educated Cecilia Gallerani, taken as a mistress by Ludovico Sforza. "La Belle Ferroniere" is probably the portrait of Lucrezia Crovelli, who succeeded Cecilia Gallerani as Il Moro's mistress. The figure is three quarters beyond a parapet *"...like a person appearing at the window for a moment. The portrait is centred on the magnetic fixity of the eyes"*. At least two recurrent themes of Leonardo's thought come out from this work: the intention of representing human beings in their dynamic (as opposed to static) attitudes (in other words something like a modern snapshot) and the way their feelings are expressed through their external countenance. At that time eyes were thought to have a particular relevance as a sort of gate to the soul, which was thought to be located in the centre of the skull. Finally Leonardo used to state that "eyes are the soul's windows", an idiom we still use today. Leonardo's most famous masterpiece is the "Last Supper" (1495–1497). He was commissioned for it by the Dominican fathers who wanted to fresco the refectory of their Santa Maria delle Grazie monastery, of which Bramante had built the cloister and the imposing entirety of the apse in the period ranging from 1492 to 1497 (Salvini, vol. II [7]). It was quite usual in those days to have monks' refectories decorated with a scene of the Last Supper of Christ. To give an idea we can quote a few of the most famous ones: Andrea del Castagno's in Sant'Apollonia in Florence (1445–1450), Taddeo Gaddi in Santa Croce in Florence (1350), and Domenico Ghirlandaio in Ognissanti in Florence (1480).

Luca Pacioli, mathematician and close friend of Leonardo, remarks on (1498) the unbelievable animation of the Apostles in the fresco. Concerning this we can use the telling and evocative description of Nardini [1]: *"...The whole scene hinges around Jesus. It is broken down into four groups, each of which reveals a precise sentiment in the expression of the faces, the gestures of the hands, the motion of the persons and even the position of the feet protruding from beneath the table"*. Also, perspective and light still play a major role in the painting, so that *"The same succession of perspective planes is scanned by light and shade.... Light plays with the most subtle 'sfumato' (soft or shaded) in the room's air and fondles the images transfiguring in lyrics the psychological exasperation of the characters, taken three by three into a wavy movement"* (Salvini [7], vol. III). However, the customary technique of frescoing required the colours to be applied quickly onto the fresh plaster, and in a definitive way, and this was not congenial to the rather hesitant Leonardo, who needed to meditate on the work done and to correct it. So he experimented with a new method of preparing the wall with three different layers of substances, which could allow him to paint in oil. While painting in this way, Leonardo realised that the preparation was not successful as many tiny cracks suddenly appeared. The consequence was that this became both one of his main concerns, and one of the major endless problems of the "Last Supper".

Anyone, who has at least heard about how Leonardo considered the art of painting as the highest and conclusive way of knowing nature, must wonder where all these concepts of "motion" of characters, three-dimensionality, perspective, and light come from. They really stem from Leonardo's conviction that painting was not a rational enterprise, like science, but a way of penetrating reality and of creation of a deeply lyrical substance. *"...by means of art man possesses the power of creating like nature does, with that same harmony and proportion nature uses for its creatures"*. If we accept the specific outworking of his experiences at Florence, it is logical to interpret this conviction of Leonardo as the strict outcome of the studies he was pursuing over that period in the field of anatomy, mathematics and geometry, and optics.

Leonardo's first anatomical studies date from the Milanese period (1487-1493), starting with the dissection of the skull, as carefully reported in his drawings (see for example Cianchi, [8]). The choice of the skull was not accidental, but rather a deliberate choice resulting from the influence of two important factors on his thinking. Verrocchio, Pollaiolo and the then general Platonic culture caused him to consider the needs of representing both the proportion of the body and the "animus", i.e. the "interior energy" showing up in a moving figure. Added to this was the accommodation of the common thought of that age that located the soul in the middle of the skull. Later on, in Rome, he would devote greater attention to the connection between the eyes and the brain to such an extent that he was the first one who drew the "Chiasma", that is the point in the skull where the optical nerves meet. The concept of bodily proportions was also connected with mathematics. Leonardo was very impressed by the remarkable work of Luca Pacioli, *Summa de arithmetica, geometria, proportioni et proportionalità*, which he ordered as soon as it was sent to press. Later, when Ludovico il Moro called Pacioli to Milan to teach mathematics, Luca and Leonardo became very close friends and Leonardo illustrated the "de divina proportione" with magnificent tables. In the other direction Pacioli taught mathematics to Leonardo, who prior to that lacked any formal training in this subject.

Among the very fruitful activities Leonardo performed in his first period in Milan (surely not exhausted in these few pages), the following are worth mentioning: the project of the lantern for the Duomo of Milan (1487), the design of settings and decorations on the occasion of the wedding of Gian Galeazzo with Isabella d'Aragon and the design of the costumes for the procession of Scythians and Tartars, for the wedding of Ludovico il Moro and Beatrice d'Este (1492). On this occasion Leonardo showed his great skill in machine construction for a spectacular theatrical display, held after the wedding. Nardini [1] reports it was "a marvellous spectacle of automation". Something similar, which could be interpreted as "Leonardo playing games with machines", is also referred to (possibly more as a legend than as a real fact) by Castelfranchi & Stock, [9]. These authors describe a mechanical lion built in honour of the King of France, Louis XII, who entered Milan in 1499. This lion moved toward the king and opened its breast with his claws as a token of submission.

The record of existence of automata lies somewhere between legend and history. For example at the end of the thirteenth century, in England, there was a rumour about two friars creating a bronze "talking head" to defend England from invaders. One of the friars was reported to be Roger Bacon (1214–1294). This is the same Bacon who asserted, well before

As we noted at the start of this chapter, Leonardo had been much influenced by his Uncle Francesco in childhood, and from that time may be traced his fascination by flight. In fact, ever since he spent much time watching birds. During his first Milanese period, he began a really systematic study of gravity and flight. But it was in Florence, at a more mature age, that these studies became more integrated, so that his codex on the flight of birds is commonly dated 1505.

At this same period (1503?) he started working on what is probably his best known masterpiece, "La Gioconda" or The Mona Lisa. So many things have been said about this painting, that we refrain from comment. As to the subject herself, Mereskovskij [4] prefers the old Florentine tradition, according to which the subject of the picture is Mona Lisa Gherardini, the third wife (still then a young girl) of Francesco del Giocondo. Mereskovskij [4] also tells how Leonardo painted this picture while he was in the house of ser Piero di Barto Martelli, commissioner of the Signoria, and a mathematician and learned man: the latter lived not far from Mona Lisa's house. While painting, it was said, Leonardo used to have musicians plying to keep the Gioconda's smile alive.

Nardini [1], on the other hand, quoting from the same tradition, probably interprets the situation more correctly: "La Gioconda is, and remains, Leonardo's great secret. This painting was not executed on commission". It is on record that Leonardo kept the painting for himself for the rest of his life.

The end of this Florentine period was marked by two very discouraging events for Leonardo. The "Battle of Anghiari" was totally lost as his new experimental technique related to the plaster base did not work; this also enraged Soderini. Thus the master isolated himself, living in Fiesole in the house of Alessandro Amadori. There he anxiously worked on his flying machine with the faithful Zoroastro. But this experiment too went wrong and the machine crashed, causing (it is said) serious wounds to Zoroastro or maybe, even, his death. So Leonardo left Florence (1506) for Milan, after being authorised by Soderini to have three months' leave, as insistently requested by the young French Governor of Milan, Charles d'Amboise. Then the French king himself insisted on having Leonardo in Milan, so that he could meet him there during his next visit to the city. Soderini could not deny a leave's prolongation for such a powerful potential patron and this was how Leonardo gained the personal acquaintance of Louis XII. Afterwards, however, he had to return to Florence (September 1507) to address the issue of a lawsuit brought by his many younger brothers with regard to the inheritance left him by Uncle Francesco. While this arduous process was in progress, Leonardo painted the "Madonna Litta" and the "Virgin of the Scales" and reorganised his notebooks, adding studies on optics, anatomy, architecture, acoustics, cosmology, hydraulics, mechanics and thermology. But, as soon as he could, he went back to Milan (1508), bringing the above Madonnas with him. In Milan Leonardo painted the "Bacchus" for the King and devoted himself to geological studies of the Lombard valleys. He deepened his anatomical studies with the greatest anatomist of the time, Della Torre from Verona, who offered to collaborate with Leonardo for the compilation of an organic treatise on anatomy. Unfortunately Della Torre died, when he was only thirty (1511), from the plague, for which he had gone to care to Riva del Garda.

Italy at that time was a sort of large battlefield, with a series of rapid changes in the political situation, as shown in the Section of historical events (before the References). At the end of 1512 the French were forced out of Lombardy, and Massimiliano Sforza (second-born son of Ludovico il Moro) occupied Milan. A little later the son of Lorenzo the Magnificent was elected Pope, raising the artist's hopes of a revival in his fortunes. Thus people like the two Sangallos, Raphael, Bramante, Signorelli and many others went to Rome and Leonardo followed this "flow" on the 24th of September 1513. He had a very warm welcome from

Giuliano de' Medici, the Pope's brother, with whom a deep friendship was instantly born thanks to their common interests in painting, mathematics and mechanics. According to Vasari they also studied alchemy together. The activities of Leonardo in the fields of geology, hydraulics, archaeology and flight also took a new lease of life. He searched for shells on Monte Mario (a hill in Rome), was in charge of planning the draining of the Pontine Marshes (1514) and of a project for the port of Civitavecchia. Actually this task (draining of the Pontine Marshes) had been assigned by the Pope to his brother Giuliano, who gave Leonardo a commission for it. The master had already drained the swamps of Vigevano for Ludovico il Moro. Utilising this experience, he worked on the new project with great enthusiasm, making a beautiful bird's-eye map of the swampy region. This map is part of the Leonardo archives in the Royal collection at Windsor, England. But around a month later Giuliano had to leave for Savoia to take a wife for political reasons, and probably Leonardo went with him. So the project was no longer pursued and, in fact, the draining of the Pontine Marshes was realised four centuries later during the Fascist period.

In his studies on flight Leonardo changed his approach and went from full-scale models to small ones and conducted experiments mainly on gliding flight and on the curvature of wings. He also painted the "Madonna with Child" for Sant'Onofrio (on a commission from the Pope) and two paintings for Giuliano: "Leda" and the portrait of a "certain Florentine woman". He had restarted his anatomical activity, after a gap of about twenty years, during the last period in Florence. This Leonardo continued by working on dissection in the Hospital of Santo Spirito. But the research had to be suddenly suspended due to an accusation of necromancy, brought by one of the German assistants, who was helping him in his mechanical works.

Several historical events occurred meanwhile and in particular Francis I, having succeeded Louis XII, had defeated the Sforza. Contemporaneously Giuliano de' Medici was dying in Florence. As a consequence of the first situation, the Pope arranged a meeting with the King, taking Leonardo and Raphael with him. There circulated a rumour that the meeting between the King and Leonardo became even affectionate. The King invited the master to follow him to France. Leonardo accepted. So it was that he left Italy for good, accompanied by Francesco Melzi and the new servant Battista de Villanis. In referring to that significant meeting, Nardini, [1], reports the King as saying: "Leonardo, mon père, …come to France, I will ask you nothing in return. It will be enough for me to speak with you sometimes, to listen to you as my great father did". Whether this is true or not it does, in any case, give an idea of the high opinion the King had of Leonardo. King Francis gave him the castle of Clos-Lucé (also called the Cloix) one kilometre from Amboise, where the King himself lived. There, in a quite peaceful atmosphere, Leonardo studied the neighbouring region to set up a water channelling project that he intended to offer the King as a demonstration of his gratitude for him. He also found a way to transform the cartoon of the Virgin in S. Anne's lap (exhibited long before at Santissima Annunziata in Florence) into a real painting and the Gioconda was finally completed. It was a fitting climax to his activity.

And it was in "Cloix" that he "returned to the Prime Mover" on the 2nd of May 1519, after writing words quoted as *"The life well spent is a long life"*.

3 Leonardo the polymath

3.1 Background

"...the Renaissance...a number of 'polytechnic' geniuses. Of these, Leonardo is undoubtedly the most emblematic and outstanding"

Alberto Lina

Table 3: Leonardo's achievements – main subject areas.

ART	PAINTING	…huge natural talent…greatest artist of his age…at the peak of his powers…perhaps the greatest living Italian artist. Mona Lisa…perhaps the most enigmatic painting in history of art [2]
	SCULPTURE	…massive bronze equestrian statue…great model in clay…instinctive understanding [2]
	ARCHITETCTURE	little real success as an architect, concept of architecture was esoteric and controversial [2] humane utopian [20] Milan cathedral [2]
	DRAWING DRAUGHTMANSHIP	…hyper-text and cutaway…500 years ahead of his time [2] ….extraordinary almost rhetorical power of his drawings [6]
ENGINEERING	MECHANICS	The most innovative aspect…is his analysis of the components of machines…accurately classified motion…screw, flywheels, springs/clockwork [6]
	MILITARY MACHINES	…rather non-innovative military engineer [6] ..following rather than leading. Knowledge of hardware of war almost without peer [2]
	HYDRAULICS	…plans for redirecting the Arno [2] …re-planning of Milan…perfectly regulated network of waterways [6]
SCIENCE	ANATOMY	…the greatest Renaissance anatomist [2] …describing arteriosclerosis of the aorta hundreds of years before…physicians [15]
	OPTICS	…possibly the first experimenter to realise the role of the lens…light and sound interrelated with water ripples…three separate elegant experiments [2]
	SOUND	…saw light and sound as behaving in a similar fashion…science of acoustics…sound and music…how sound could reach us [2]
	MATHEMATICS	…weakness as a mathematician…tutelage of Pacioli (appreciation of) [2] …no investigation can be called scientific unless it admits of mathematical demonstration [21]

Thirdly, there is the issue of Leonardo's flying machines. "Leonardo's understanding of aerodynamics was actually amazingly sophisticated [2] and we are now in the situation where replicas of Leonardo's designs are being constructed by enthusiasts.

Table 3, then, demonstrates an almost unbelievable range of interests. The consideration of just three additional points, as above, serves only to show how difficult it is for us to appreciate that range. Our sense of awe at Leonardo could be summarised as follows: how is it that a world class painter can be at the same time a pioneer anatomist *and* a prodigiously inventive engineering designer…besides undertaking all the other activities? Is there *any* answer to this question?

Paolo Galluzzi [6] does attempt to give a partial answer, based on a rationale of Leonardo's background as in the previous section. Like Alberto Lina, Galluzzi places Leonardo in the context of a collective development lasting several decades …the practical potential of technology …enthusiastically shared by many "artist–engineers" of the fifteenth century. In fact Galluzzi's entire Introduction could be read as an expansion of that quote. He continually returns to the theme of Leonardo himself.

In Figure 1 we have extended Galluzzi's approach to show the various connections between the main subject areas studied by Leonardo. Firstly, with a general artistic capability, he has a (2-dimensional) major activity in painting with a 3-dimensional counterpart in sculpting, and a further attainment in draughtsmanship, again in 3 dimensions. These three subjects form a discrete group. Then, because of the workshop (bottega) concept of training artist-engineers, there is a "horizontal axis" connecting art with engineering and mechanics. These latter, as we have seen, were Leonardo's consistently principal source of income. (We have removed, however, any distinction between what used to be termed "military" and "civil engineering"). The architecture and hydraulics arise naturally as an extension of his professional activity.

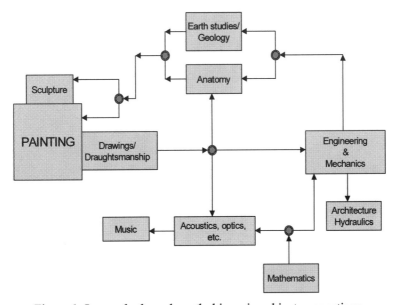

Figure 1: Leonardo the polymath: his main subject connections.

Mathematics forms a separate subject "input" arising from his association with Pacioli. While, as we have seen, Leonardo lacked an inherent grasp of this, at the same time his appreciation of its significance had relevance both to the mechanics and his "scientific" studies of optics/acoustics. Now it was Leonardo's intensively inquisitive nature, combined with his professional training, that formed the chief stimulus to this combined optics/acoustics studies. The personal "mix" of a superior ability to think and draw in 3 dimensions, of being extremely industrious when focused, and writing up his observations, is what enables White to term Leonardo "the first scientist". In fact, it might also be said that such has become our faith in Leonardo's multi-faceted attributes that somehow we *expect* him to be mathematically adept and are left frustrated on realising that this is not so!

The lower part of Figure 1 is completed with "music". In the same way that today "physics" and "music" are sometimes combined academically, so Leonardo's acoustics studies extended to music in general and to the design of musical instruments in particular. Were Figure 1 3-

What really marked the revolution in scientists' attitude was the confidence Galileo placed in an apparatus built by technicians, accompanied by progress only by way of practice. This was an approach essentially ignored by official science. In terms of today's research-paper-oriented components, we could say Galileo's approach comprised the "equipment" and "experimental method" sections.

It was *the seventeenth century* which marked the real origin of the scientific revolution. Here we simply name but a few of the main well known contributors to this revolution, such as Francis Bacon, Renè Descartes, Blaise Pascal and Isaac Newton. A great impulse was given to medicine and biology by scientists like William Harvey (1578–1657), who discovered the principles of blood circulation. So were constructed the fundamentals of modern physiology as experimental medicine. Again there were Marcello Malpighi (1628–1694), Robert Hooke (1635–1703), Antony van Leeurwenhoeck (1632–1723) and others, who are known as the classical microscopists. While Galileo explored the infinitely large, provoking a sort of "scientific earthquake" in a field widely discussed for centuries, they initiated a kind of "silent revolution" in a real new, unexplored, field

All this happened around a century after Leonardo's death (1519) and subsequent to such key developments as the Copernican revolution and the fundamental anatomical work of Andrè Vesale in Padua (1514–1564). Bearing these connected events in mind will help the reader to avoid making anachronistic errors in evaluating Leonardo's work. Another aspect which could help to keep things in the right perspective is Rossi's outline [10] of the situation of Science and Technology in those times, which we will now follow. From the thirteenth to the sixteenth century, any Italian university had a pyramidal structure: arts (arts might include mathematics, moral philosophy and humanistic matters) as a lower level faculty and theology, law and medicine as superior faculties. No one might teach mathematics in a university, for example, without already having a degree in theology, law or medicine. Passing from teaching mathematics to teaching philosophy was considered a promotion for a professor. Excellence in mathematics was insufficient for what we call today career progress; excellence had to be demonstrated in several other fields as well. Therefore there was no real incentive to devote systematic attention to "scientific subjects".

The importance of mathematics, then, was placed at a secondary level within university curricula. Even in the sixteenth century there was an average of one mathematician for each twelve teachers of medicine in the major universities. Table 4 summarises the overall situation in four of the most important European universities. While universities had a strong influence on contemporary culture they were absolutely unable to promote innovation. The strong criticism against universities of Francis Bacon and Renè Descartes testifies to this. As late as 1650 Thomas Hobbes (in *The Leviathan*) stated that, in universities, philosophy only coincided with Aristotle's thought, geometry was not even taken into account and physics was just empty talk. It reinforced what John Hall had written to university students one year earlier, stressing the lack of the teaching of chemistry, anatomy and experimentation: "*It is as if students had learned three thousand years ago the whole Egyptian culture and since then they had ever slept like mummies to wake up only today*".

In addition we must stress how slow then was what we now call information transfer. We are so used to almost real time communication (or at least to the rapid diffusion of journals and books), that we are psychologically inclined to undervalue this point. To give an example, it took three centuries for the diffusion of the Indian numerical system (which Western merchants had learnt from the Arabs) in Europe, after Leonardo Fibonacci wrote his historic study in 1202.

Table 4: List of funded chairs, by subject, in different European universities, from Ben David, 1975 as reported in Rossi, [10]. (a): astrology, natural philosophy, physics; (b) arithmetic and geometry, astronomy; (c) mathematics; (d): natural philosophy, geometry, astronomy; (e): arithmetic and astrology, physics and natural philosophy.

	1400	1450	1500	1550	1600	1650	1700
Bologna							
Science	3(a)	-	2(b)	2	2	2	2
Medicine	11	2	3	3	5	5	3
Others	33	9	16	16	20	22	23
Paris (Sorbonne and Collège de France)							
Science	-	-	-	2(c)	2	2	2
Medicine	-	-	-	2	3	3	3
Others	-	-	-	8	12	18	20
Oxford							
Science	-	-	-	-	-	3(d)	3
Medicine	-	-	-	1	1	2	2
Others	-	-	-	15	15	20	20
Leipzig							
Science	-	-	-	-	2(e)	2	2
Medicine	-	-	-	-	4	4	6
Others	-	-	-	-	17	17	23

Outside the university the daily life of society was progressing on its own, with all its practical needs and following an almost entirely separate evolution. In 1400 and 1500 there was a very large diffusion of technical treatises, in some cases becoming real handbooks. They were authored by engineers, artists and artisans of the stature, for example, of Brunelleschi (1377–1446), Ghiberti (1378–1455), and Piero della Francesca (1406–1446). For the actual period of Leonardo's lifetime we can list the following publications together with their subjects.

- Architecture – Works by Leon Battista Alberti (1404–1492), whom Leonardo personally knew, Francesco Averlino nicknamed "Filarete" (1416–1470), Francesco di Giorgio Martini (1439-1501), who among other things was one of the of the "Sienese Engineers" (see below). These devoted much attention in addition to machines.

- Fortifications – Albrecht Durer (1471–1528), his treatise-type publication (1525) being after Leonardo's death.

- War and military machines - Konrad Keyser (1366–1405) and Roberto Valturio who both produced treatises.

- Descriptive geometry – again Durer.

For a man of such prodigious talent it is appropriate that even from our limited viewpoint, we see his context as covering some quarter of a millennium. When we turn to Leonardo

Codex Atlanticus, f. 1063. "Without notes by Leonardo. None of Leonardo's drawings shows the whole boat but just some detail or other. The paddles or blades are moved by large man-powered cranks."

Figure 5: A selection of Leonardo's designs: Paddle Boat.

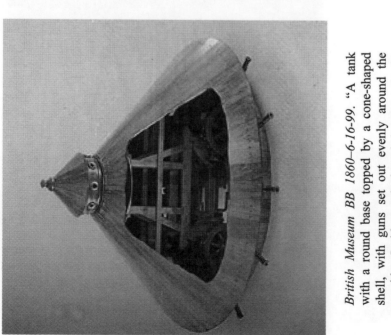

British Museum BB 1860–6-16-99. "A tank with a round base topped by a cone-shaped shell, with guns set out evenly around the outside. Inside toothed and lantern wheels, moved manually, allow the tank to be driven in any direction."

Figure 4: A selection of Leonardo's designs: Tank.

Codex Atlanticus, f. 133 v. "The drawing of this bicycle among Leonardo Papers was far too unexpected not to take scholars by surprise and raise a few queries. It came to light during restoration work on the Codex Atlanticus, after taking apart two sheets that had been pasted to mountings by Pompeo Leoni at the end of the 16th century. The drawings, being on the reverse side of a sheet Leoni had torn in half, had remained invisible for more than three hundred and sixty years and nobody could obviously have tampered with the sheet in that period. According to Augusto Marinoni it must have been drawn by Salai, Leonardo's young pupil and model; his name is, in fact, the only word written on the sheet. He was clearly copying one of his maestro's drawings. The transmission system with the chain for square-shaped teeth (rounded off on the reconstructed model) most certainly stems from the Vincian drawings in Codex Madrid I, f. 10 that nobody knew about before 1966" (from [19]).

Figure 6: A selection of Leonardo's designs: Bicycle.

many of these aspects. For example he says Leonardo's energy was mainly devoted "to scientific and technological pursuits". When added to the artistic and draughtsmanship capabilities, Galluzzi's assessment of the *range* of Leonardo is the same as White's. Moreover Galluzzi agrees that Leonardo was "the prophet of automation" and supports Benvenuto Cellini's description, "a very great philosopher". Finally Galluzzi discusses Leonardo's "faith in the development, through technology, of unheard powers in man: to fly like a bird, to live underwater like a fish". He mentions that "technological dreams often have a prophetic tone" – so in the context of flying machines and submarine activity Leonardo is a visionary prophet.

Turning from Leonardo the artist-engineer to Leonardo the anatomist-dissector, Galluzzi does not make value judgements, nor does he seek to place Leonardo in the corresponding *anatomical* context of genius. Of course, by removing anatomy it immediately degrades Leonardo's achievements, and that makes Galluzzi's assessment less valuable. White, on the other hand, collates all Leonardo's activities, and for him there is absolutely no doubt as to Leonardo's genius. In fact, he is as enthusiastic as Galluzzi is reluctant. For White, Leonardo is "a supremely gifted artist" who also had a "prodigiously inventive mind".

White is not alone. The almost exhaustive studies of Theodore Cook on spirals in nature, science and art, [21], were inspired by "the manuscript of Leonardo da Vinci". For Cook "the genius of Leonardo was of so universal a quality". Now Cook dated his preface as March 28th, 1914, and it is telling that approaching a century later, the surgeon and medical historian Sherwin Nuland has the same unreserved admiration. He describes Leonardo as "a man whose name has become synonymous with genius" [10].

Part of the character of genius, we feel, is its capacity to inspire highly, whether the near-idolatrous fascination of Nuland, or the bridge-building zeal, working with genius, of Sand [24].

While fully accepting the context of the studies of Galluzzi, and not ignoring the reservations of authorities like Stephen Jay Gould (as discussed by White, [2]), we feel the evidence for the first-rank holistic genius of Leonardo is more than sufficiently compelling.

8 Conclusion

"...a bold attempt at unifying Nature under a small number of universal laws"

Paolo Galluzzi

In this section, we seek to evaluate Leonardo's life in the context of our Book Series "Design and Nature". For this he has proved to be an excellent subject. As far as we can judge, he was forever pursuing parallel studies in both the natural and engineered worlds. His insight, creativity and drive were outstanding whether in art, in anatomy, in light and sound, in mechanics of machines or in engineering in general. To say this much is hardly novel, but we believe he was much more of a consistent person undertaking [much more] integrated activities than modern commentators allow.

By seeing Nature as a whole, and by endeavouring to unify Nature "via universal laws", [6], essentially means that he pursued the rationale for our Book Series about five hundred years earlier.

Firstly there is the question of his inconsistent moral attitudes, both in relation to anatomical studies, and to warfare and human suffering. Tending the dying patient and then dissecting the body, is closest to the current medical ethics situation of transplant procedures. In the UK, the possible "conflict of interest" has been recognised by having different medical teams for care, and for organ transplant. Now admittedly, Leonardo's work (in anatomy) could barely be equated in a local sense with saving a human life through organ donation, but in the longer term it was arguably as or more beneficial to humanity. Also Leonardo had to battle with continual criticism of semi-necrophilia, so that the work had to be done as secretly as possible. One feels that most lesser mortals would have given up. Leaving aside his enormous inquisitiveness, he did have a clear basis for his anatomical work in his holistic view of all of life, whether natural or man-made. The picture is, we feel, more favourable to Leonardo than at first appears.

Further, there is the apparent contradiction between his turning away from submarine warfare invention and vegetarianism on the one hand, and working as a very pro-active military engineer on the other. The ethical issues of waging (or not waging) war are beyond this chapter, and complicated in Leonardo's case by his earning a living as an effective "participant" and by the Papacy being at this time a truly political and military force. Also, Leonardo's artistic status made him an international figure.

The patronage of his work meant that he was known to the political leaders often personally. It is conceivable that Leonardo viewed submarine warfare as an ethical bridge too far, akin to current outlawing, say, of germ warfare. If so, his vegetarianism and generally softer side became his true self, with the other features being a necessary consequence of the times in which he lived. This does not answer the question of indifference to suffering raised in the Introduction, and, for this moral issue, the scientific ambition which drove his life could possibly be the answer. Again, current ethical issues related to genetic modification have tended to remove policy responsibility from the actual scientific workers themselves – leaving them the almost headlong dash to complete the genome! – to overarching public responsibility committees of quasi-legal status.

We tread somewhat easier ground when we survey his work and achievements. There is no question but that his restless brain searched multi-dimensional problem space, much like the Genetic Algorithm design procedures assessed in a Chapter in this Volume. However, as we have seen, the bottegas themselves promulgated a multi-disciplinary approach and Leonardo's being a (sort of natural plus engineered) "One World Man" was in the same mould. Yet we *can* detect a unity on these lines. In seeking to perfect his artistic capabilities, he studied muscle structure effects on human and animal surface configurations, especially faces, and this led him on to anatomy. In his parallel engineering work his mechanics of machine achievements led him to view that anatomy in terms of mechanical behaviour.

The current academic movement towards inter-disciplinarity is a kind of seeking for the unity that our individual departmental structure tends to obscure. For Leonardo that unity always existed, from the bottega onwards. Let us take one present-day example of particular relevance to his work. The subject of Computational Fluid Dynamics has been transformed over the last 10-15 years from two-dimensional steady cases to three-dimensional time-dependent flows with irregular geometries. The old (typically-engineering) graphical presentation of results is being transformed to virtual-reality-like methods which display the very much larger amounts of data being generated. In parallel with CFD, new non-invasive methods of (usually optical) measurement often require tomographical reconstruction of fluid structures. The tomographical procedure is inherent in fully processing the matching clinical methods of magnetic resonance and ultra sound used so much in medicine today. We are living much more in a three-dimensional world, and this may help us to appreciate Leonardo's sure grip of being able to think in this way in both his anatomy and machines. His layering description of human anatomy on the one hand, and his almost casual ability to conceive three-dimensional machine systems (witness the illustrations of present-day reconstructions in Galluzzi, [6]), on the other, speak for themselves.

[13] *The Complete Family Encyclopedia*, Frase-Stewart Book Wholesale Limited/Helican Publishing: London, 1992.
[14] Millon, H.A. & Lampugnani, V.M. (eds.), *The Renaissance from Brunelleschi to Michelangelo. The Representation of Architecture*, Rizzoli: New York, 1997.
[15] Nuland, S.B., *Leonardo da Vinci*, "Lives" Series, (General Editor J. Atlas), Weidenfeld & Nicolson: London, 2000.
[16] Darwin, C., *The Origin of the Species*, Wordsworth, Ware, UK, 1998.
[17] Bailey, M. (General Editor), *The Folio Society Book of the 100 Greatest Paintings*, The Folio Society: London, 2001 (individual essayist on Leonardo, M. Kemp).
[18] White, M., *Isaac Newton the Last Sorcerer*, Fourth Estate: London, 1997.
[19] *Il Museo Leonardiano di Vinci*, (museum catalogue), Città di Vinci, 21 October 1986, New Revised Edition, 1994.
[20] Grafton, A., *Leon Battista Alberti*, Allan Lane/The Penguin Press: London, 2001.
[21] Cook, T.A., *The Curves of Life*, (originally published by Constable: London, 1914), Dover Publications: New York, 1979.
[22] Cooper, G. & Altman, W., *Incentive for Invention*, Business file, page A, *Sunday Telegraph*, London, 29 September 2002.
[23] Osborne, H. (Ed.), *The Oxford Companion to Art*, Oxford, 1970.
[24] Boyes, R., *Norway Builds da Vinci's 500-year-old Bridge*, The Times, 1 November 2001.

CHAPTER 13

The evolution of land-based locomotion: the relationship between form and aerodynamics for animals and vehicles with particular reference to solar powered cars

D. Andrews[1], A. Shacklock[2] & P.D. Ewing[3]
[1]Department of Mechanical Engineering, Imperial College, London, UK and Department of Architecture and Design, London South Bank University, UK.
[2]Singapore Institute of Manufacturing Technology, Singapore.
[3]Department of Mechanical Engineering, Imperial College, London, UK.

1 Introduction

In his publication 'On The Origin of Species' Charles Darwin argued that the diversification of species occurred through natural selection as both animals and plants adapted to their environments and developed various physiological, anatomical and behavioural traits to facilitate maximum exploitation [1]. For animals these traits include specific means of locomotion whether for feeding, mating or escape from predators, and locomotion is therefore essential for survival.

Locomotion is allied to the development of tools, which was once believed to be discreet to man (*Homo sapiens*) although experts have subsequently witnessed innate and learned use of tools among bird and animal species both in captivity and the natural environment. Egyptian Vultures, for example, throw stones to break the hard shells of ostrich eggs while the Woodpecker Finch, a resident of the Galapagos Islands, uses cactus spines to extract grubs from tree branches [2]. Similarly primates (such as monkeys, gorillas and chimpanzees) use tools for feeding purposes and wild chimps have been observed trimming blades of grass to make tools for the extrication of termites from mounds and honey from beehives, while captive chimps have similarly learned to use sticks as a means of reaching bunches of bananas [3]. However, man uses the most extensive and diverse range of tools and a plethora of gadgets and appliances have been developed to facilitate task completion and automate numerous activities including locomotion.

Although *Homo sapiens* can swim, without tools (or auto-locomotion), man's performance in both water and the air is not as good as that on land as reflected in the evolution of human

optimisation within the animal world, which are then compared with human locomotion. The mechanical locomotion of man on land is investigated through the development of the bicycle and its aerodynamics, followed by a review of the forms of contemporary mass-market cars. In the final section we discuss the evolution of the form of solar powered cars. In the context of limited global power and energy supplies, we conclude that this type of vehicle adheres to the laws of nature (and aerodynamics in particular) *more closely* than internal combustion engine (ICE) vehicles.

2 Aero and hydrodynamics in the animal kingdom for high-speed species

As previously stated, Darwin argued that plant and animal species evolved through a process of natural selection determined by survival of the fittest. Creatures developed specific attributes, including means and speed of locomotion, that enabled them to take full advantage of the environments in which they lived and thus to perpetuate the species.

The fastest animals on land, in the air and water are the Cheetah, the Peregrine Falcon and Sailfish respectively. Although a horse can outrun a cheetah over long distances, the cheetah can reach speeds of 100 km/h over distances up to 1 km, accelerating from 0 to 70 km/h in 2 seconds while the Sailfish can swim at speeds as great as 110 km/h over 100 metres and the top speed of the Peregrine Falcon exceeds 300 km/h when diving at prey. In addition to specific physiological characteristics (such as lung and heart capacity), each of these creatures has developed particular anatomical attributes (skeletal form and muscle arrangement and strength) and body forms that enable them to attain these speeds.

2.1 The Cheetah

A member of the cat family, the Cheetah *(Acinonyx jubatus)* (Figure 2) is built for flight rather than fight. It protects its food and young by tricking or scaring predators away but flees from packs of hyenas and other big cats. Cheetahs have excellent binocular vision and hunt by day and on moonlit nights when they stalk prey until close enough for a short high-speed chase. The quarry is then attacked from the side, knocked to the ground and seized by the throat until dead.

Figure 2: The Cheetah.

Figure 3: Female and male Peregrine Falcon.

In addition to a small head, short ears, a long sleek body and legs and long tail for good balance at high speed, the Cheetah has partially retractable claws, which facilitate traction, quick turns and acceleration. Cheetahs differ from horses because the horse's spine remains relatively rigid when galloping, and so the legs are the predominant means of propulsion whereas the cheetah's flexible spine curls like a whip and thus ensures rapid acceleration. Cheetah anatomy and behaviour are therefore ideally suited to the environment in which it lives, to the means by which it hunts and to its survival in the natural world [9, 10].

2.2 The Peregrine Falcon

At 170 km/h, the White-throated Needle-tailed Swift (*Hirundapus caudacutus*) is believed to be the fastest bird when flapping its wings in flight. The Peregrine Falcon (*Falco peregrinus*) (Figure 3), however, exceeds this speed and can reach speeds of 300 km/h when diving for prey (stooping). These long-tailed raptors exploit aerial height and thus gravitational force when launching an attack and frequently dive from 300 m. The adoption of a streamlined body form (where the bird's feet lie against the tail and the wings are half-closed) minimises drag and, combined with gravitational force, enables them to achieve such high speeds. When stooping, in order to disable or kill medium-sized and large prey, the bird delivers a fierce blow with a half-closed foot. If the quarry is too heavy to carry, it is allowed to fall to the ground and the bird lands beside it to feed while lighter and smaller prey (such as swallows or sandpipers) are caught in mid-air or are struck down for later retrieval. Falcons specialise in hunting in the open and benefit from extremely acute eyesight. They are therefore ideally adapted to hunting at dusk and dawn and so have an advantage over their prey as does a system of baffles in their nostrils that aids respiration during high-speed dives. The combination of body form, a body length of 34–50 cm, a wingspan of 80–120 cm and a weight of 0.5–1.5 kg optimises these birds for the environment in which they live.

It is appropriate to note at this point that although much aerospace engineering and design does not adhere closely to the laws of nature, some engineering and design solutions derive directly from the natural world. For example, when faced with airflow problems in high-speed jet engines, aeronautical engineers referred to this particular biological feature in Falcons and subsequently developed baffles to correct these problems [11, 12].

2.3 The Sailfish

Istiophorus platypterus or Sailfish (Figure 4) belong to the billfish group, which include Marlin and Spearfish. The bill is not used for feeding but to aid locomotion as the bill length minimises turbulence before the maximum cross-section of the body area is reached. Atlantic Sailfish grow to approximately 2.4 metres in length and can weigh as much as 50 kg. The combination of size, torpedo body shape and narrow caudal peduncle (the stalk-like part of the body to which the crescent-shaped, rear fins are attached) enable Sailfish to swim at speeds of 110 km/h. Although Sailfish cannot sustain such speeds for more than 100 metres, these rapid forays enable them to out-swim prey such as tuna. The large dorsal fin is folded into a groove in the Sailfish's back to minimise drag, improve hydrodynamic performance and thus energy use when necessary. These fish frequently hunt in groups of thirty-five to forty and use their dorsal fins to create a wall around shoals of smaller fish. Each Sailfish darts into the middle of the corral to feed and then returns to the circle until the next feeding opportunity [13, 14] and it is evident that these animals have also evolved in order to capitalise on the environment in which they live.

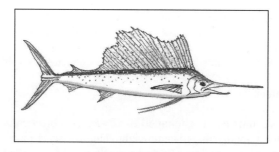

Figure 4: The Sailfish.

2.4 Summary

Since the Industrial Revolution in particular, human activity has altered numerous ecosystems and, in biological terms, the evolution of *Homo sapiens* has subsequently influenced the adaptation and survival of many other species. However, most species evolved many millennia prior to human intervention. Like the Cheetah and Peregrine Falcon, Sailfish developed anatomical and physiological attributes that are perfectly suited to the particular habitat in which they live while enabling them to capitalise on the resources therein. Food (energy) intake and energy expenditure are also perfectly balanced for the survival of individuals and thus the species. This physiological balance has developed both as a result of, and has contributed to, the evolution of each animal's form by minimisation of aero and hydrodynamic drag, which consequently permits these animals to move at the high speeds necessary for their survival.

3 Human locomotion

3.1 Human evolution

A member of the primate genus, *Homo sapiens* differs from other apes in that they walk erectly on two feet and are therefore classified as bipedal hominids. The earliest examples of bipedal hominids (*Ardipithecus ramidus* and *Australopithecus anamensis*) date back 4 million years and were discovered in Ethiopia and Kenya respectively [15]. Evidence suggests that *Homo sapiens neandertalensis*, the earliest ancestors of modern man, inhabited parts of Africa at least 120,000 years ago, migrating to Europe and Asia 50–40,000 years ago, and that *Homo sapiens sapiens* (modern man) evolved approximately 30,000 years ago [16]. Walking upright on two feet is advantageous because in addition to increasing the agility and speed of primates when running, it allows the forelegs or arms to be used for purposes other than locomotion and for the production of the complex sounds necessary for speech. These factors all contributed to the evolution of, and are evident in, modern man.

3.2 Gait patterns

Human locomotion comprises two standard gait patterns, namely walking (where one foot is on the ground at all times) and running (where both feet are off the ground simultaneously at some point). Human anatomy and physiology have evolved so that both of these activities are energy efficient, integrated cyclical patterns of movement. As with all land mammals, human locomotion is dependent on the interaction of the nervous system, including brain, spinal cord, peripheral nerves, with the skeleton and articulating joints, muscle tendon complexes and blood supply [17, 18].

3.3 Walking

Walking minimises both energy expenditure and the excursion of the centre of gravity and so a 'normal' speed of 6.3 km/h only requires 10.5 kJ per minute or 630 kJ per hour. The pendulum like motion of walking contributes to energy efficiency as the walker rolls up and over the high midpoint of a step on near-straight legs, then rolls down using gravity [19].

3.4 Running

Running involves longer, freer steps than walking and as the muscle-driven legs bend and unbend rapidly, energy is stored in the tendons and ligaments. These tendons and ligaments also form linkages to bind the muscles to the bones, and as the tendons stretch and relax the energy peaks are smoothed and thus overall effort is reduced, while the resilience of tendons enables the runner to bounce elastically [20].

Like the cheetah, human running speed is influenced by energy expenditure and the highest speeds can only be sustained over short distances. Performance can of course be enhanced through training and in 1996, the sprinter Michael Johnson averaged 37.267 km/h over 200 metres [21]. Average maximum running speed is usually about 32 km/h over short distances although the record for completion of a marathon (41.92 km or 26.2 miles) stands at 2 hours 5 minutes and 42 seconds and average speed is therefore 20.8 km/h [22].

In addition to training, variations in achievable maximum running speed and endurance among humans and other animals are dependent on body weight, musculature and strength, length of limbs, lung capacity and heart rate. At an average of 462 kJ per mile a marathon runner therefore uses approximately 12,000 kJ during a race [23]. However, the speed at which energy is used increases concurrently with running speed. That, in conjunction with oxygen supply to muscles and variations in muscle recovery rate (which are influenced by the speed and frequency of muscle use) have an impact on potential achievable distance [24]. Although maximum sprint and distance running speeds have increased as a result of human growth, dietary changes and developments in sports science, mathematical modelling currently predicts that human anatomy and physiology will ultimately limit maximum achievable running speeds. Consequently, even though Michael Johnson covered 100 m in 9.66 seconds, it is unlikely that men will be able to cover this distance in less than 9.37 seconds and in women, less than 10.15 seconds [25].

3.5 Summary

It is evident that human body form differs significantly from that of the cheetah, peregrine falcon and sailfish and does not appear to be 'built for speed'. Although human locomotion is energy efficient, being upright, the human body is not as aerodynamically efficient as the forms of the other animals (a detailed explanation of which is found in sections 6 and 9.5). Although incapable of out-running a cheetah or similar predator, the lack of speed and aerodynamic efficiency is of less significance to human beings than other creatures. In the case of humans, the development of other physical and intellectual skills (including use of tools and speech) ensured the survival and perpetuation of the species.

4 Assisted locomotion - the bicycle

Bipedal hominids travelled by foot for millions of years until the introduction of animal-assisted transport 5000 years ago and so on this time scale, the introduction of human-powered and automotive transport approximately 100 years ago is comparatively recent. Drawings for a chain-driven two-wheeled machine have been linked to Leonardo da Vinci and are dated around 1493 but there is no evidence of construction [26]. It is thus fair to say that the bicycle evolved during the latter half of the nineteenth century and its development is therefore concurrent with that of the motor car.

4.1 Development of the bicycle

In addition to being an artist, engineer and inventor, Leonardo da Vinci was a visionary genius, as is evident when the evolution of the modern bicycle is considered. Although Leonardo's drawings of the chain-driven two-wheeled machine solved the majority of mechanical problems, at the time of the earliest bicycles in the 1790s, Leonardo's work remained undiscovered. The Romans had similarly exploited the conversion of leg-muscle power to rotary action for the operation of treadmills [27], but this technology and human-powered locomotion were not linked and consequently it took about eighty years for the modern bicycle to evolve from its initial invention.

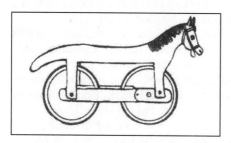
Figure 5: 'Célerifère' type cycle.

Figure 6: Macmillan 'Hobby-horse'.

In 1791 the reputedly eccentric Comte de Sivrac was seen in Parisian parks astride a 'Célerifère', renamed the 'Velocifère' in 1793 (Figure 5). This two-wheeled 'horse' resembled children's toys and was pushed forward by the feet, but lack of a steering mechanism proved awkward because the contraption had to be lifted to change direction. In 1817 Karl Drais von Sauerbronn patented a 'running machine' with a steering mechanism but that too was propelled by pushing the feet on the ground. Although these 'Draisiennes' were popular, they were uncomfortable to ride and consequently means of lifting the feet off the ground were sought. One solution was the use of treadles that mimicked walking motion and in 1840 Kirkpatrick Macmillan, a Scottish blacksmith, developed such a treadle-driven 'hobby-horse' (Figure 6).

Macmillan is commonly credited with inventing the real 'bicycle', although the term 'bicycle' did not actually appear until 1869 when it was used in a British patent and subsequently replaced all other names. However, the treadle mechanism also required considerable effort, and further mechanical experiments were undertaken resulting in rotary action pedals. These initially drove the front wheel, and the first patents for such velocipedes were issued in 1866 and 1868 to Pierre Michaux in France and to Pierre Lallement in America respectively.

Figure 7: 1870 'Ordinary'.   Figure 8: Lawson's 'Bicyclette' 1879.

The earliest bicycles were made from wood, but iron was rapidly adopted as the preferred manufacturing material and the subsequent addition of rubber rings around the metal wheels improved performance by preventing slipping and reducing noise. The introduction and patenting of wire spokes in 1869 then led to the production of lighter and more rigid wheels and the first lightweight all-metal 'Ordinary' bicycle (later known as the 'Penny-Farthing') was patented in 1870 (Figure 7). The directly driven large front wheel eradicated the need for gears and even though somewhat unstable, these high-wheeled bicycles quickly became popular. Derived from the 'Ordinary' and with a large front and smaller rear wheel, the rear-wheel chain driven 'Bicyclette' was patented by H.J. Lawson in 1879. Even though more stable than its predecessors, this bicycle was deemed cumbersome by both the public and trade and was a commercial failure (Figure 8). However in 1884, the 'Rover', manufactured by John Kemp Starley and William Sutton, was launched at the Stanley Exhibition in London and with its identically sized wheels, coupling-rod steering, chain-driven rear wheel and diamond-shaped frame, the Rover established the modern bicycle configuration (Figure 9).

 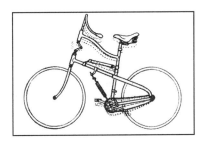

Figure 9: The 'Rover' 1884.   Figure 10: The adjustable 'Whippet'.

Further technological development led to the manufacture and patenting of the 'Whippet Safety' bicycle by O. Macarthy in 1885 (Figure 10). The sloping backbone joined the rear axle to the steering head and the front fork was connected to the rest of the bicycle by hinges and springs. All parts with which the rider had contact (the saddle, handlebar and cranks) therefore had an elastic connection to the rest of the machine. The 'Ordinary' bicycle and tricycles continued to be popular but vibration was significantly reduced with the invention and patenting of the pneumatic tyre by John Boyd Dunlop in 1888, and the Safety bicycle soon predominated [28–30].

As with the use of horses and carriages, the bicycle increased travel speeds and, although initially a luxury and leisure product, it rapidly became a popular and populist means of transport and there are now at least 1 billion on roads globally, 400 million of which are found in China [31].

4.2 The aerodynamics of the upright bicycle

To date the bicycle has proved to be the most efficient means of human-powered transport and 95% of muscle energy expended is transferred to forward movement through the crank, while the remaining 5% tends to be lost as heat. In addition to changes in terrain, cycle speed has always been subject to a number of variables including weight of the bicycle and rider, rider fitness, mechanical friction (bearings and gear train), rolling resistance of tyres (which currently varies from 0.004 to 0.14 according to tyre type), and most significantly, wind speed, air resistance and resultant aerodynamic drag.

Figure 11: Racing cyclist 2002 – carbon-fibre cycle showing standard wheels

Figure 12: Racing cyclist 2002 – carbon-fibre cycle showing carbon-fibre wheels

Aerodynamic drag increases exponentially with speed, and at speeds above 24 km/h, the energy required to overcome air resistance greatly exceeds that of the rolling and mechanical resistance of the bike: for example when speed is increased from 12 km/h to 32 km/h, mechanical resistance increases 2.25 times, rolling resistance 3.63 times and air resistance 18 times [32]. Consequently greater effort does not result in an equivalent increase in speed because power is approximately equal to velocity cubed.

Most adults deliver 75W (0.1HP) when travelling at 19 km/h while a well-trained cyclist can produce 187.5W (0.25HP) to 300W (0.4HP) at speeds between 32 and 38.5 km/h. While world champion cyclists can generate 450W (0.6HP), average road speeds are only 43 to 48 km/h [33] because of an exponential increase in air resistance-drag, and so various means of reducing drag have evolved.

Cycle racing has been practised since the invention of the bicycle and many records have been set in a variety of conditions. These include road racing where cyclists are subject to natural weather conditions (such as the Tour de France) and velodrome track events with more controlled conditions. Although the earliest bikes adopted an upright sitting position, drop handlebars and the forward body position associated with racing cyclists (Figures 11 and 12) were introduced in the 1890s. An upright body position accounts for a drag coefficient of 0.9–1.0 while research shows that adoption of the racing position can reduce drag up to 31% [34]. In addition to changes in cycle technology such as the use of composite materials and tri-bar handlebars, developments in clothing and helmet technologies have also helped to reduce aerodynamic drag and subsequently increase speed. Thus in 1996 Chris Boardman set a new world record when he covered a distance of 56.375 km in one hour in Manchester Velodrome.

Road racers in particular exploit 'drafting' or group cycling where riders to the rear of the group are protected from wind and air resistance by those in front and can therefore conserve energy. Drafting is also exploited in events such as motor pacing where riders utilise the wake of specially designed vehicles, the speed record for which was set by John Howard in 1985 and stands at 243.5 km/h [35].

4.3 Recumbent bicycles

Adoption of a reclined or recumbent body position has also proved successful in reducing drag. The earliest recumbents had four wheels but cornering proved difficult at speed and so Charles Mochet developed a two-wheeled version, the 'Velocar'. The 'One Hour' upright bicycle record stood at 44.25 km/h for about twenty years until 1933 when Francis Fauré reached a speed of 45 km/h on a Velocar. Even though this machine weighed 11kgs more than contemporary upright bicycles, the addition of fairings further increased speed so that in 1939 Fauré finally achieved his ambition and became the first person to exceed 50 km/h on a human-powered vehicle.

Figure 13: Jaray treadle-driven recumbent 1920.

Figure 14: 'Greenspeed' solar-electric assisted recumbent 1999.

Despite this success, in 1934 cycle racing organisations stated that recumbents were not true bicycles because they were said to have 'special aerodynamic features' and so Fauré's record was not recognised. Notwithstanding the fact that recumbent bicycles were invented shortly after the 'safety bicycle' (Figure 13), they have never proved as popular as their upright counterparts. Nevertheless, members of the Human Powered Vehicle Association continue to build and race such machines and to push both technical and human boundaries. Racing recumbents are sometimes enclosed to further minimise drag (Figure 14), the current record for which was set by the Blue Yonder team and stands at 128.8 km/h or 5.5 seconds for the 200 m sprint [36, 37]. Records demonstrate that although the drag of recumbent cycles differs according to design, it is invariably lower than that of an upright bicycle. Less effort is therefore required for propulsion and the heart-rate of a cyclist on an upright bicycle travelling at 28.8 km/h is the same as that of a cyclist on a recumbent travelling at 30.4 km/h. A cyclist on an upright bicycle, however, would expend considerably more energy and effort to achieve this 5.5% increase in speed because of the increased air resistance.

4.4 Maximum achievable speeds

The combination of many factors including developments in sports sciences, training programmes and clothing and helmet technologies have all contributed to faster cycling records, but the introduction of lenticular (convex) wheels and aerodynamically improved frames in particular led to an overall improvement in performance of 4% [38]. However, like running, human anatomy and physiology will eventually limit maximum bicycle speed.

In 1995 mathematical models predicted that a hypothetical 'super-athlete' cycling in perfect conditions (namely at sea-level with a relative humidity of 60% and temperature of 20° C) could achieve an average one-hour speed of 56.9 km/h [39], which is only 0.28 km/h faster than Chris Boardman's 1996 record. Additional models showed that, unlike running, cycling at altitudes up to 4000 m above sea-level could increase overall speed because the effect of reduced air pressure, and therefore air resistance and energy requirement, outweigh the lower oxygen levels and resultant energy production. Consequently at an altitude of 2230m, the 'super-athlete' would be able to cover a distance of 58.7 km in an hour while at 3500 m above sea-level the distance covered in one hour would be 61.6 km. A final model showed that if air pressure is lowered to the extent that the oxygen content of the blood is reduced, the 'super-athlete' could reach speeds up to 98.14 km/h [40].

As yet these figures remain hypothetical, but the models illustrate the impact of air resistance on cycling in particular while comparisons between upright and recumbent bicycles show that a reduction in aerodynamic drag can significantly reduce energy expenditure and thus contribute to system efficiency.

Table 2: Variations in energy efficiency.

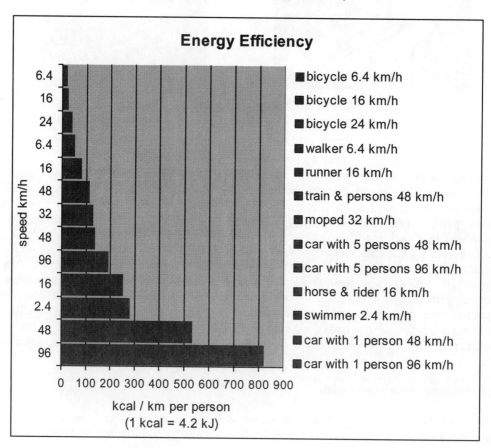

4.5 Summary

It is pertinent to note that despite poor aerodynamic efficiency (the drag coefficient when walking and cycling upright is approximately 1.0), energy use in the human body is far more economic than that of an internal combustion engine car. For example, using 1 litre of petrol (equivalent to approximately 32,600 kJ of energy) [40], a typical car travels about 11 km. In comparison, at average walking pace, a human would travel at least fifteen times that distance using the same amount of energy, and because cycling is even more energy efficient than walking, at 24 km/h a human on a bicycle could travel more than 200 km [41].

Nevertheless, all of these modes of transport dramatically increased man's potential travel at speed over distance from an average of 16 km/h in the horse-drawn coaches of the 1830s and early automobiles [42] to an average of 24 km/h for bicycles while contemporary cars frequently exceed 160 km/h. Use of carts, carriages and cars also increased load and passenger carrying capability and their appeal was further enhanced because they require far less human effort than bicycles for operation.

5 Assisted locomotion – the automobile

5.1 The popularity of the car

Invented little more than one hundred years ago, the automobile or motor car has proved to be one of the most successful products of the twentieth century; there are now approximately 600 million cars on the road globally [43] and the World Bank predict a further increase to one billion by 2030 [44].

The success of the motor car has transformed human existence in many ways, including perception of time and distance, with the average urban road speed limit being nine times greater than average walking speed and the potential to travel at 190 km/h commonplace. In addition to being an aid to liberation, an emblem of social status and a costume for many drivers, the car also influenced the design of urban and rural settlements and has thus contributed to changes in employment and leisure patterns. This has in turn encouraged perceived dependence on the car, the impact and results of which are both positive and negative and have been repeatedly debated since its invention [45]. Despite the various negative impacts, a world without vehicles for personal transportation is now unimaginable and the car in some form or other is undoubtedly here to stay.

5.2 The future of fossil fuels for transportation

At the dawn of the motor age, in addition to petrol and diesel, steam powered and more particularly electric vehicles were comparatively common. The high energy density of petrol and convenient and rapid refuelling ensured that the fossil fuel internal combustion engine vehicle predominated by the First World War [46]. During the past thirty years, however, several factors have resurrected interest in so-called alternative fuels and energy sources. Oil is a finite resource and, although estimates vary about the date at which either supplies are expected to run-out or increased extraction costs make oil economically unviable, expert opinion currently predicts that supply will end by 2050 [47]. The oil crises of 1956 and early 1970s and the 1991 Gulf War highlighted the reliance of many countries on imported oil. Consequently, in conjunction with the high cost of importation, several nations are now seeking greater fuel autonomy [48]. In addition to political and economic factors, environmental concerns have also contributed to incremental changes in mainstream automotive engineering and design, specifically more fuel-efficient internal combustion engine models and the development of alternatively powered vehicles.

5.3 Fossil fuels and the environment

The combustion of fossil fuels produces a variety of emissions, which have escalated with increases in car ownership and use. Emissions include oxides of nitrogen and sulphur, hydrocarbons, halocarbons, CO, CO_2 and PM_{10} particulates, all of which have been shown to have a detrimental impact on the environment at local, national and/or international levels [49]. Many experts believe that the escalation in 'greenhouse gases', and CO_2 in particular are linked to climate change. In order to slow (and ideally to arrest) climate and other changes therefore, a significant reduction in these emissions is vital be it from industry, energy generation and usage including transport. Since 1979, the United Nations Framework on Climate Control has negotiated a number of protocols advocating reductions in pollutants and emissions, the outcome of the 1997 Kyoto meeting being a policy to encourage a global decrease in CO_2 emissions [50]. In addition to this global initiative, regional and national programmes have been established. In the USA they include that drawn up by CARB (the Californian Air Resources Board) and PNGV (Partnership for New Generation Vehicles). CARB proposed a stepped programme introducing zero-emission vehicles (ZEVs) to the Californian market [51], while PNGV was established to accelerate the development of lighter-weight, more fuel-efficient, cars by major manufacturers [52]. The EEC has devised similar legislation and from 2008 for example, all new cars must emit no more than 140 gm CO_2 per km [53] and by 2015 at least 95% of every end of life vehicle (ELV) must be recycled including the recovery of energy through waste incineration [54]. There is already evidence that both of these directives are influencing aspects of vehicle design including developments in engine technology and overall vehicle design and assembly.

5.4 'Alternative' and emerging technologies

Although research relating to alternatives to petrol and diesel has been on going since the invention of the car, and several (LPG (liquefied petroleum gas), CNG (compressed natural gas) and alcohol (methanol and ethanol)) are currently available [55], these fuels still produce emissions. There are, however, some fuel and energy sources that produce zero emissions during vehicle propulsion. These include compressed air, hydrogen fuel cells and electricity stored in batteries. However, if the means of producing the fuel or energy itself produces emissions, then one type of pollution may simply be exchanged for another, which is certainly true in the case of electricity: when coal fired power stations are used for power generation, pollution is shifted from a mobile line source (the vehicles) to a point source (the power station) [56]. If sustainably generated electricity (such as use of water and wind to operate turbines or sunlight via photovoltaic cells) is used, then these emissions can be significantly reduced.

At present most alternatively powered vehicles tend to be one-off prototypes, concept and development models, or manufactured in limited numbers with few available on the mass market, and the overwhelming majority of vehicles remain steel-bodied and propelled by internal combustion engines. Fuel consumption has been reduced through weight reduction as a result of developments in materials, engine technologies and management systems and lower road resistance (which comprises rolling resistance of tyres and aerodynamics).

The results of these technological developments (including lower fuel consumption) may be apparent to consumers but the components themselves tend to be out of view and of far less importance than the visible consumer-vehicle interface, the vehicle body. Body design provokes both conscious and unconscious emotional response and can influence the commercial success of a particular model. Consequently, changes in automotive styling have tended to be gradual rather than dramatic, as manufacturers exercise caution as opposed to launching vehicles that represent significant stylistic departures from immediate antecedents. Body design also has a significant impact on vehicle aerodynamics, which are therefore subject to taste, style and fashion [57].

6 Vehicle aerodynamics

Air affects moving solids by flowing around the body. A moving vehicle is affected in three ways: by flowing around the vehicle, by flowing through the body, and within the processes of the machinery itself. Airflow around a moving vehicle causes aerodynamic forces in longitudinal, vertical and lateral directions, and because air resistance opposes forward movement it influences both the amount of power required to propel the vehicle and the maximum attainable speed. Aerodynamic drag can thus be defined

$$D = C_d A \frac{\rho V^2}{2} \qquad (1)$$

where C_d is the non-dimensional drag-coefficient, A is the projected frontal area of the vehicle, ρ is the density of the surrounding air and V the vehicle speed.

This may be described graphically as in Figure 15 which shows that, while overall road resistance increases significantly with speed, aerodynamic drag increases far more dramatically than rolling resistance. Hence for an average ICE passenger car, at approximately 67 km/h, aerodynamic drag and rolling resistance are equal. At 100 km/h, however, aerodynamic drag accounts for 80% of road resistance and rolling resistance only 20% [58].

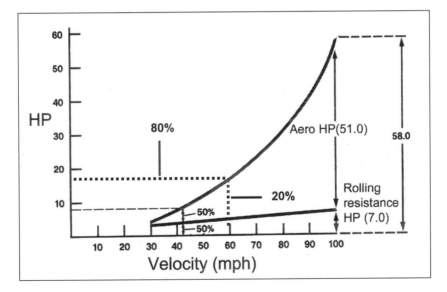

Figure 15: The increase in rolling-resistance and aerodynamic drag with vehicle speed for a typical internal combustion engine car.

Although the significance of aerodynamic drag and its impact on fuel consumption has long been appreciated within the automotive industry, it has become an increasingly influential and important factor within the automotive design process during the past twenty years as typified by the cars in Figures 16 and 17. Between 1980 and 1990 the Volkswagen Group achieved a 30% reduction in the non-dimensional drag coefficient (C_d) of their most popular models equivalent to a 10% reduction in fuel consumption [59]. Continuing research now means that the drag coefficient for the Golf has been further reduced from approximately 0.34 in 1984 to 0.31 in 2001 contributing to even greater fuel economy [60].

Figure 16: Golf Mk 1 1980.

Figure 17: Golf Mk 4 2001.

Other manufacturers including Fiat have also reduced the drag coefficient of populist models: for example the 1960's 850 had a C_d of 0.42 and 1970's Strada (also known as the Ritmo) had a C_d of 0.4 while by 2000, the drag coefficient of the Punto (Figure 18) was only 0.34 [61].

Figure 18: Fiat Punto 2000.

7 Automotive styling

7.1 Design constraints

As noted in the equation (1) for drag, A is the projected frontal area of a vehicle (Figure 19) and is therefore crucial to vehicle aerodynamics and influenced by styling, aesthetics and ergonomic considerations because, in addition to fulfilling a variety of psychological and societal functions, the car is essentially a means of transport and is designed to carry two, four or more passengers and luggage. Driver and passenger seating position and ease of access (ergonomic factors) consequently influence vehicle shape (Figures 20 and 21) while the dimensions of vehicle interiors have evolved to accommodate the physical growth of the general population as well as heightened expectations of comfort [62].

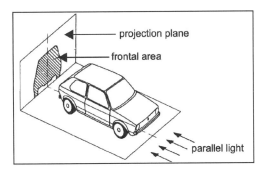

Figure 19: Frontal area A.

Although automotive manufacturers refer to the non-dimensional co-efficient of drag (C_d) in publicity and other material, it is the product of drag coefficient and the frontal area ($Cd.A$) that must be considered when calculating drag. Assuming that the average frontal area of the above vehicles is ~2.4 m², the $C_d A$ for the three Golf models would be approximately 1.2 m², 0.81 m² and 0.74 m² and that of the Fiat 850 1 m², the Strada 0.96 m², and the Punto 0.76 m².

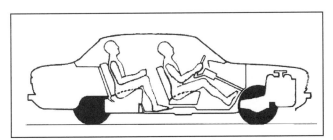 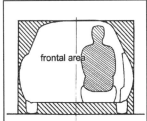

Figures 20 and 21: Design constraints for passenger vehicle.

7.2 Aerodynamics and early vehicle design

The earliest automobiles, known as horseless carriages, were just that (Figure 22) but in 1899 a torpedo-shaped electric car, 'Le Jamais Contente' (Figure 23) built by Camille Jernatzy, became the first automobile to exceed 100 km/h. The body was streamlined and appears to be the first design to adopt aerodynamic principles.

Figure 22: Bersey Electric Cab 1897.   Figure 23: 'Le Jamais Contente' 1899.

Aerodynamics research was undertaken during the 1930s but the majority was confined to half-body models, some scale and others full size. The recorded non-dimensional drag coefficients (C_d) of these scale models were extremely low (ranging from as little as 0.13 to 0.24) in comparison with many contemporary standard US production cars (0.61 to 0.74) [63], but very few of these vehicles were manufactured due in part to the limitations of contemporary manufacturing processes and resultant cost.

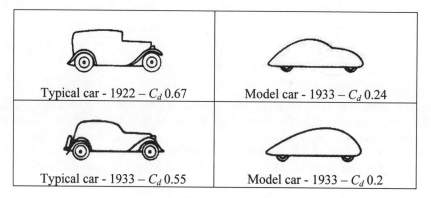

Figure 24: Actual and optimised vehicle aerodynamics

Although experimental models like the Rumpler Trumpfenwagen (Figure 25) were produced, the majority of manufacturers believed them to be too radical for the market at that time. The restricted interior space of more aerodynamic forms did not lend itself to saloon models, but variations were adopted for sports and racing cars. Following the First World War and exploiting construction techniques used in the aircraft industry for example, the Italian Ugo Zagato gained a reputation for the design of lightweight sports cars with comparatively low drag coefficients, and the Zagato company produced and continues to produce sports car designs today [64].

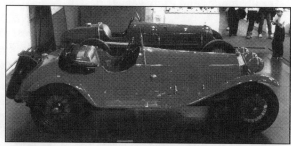

Figure 25: Rumpler Trumpfenwagen 1922.

Figure 26: Alfa Romeo Gran Sport (Zagato) 1930.

The earliest cars were adopted as status symbols because they were novel and extremely expensive (Figure 27). However, by introducing the assembly line production process and limiting the number of available models, Henry Ford transformed the automotive manufacturing process. Ford opened up a new market producing cheaper vehicles, millions of which were then sold to 'middle' and 'working' class consumers [65]. As the number of Ford Model T (Figure 28) sales grew and these comparatively small, cheap cars became ubiquitous, they were increasingly perceived as inferior to other large, more exotic vehicles, which remained symbols of social status for those who could afford them and objects of desire to those who could not.

Figure 27: Rolls-Royce Alpine Eagle,1910: £985.   Figure 28: Ford Model T 1915: £135.

Conspicuous consumption and thus vehicle size endorsed status in this era when oil was considered plentiful and supply infinite and high fuel consumption was not seen to be problematic. As vehicle speed was increased through use of larger engines, many vehicles only travelled about 6 miles on a gallon of fuel [66]. In addition to the use of large engines, during the 1950s many American cars in particular borrowed motifs like chrome streamline trim and 'fins' from the visual language of aeronautics to suggest the potential to travel at speed (Figure 29). These were, however, illusory and in no way contributed to actual vehicle performance and the drag coefficient of many of these cars was very high. But this was irrelevant to drivers who desired and were supplied with cars that, through form and design detail, connoted speed [67].

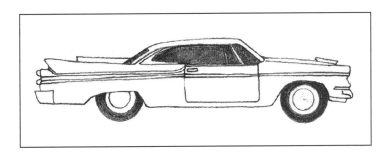

Figure 29: Swept-wing Dodge 1957.

7.3 Consumer expectation and contemporary vehicle design

Although aerodynamic performance has improved and is very much part of the contemporary automotive design process, there are several reasons why manufacturers still do not offer aerodynamically optimised vehicles to the mass market. Aerodynamically optimised vehicles would be virtually indistinguishable from one another, and because consumers demand a level of design individuality, would not be deemed acceptable. Although the overall form of many contemporary car bodies is very similar, relative design individuality is achieved through cosmetic detail such as use of colour and graphics, by sculpting and detailing body form, and the addition of external accessories including wing mirrors and radio aerials. The latter all affect aerodynamics and performance as illustrated below (Figure 30). Increased demand for comfort and inclusion of internal features such as air conditioning also have an impact on vehicle performance as air flows through the car interior [68].

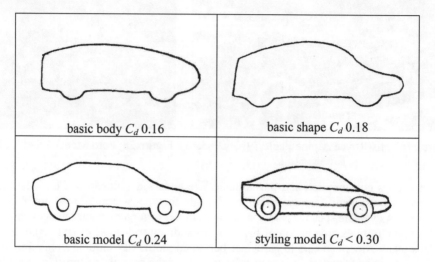

Figure 30: The development of a low-drag car body from 'aerodynamic ideal' to styling model.

7.4 Contemporary vehicle nomenclature

Contemporary ICE passenger vehicles can be categorised as follows: sports utility (large urban vehicles with off-road capability), saloon (family and executive models), sports (two-seater) and small (four seater cars for urban use). It is no coincidence that the nomenclature of models alludes to emblems of strength and adventure (Mitsubishi Shogun (Figure 31), Land Rover Discovery (Figure 32) and Freelander), holiday destinations, leisure activities (Seat Ibiza (Figure 33), Fiat Marbella, VW Polo) and energy and style (Lotus Elan (Figure 34)).

Figure 31: Mitsubishi Shogun.

Figure 32: Land Rover Discovery.

Figure 33: Seat Ibiza 1987.

Figure 34: Lotus Elan 1970.

Allusions to the natural world and in particular to animals are equally common, for example, the VW Sirocco (Figure 35) Alfa Spyder (Figure 36) and Ford Mustang. Several luxury and sports car manufacturers adopted the names of big cats (Cougar, Puma and Jaguar (Figure 37)) and snakes (Dodge Viper and AC Cobra (Figure 38)). These labels aid model identification and are used specifically to provoke association with animal traits and thus connote beauty, speed, danger and excitement.

Figure 35: VW Sirocco 1984.

Figure 36: Alfa Spyder.

Figure 37: E-type Jaguar 1962.

Figure 38: Shelby AC Cobra 1965.

Some of these vehicle forms suggest the potential for speed and power be it through zoomorphic reference or allusion to the established formal language of the sports car, and thus fulfil consumer expectations because there is a demand for vehicle appearance to reflect performance potential, the exceptions being marques associated with affluence, success and the establishment such as Rolls-Royce and Bentley (Figure 39) [69].

Both the marque and vehicle form imply dignity and understatement even though the maximum achievable speed of such cars is 225 km/h and therefore comparable with that of the Shelby AC Cobra and Lotus Elise sports cars.

Figure 39: A typical Bentley Limousine.

7.5 The influence of aerodynamics on contemporary car styling

Although contemporary vehicle body form varies, and most correlate with the aforementioned categories (sports utility, saloon and sports and small), the non-dimensional drag coefficient (C_d) of the majority of models is surprisingly similar. The BMW X5 SUV (Figure 40) for example has a C_d of 0.36 [70]), that of the Ford Focus saloon (Figure 41) ranges from 0.3 to 0.36 according to model type [71], that of the Porsche Boxster (Figure 42) is 0.31 [72] while that of the Lotus Elise (Figure 43) is estimated to be 0.36 [73]. There is, however, a marked difference between the C_dA of the Bentley Limousine, BMW X5 and other models listed in Table 3. Although the C_dA of the Lotus Elise is marginally higher than that of the Boxster, lower fuel consumption derives from use of an aluminium chassis and composite body panels so that the Elise weighs 542 kg less than the Boxster.

Figure 40: BMW X5 2002.

Figure 41: Ford Focus 2001.

Figure 42: Porsche Boxster 2001.

Figure 43: Lotus Elise 2002.

The C_d of many 'small' models differs little from that of many larger vehicles, being 0.316 in the Toyota Yaris [74] (Figure 44), 0.32 in the Hyundai Amica [75] (Figure 45), and slightly lower at 0.28 for the Nissan Micra (Figure 46) [76].

The evolution of land-based locomotion 275

Figure 44: Toyota Yaris.

Figure 45: Hyundai Amica.

Figure 46: Nissan Micra

In addition to the lower weight of these cars, lower fuel consumption derives from differing engine size and technologies and smaller frontal area (A) as is evident in Table 3.

Table 3: Comparative vehicle weights, coefficients of drag, and fuel consumption.

Model	Weight kg	C_d	A m^2	$C_d A$ m^2	Petrol consumption
Bentley Limousine	2600	0.45	3.54	1.59	15.1 mpg
BMW X5	2120	0.36	3.23	1.16	23.3 mpg
Porsche Boxster	1312	0.31	2.29	0.71	29 mpg
Lotus Elise	770	0.36	2.16	0.77	38.5 mpg
VW Golf 2001	1316	0.31	2.46	0.76	40.4 mpg
Fiat Punto	1000	0.34	2.45	0.76	50 mpg
Hyundai Amica	918	0.32	2.36	0.75	44.8 mpg
MCC Smart City Coupe	720	0.25	2.31	0.57	58 mpg

It would be incorrect to assume that interest in small, fuel-efficient cars is a new phenomenon. During the post-war economic boom in Italy, in addition to producing large and exotic models, Fiat responded to the market for small cars with the Topolino followed by the 600 in 1955 and then the iconic Fiat 500 in 1957 (Figure 47). Just as Ford produced the Model T, VW the 'Beetle' and Citroen the Deux Chevaux 'people's cars', Fiat launched this small, comparatively inexpensive and economical vehicle to aid mobility of the increasingly affluent working class population [77]

Figure 47: Fiat 500.   Figure 48: Isetta 300.

The launch of this model coincided with the 1956 Suez oil crisis when several other manufacturers also developed small cars to satisfy consumer demand for more fuel-efficient vehicles. These included one and two-seater 'bubble cars' (Figure 48) and, with its radical transverse engine designed by Sir Alec Issigonis, the Mini (Figure 49) launched in 1959.

Figure 49: Morris Mini 1959.   Figure 50: MCC Smart City Coupé (Smart Car).

These small cars were particularly suited to urban use being easy to park in small spaces and ideal for use on narrow roads in cities built prior to the automotive era. One of the latest additions to this group of 'micro-cars' is the Mercedes Swatch Smart Car (Figure 50). It differs from the majority of models in this group, being a two-seater and only 2.5 metres long. Currently the most fuel-efficient mass-market ICE car, the Smart Car body is comparatively light at 720 kg and with a comparatively low C_d of 0.25, and C_dA of 0.57 m^2: the average combined mileage is 58 mpg for the petrol model while the diesel model exceeds 80 mpg [78, 79]).

7.6 Summary

It appears that, in the case of internal combustion engine vehicles, even though the mainstream automotive industry has responded to government and other legislation and is producing a variety of more fuel efficient, lower emission vehicles, there is still a market demand for large vehicles. The kerb weight of vehicles within the four major categories (sports utility, saloon and sports and small) varies considerably and significantly influences fuel consumption, as does engine capacity which ranges from 8 litres in the Dodge Viper sports car and 5.4 litres in a Rolls-Royce Silver Seraph to a mere 599 cc in the Smart Car.

Developments in vehicle aerodynamics have resulted in lower non-dimensional drag coefficients C_d, and have, in conjunction with developments in engine technologies, contributed to improved fuel economy. The frontal area (*A*) of vehicles influences overall aerodynamic drag so that smaller vehicles with a lower C_dA and similar engines tend to be more fuel-efficient. However, the design and dimensions of the frontal area are constrained by ergonomic considerations, fashion and consumer demand for stylistic similarities at a given time.

8 The future of automotive aerodynamics

8.1 Concept cars

Political legislation has forced the majority of car manufacturers to reduce exhaust emissions through improved fuel economy and the result of one initiative in particular, namely PNGV (the Partnership for New Generation Vehicles), has produced some interesting concept design models.

Established in the USA in 1993, one of the primary aims of the PNGV was to encourage major automotive manufacturers to build passenger vehicles that consumed one third of the amount of fuel consumed by mass market cars at that time. To date GM, Chrysler and Ford have all produced concept vehicles, the ESX 3 (Figure 51), Prodigy and Precept respectively. In addition to use of lightweight materials, and utilisation of new technologies including ICE-electric hybrid propulsion and on-board cameras that eradicate the need for wing mirrors, these models all exploit low aerodynamic drag as a means of improving fuel economy.

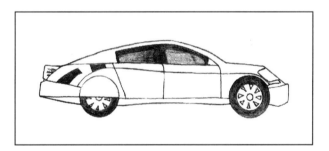

Figure 51: GM ESX 3 concept car.

The C_d values for these cars are 0.16, 0.22 and 0.2 and the values for C_dA are estimated at 0.38 m², 0.52 m² and 0.48 m² respectively but, at the time of writing, they remain as concepts [80]. European manufacturers have also produced concept models such as the Fiat Ecobasic. Weighing a mere 750 kg and with a steel space frame and plastic body panels, like the Smart Car, it is said to be almost entirely recyclable. The low weight in conjunction with a C_d of 0.28, C_dA of 0.64 m² and diesel injection engine contribute to an average combined fuel consumption of less than 3 litres per 100 km but again is not yet available to the general public [81].

8.2 ICE - electric hybrid vehicles

Hybrid technology combines an internal combustion engine with an electric motor and was developed to minimise fuel consumption and thus emissions without limiting travel range. Two such low-emission vehicles were launched commercially in 2000, namely the four-seat Toyota

Prius (Figure 52) and two-seat Honda Insight (Figure 53). Prius construction is more traditional than that of the Insight, which utilises plastics extensively and so at 773 kg is considerably lighter than the Prius at 1200 kg. The C_d values of these vehicles is 0.29 and 0.25 and C_dA values of $0.7m^2$ and $0.57m^2$ respectively. Combined with the difference in weight, the difference in drag contributes to the difference in combined fuel consumption at 57.6 mpg (4.9 l/100 km) for the Prius [82] and 83.1 miles per gallon (3.4 l/100 km) for the Insight [83]. Despite the fact that the Prius is considerably bigger than the Smart Car however, hybrid drive technology produces a comparable fuel consumption to that of the Smart Car at 56.5 mpg.

Figure 52: Toyota Prius 2001.   Figure 53: Honda Insight 2001.

Like internal combustion engine vehicles these cars can be refuelled easily and have far greater potential range than pure electric vehicles. It is possible that, although lower than that of many internal combustion engine vehicles, the aerodynamic performance of these models has also been compromised in favour of appearance and the demand for passenger comfort.

8.3 Electric vehicles

Electric vehicle (EV) technology is at least as old as that of internal combustion engine vehicles, although the ICE vehicle predominated by the First World War due in part to limited range and prolonged battery charging time of EVs. However several manufacturers currently produce electric vehicles although availability varies from country to country. Zero-emission electric vehicles are either conversions (ICE models with a replacement electric motor) or 'ground-up' designs. Although range is always limited by battery storage capability, it varies according to battery type as evident in the Solectria Force. This four-seat model is supplied with either lead acid, nickel cadmium or nickel metal hydride batteries which provide respective ranges of 80 km, 136 km and 170 km [84].

The Peugeot 106E (Figure 54) is the most common European EV. A steel-bodied ICE-electric conversion, with body weight 1087 kg, and batteries limit range to 80 km [85] even though the nickel cadmium (Ni-Cd) batteries can store approximately 1.5 times more energy than the ubiquitous lead-acid batteries used by many other EVs [86].

In addition to the Solectria Force, several other 'ground-up' electric vehicles are in production including the Nissan Hypermini (Figure 55). The plastic bodied Ford Think (Figure 56) was due to be launched commercially in 2002 but has now been shelved. Nevertheless at 960 kg it is considerably lighter than ICE-conversion models. Like the Peugeot 106 it uses Ni-Cd batteries; maximum range from one charge is limited to 85 km and is therefore virtually the same as that of the Peugeot 106 [87]. The higher weight of the Peugeot 106 is more than likely offset by the slightly higher battery reserve and a lower coefficient of drag.

The evolution of land-based locomotion 279

Figure 54: Peugeot 106 Electric 2000.

Figure 55: Nissan Hypermini 2002.

Other 'ground up' electric vehicles were specifically designed to exploit low drag coefficients. Launched in 1991, the GM EV1 (Figure 57) was one of the first American production models. This was superseded by the Impact, the kerb weight of which is lower than that of the above conversions at 1000 kg. The change from lead acid to NiCd batteries increased range on one battery charge from approximately 80 km for earlier models to over 125 km. With a listed drag coefficient of 0.19 however [88], this vehicle evidently benefits from and is closer to an aerodynamic ideal than many other vehicles.

Figure 56: Ford Think 2000.

Figure 57: GM EV1.

8.4 The Hypercar concept

In 1991 The Rocky Mountain Institute in Colorado (RMI) proposed an ultra-lightweight, 'Hypercar' (then called the 'supercar') concept at a National Academy of Sciences hearing. This project has subsequently evolved, due in part to PNGV, and now the Hypercar Centre and Hypercar Inc. are actively promoting the commercialisation of such vehicles. The RMI argues that the only way to produce a significantly innovative vehicle with superior fuel and energy economy is to design a vehicle where

> *Everything* [is] *considered simultaneously and analysed to reveal mutually advantageous interactions (synergies) as well as undesirable ones* [89].

The Hypercar concept (Figure 58) is therefore an example of 'Whole-System Design' and examines all aspects of auto-technology and the relationship(s) between all components. RMI states that in order to be considered a 'hypercar' the vehicle must incorporate ultra-light construction, efficient accessories and low-drag design. Original research stated that the Hypercar should have a C_d value of 0.2, a frontal area of 1.9 m² and thus a C_dA value of 0.38 m², which is considerably lower than other vehicles on the road. This model uses a hydrogen fuel cell, which emits water, although this technology is not yet commercially available in private cars. Although on-board production of hydrogen from a fossil fuel source would produce some emissions, hydrogen could be stored on board in tanks, which if produced sustainably (such as through use of solar or wind generated electricity for hydrolysis) would eliminate 'up-stream' emissions.

The proposed production model is a four-seat family vehicle for urban and rural use and tailored to the American market [90]. This concept is not only a radical departure from the current automotive manufacturing paradigm, but as an example of design optimisation, is a solution that adheres closely to the laws of nature.

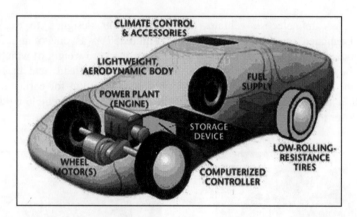

Figure 58: The Rocky Mountain Institute 'Hypercar' concept vehicle.

8.5 Summary

The range of both ICE and hybrid vehicles on one tank of fuel invariably exceeds 500 km and refuelling can take as little as 5 minutes, while battery storage capacity limits electric vehicle range to an average of 80 km and a full recharge can take up to 8 hours. These factors assume that battery type and component technologies are crucial to the future design development of EVs as well as improving performance of hybrid vehicles. In conjunction with vehicle weight and future developments in materials, aerodynamics and low drag coefficient significantly influence performance as is evident above. It is likely that in response to legislative and market demand for more fuel-efficient vehicles and lower-polluting vehicles, in addition to the development of more efficient engines and new fuel systems, emphasis on vehicle aerodynamics and subsequent reduction of drag coefficients will assume even greater importance within the automotive design process.

9 Solar powered cars

9.1 History

The motor car was invented by individual trailblazing engineers and its manufacture has subsequently become the largest industry in the world [91]. It would be incorrect to assume that all alternative personal transportation solutions evolved within corporations or that legislature initiated their development. The inverse may well be true, as individual pioneers have sought not to offer incremental but radical and holistically designed alternatives to the established personal transportation paradigm and thus perpetuate the maverick spirit from which the car originated [92, 93].

During the past twenty years an increasing number of such individuals have proved that sunlight alone can propel vehicles for prolonged periods and over considerable distances. Because solar-powered cars are essentially electric vehicles with integrated on-board charging capability, a considerable amount of information and expertise developed through their design and construction has been transferred to more conventional electric and other road vehicles including the GM EV1 and Impact and Solectria Force [94]. Nevertheless, solar car design and racing was initiated and is continuing to evolve outside the mainstream automotive industry.

The photoelectric effect was discovered in 1839 and the first photovoltaic cells were exploited on space satellites in 1954. As efficiency improved and cost decreased, use of passive solar and photovoltaic installations has increased for heating water and other electrical applications in buildings, while the first significant use of photovoltaic cells for transportation was to power a plane across the English Channel in 1981 [95].

Inspired by this event and regarded by enthusiasts as the father of solar-powered car racing, Hans Tholstrup built and drove the first solar car 'Quiet Achiever' from west to east Australia in 1982. Tholstrup's success encouraged other engineers and inventors in Europe and America and various races and challenges were established. These include the Swiss Tour de Sol, the GM Sunrayce U.S.A. (now the American Solar Challenge), the trans-Australian World Solar Challenge and Japanese events such as the World Solar Rallye. Events differ as some are 'staged', some 'first past the post', and others the maximum distance covered within prescribed time limits, but all were organised as a means of demonstrating the potential of solar power for vehicle propulsion and of 'forcing technology forward' [96].

9.2 Solar car design principles

The principles of solar car design are very simple, the aim being to maximise the power obtained from the sun while minimising both losses in conversion and the power required to drive at a given speed. As the power required increases with speed, the nominal speed of a car is determined by the point at which input power balances the power required to overcome losses. A simple analysis of the governing power equation will explain this concept and highlight the key areas that influence solar car design.

The power available from the sun can be expressed as the solar flux S (100 W/m^2). Under typical racing conditions, this is approximately 1000 W/m^2 but not all of this power can be converted to electrical power. The efficiency of the solar cells within the panel (array) determines the proportion of solar power that can be converted into electrical power. Hence a commercial grade solar cell could have an efficiency of 16%. This means that with an array of solar cells of total area of 8 m^2, the output power is given by

$$P = A_{pv} \times S\eta = 8 \times 1000 \times 0.16 = 1280 \text{W} \qquad (2)$$

where P is power, A_{pv} is the area of the array, S is solar flux and η is efficiency but it must be remembered that solar flux changes with atmospheric conditions and that this equation assumes that the sun is directly above the solar cells.

The mechanical power required by a car is equal to the product of opposing forces and velocity. As previously stated, the forces in this simple analysis can be divided into rolling resistance and aerodynamic drag. If the car is moving on a slope, gravitational forces will also have to be taken into account. We have already seen that aerodynamic drag is given by

$$D = C_d A \frac{\rho V^2}{2} \qquad (1)$$

where A is the frontal area of the car, C_d is the drag coefficient and ρ is the density of air while rolling resistance is expressed as

$$Fr = C_{rr} Mg \qquad (3)$$

Combining these two forces and multiplying by velocity gives an expression for the mechanical power as a function of velocity

$$P = C_{rr} Mgv + \frac{1}{2} C_d \rho A v^3 \qquad (4)$$

Before balancing input and output power, losses within the system must be accounted for. These can be attributed to the efficiency of the electrical system η_e, the efficiency of the motor η_m and the efficiency of the mechanical transmission to the drive wheel η_t. This results in the power equation

$$(A_{pv} S \eta_{pv}) = \left(mgC_{rr}v + \frac{1}{2} C_d \rho A v^3 \right) \frac{1}{\eta_e \eta_m \eta_t} \qquad (5)$$

where A_{pv} is area of array, S is solar flux and η_{pv} is efficiency of the array.

This equation reveals the guiding principles of solar car design. The efficiency of the solar cell is very important and a large area of solar cells is needed to generate modest amounts of power. In the previous example, the car designer only has 1280 W available. It is therefore imperative to minimise mass, rolling resistance, drag and frontal area. The efficiency of the electrical systems, motor and transmission must be as close to 100% as possible. The issue of car mass becomes even more critical when driving up an incline. This is best illustrated by way of a simple example. Consider a car of mass 500kg on a slope of 5 degrees. The weight component down the slope is 500.9.81.sin(5) = 428 N. The power required to overcome the weight component when travelling at 2 ms^{-1} (7.2 km/h) up the slope is given by

$$P = Fv = 428N \times 2ms^{-1} = 956 \text{ W} \qquad (6)$$

This result is comparable to the total output power of the solar array from the previous example.

The design of a solar car is very different from conventional car design in that a 'normal' car carries its energy source with it and is thus able to produce large quantities of power when required so that a small family car may have 60 kW of available power and therefore considerably more than a solar car.

The principal components of solar cars (Figure 59) have remained unchanged since 1982, although the development of various constituent technologies has resulted in improvements in vehicle performance. A solar car is comprised of a lightweight aerodynamic space-frame or monocoque chassis and body shell, suspension, a panel (array) of photovoltaic cells, batteries, a variable speed motor with motor controller and drive wheel. To be competitive, a solar car must operate at the limits of efficiency in all areas of its design and the means by which this is achieved are discussed in the following sections.

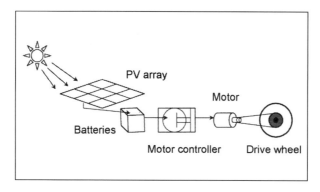

Figure 59: Principle solar car components

Figure 60: Mad Dog II, (South Bank University's Award Winning Car) and others In the World Solar Rallye, Akita, Japan, 1998.

9.2.1 Solar car dimensions

Electricity is generated via the array and a larger array produces more energy but it is unfeasible to increase the array size indefinitely and so both vehicle and array size are determined by the International Solar Car Federation (ISF) rules. 'Classic Class' cars like Mad Dog III (Figure 61) have dimensional limits of 6 metres length, 2 metres width, 1.6 metres height and a maximum array size of 8 m^2.

Figure 61: Mad Dog III 2000.

Figure 62: Aurora 1996.

Single-seat cars that conform to more recent ISF5000 regulations like Aurora (Figure 62) and Sunswift (Figure 63) have corresponding limits of 5 metres × 1.8 metres × 1.6 metres, most of the upper surface of which is covered with PV cells, while two-seat cars, (e.g. Honda Dream II (Figure 64)), are the same size as 'Classic Class' cars, but a 12 m^2 array is permitted.

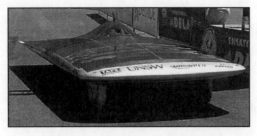

Figure 63: Sunswift 2001.

Figure 64: Honda Dream II.

9.2.2 Chassis and body shell

Lightweight materials like carbon fibre are used for the body shell, space-frames are constructed from aluminium and monocoque chasses are constructed from materials such as Fibrelam. Manufactured by Hexel Composites, this honeycomb sandwich material is used for aircraft floor panels because it is light, strong, rigid, has a high tolerance to extremes of temperature and is highly resistant to corrosion and fatigue [97]. Some solar cars also incorporate fibreglass while others like the Japanese car 'Jona Sun' (Figure 65) use more traditional and recyclable materials including bamboo and paper. Whatever materials are used, designers strive to keep weight to a minimum and so the average overall weight of a solar car is approximately 200 kg plus driver (80 kg) and batteries. All racing teams must either carry a driver or driver and ballast equivalent to at least 80 kg and so use of a light driver has no advantage whereas use of a driver over 80 kg will prove disadvantageous.

Figure 65: Jona Sun 1999.

Figure 66: Nuna 2001.

9.2.3 Photovoltaic (PV) cells

The physical composition of photovoltaic cells varies and influences efficiency, which ranges from 7% for Crested Ibis amorphous silicon cells up to 28% for gallium arsenide space grade cells. Although extremely expensive, they were used on GM Sunraycer, the 1987 World Solar Challenge winner and the Dutch car 'Nuna' (Figure 66), winner of the 2001 WSC. Many other teams including 'Mad Dog' (the only British solar car team) use less expensive monocrystalline silicon cells. Efficiency ranges from 14 to 21% but the cells on 'Mad Dog III' for example were treated with an anti-reflective coating resulting in an overall efficiency of 16.5%. Lamination also protects the delicate and brittle PV cells, and companies with the requisite expertise such as Gochermann Solar Technologies use special processes that both protect the cells and allow them to be curved over increasingly complex solar car body forms.

9.2.4 Batteries

A solar powered car needs a battery to act as an energy store. The battery should act as a reservoir that can be drained when power demand is high and is refilled when energy supply is high. It was shown previously that a slight incline causes a dramatic impact on power required, and that changes in the received sun power caused by cloud cover or shade from buildings and trees can also reduce the speed of the car. Ideally solar cars should be driven at a constant speed, and a solar car is usually configured to drive at optimal speed given the prevailing conditions in order to match the motors' peak efficiency speed with driven wheel speed.

Batteries have poor energy density in comparison with petrol and diesel and so high battery mass is required to propel a vehicle over distance. A large battery capacity incurs the penalty of a large mass, which in turn causes the undesirable forces attributable to rolling resistance and weight components when climbing hills. The optimum type and size of battery is related to race strategy, local conditions and car performance.

Battery charge and discharge characteristics, energy storage capacity (and therefore weight) all vary according to type and physical composition. Invented in 1854, lead-acid (Pb-acid) batteries were the first type to be used in electric vehicles and, with an energy density of about 35 Wh/kg, are also used in current ICE models to operate the starting motor.

Other more efficient batteries have subsequently been developed and are used in solar cars, one type being nickel cadmium (Ni-Cd) with an energy density of 52 Wh/kg. With rapid recharge capability and an energy density of 80 Wh/kg, nickel metal hydride (NiMH) batteries appeared promising but problems relating to high self-discharge rates make them comparatively unstable although they have proved successful in certain solar cars.

Lithium-ion (Li-ion) batteries are the latest type to be used in solar-powered cars and, with an energy density equivalent to 3½ times that of Pb-acid batteries are proving highly successful. This is due in part to greater energy density and also because charge and discharge rates are approximately 10% more efficient than those of Pb-acid batteries. However, present high costs prohibit widespread application for electric road vehicles although increased sales could reduce production and purchase costs [98, 99].

As previously mentioned, vehicle dimensions are restricted by ISF regulations as is battery weight. Because of variations in energy density, maximum permitted weight of batteries and therefore energy storage capacity is determined by type to ensure parity. 'Stock Class' cars can use a maximum of 165kg Pb-acid batteries while cars in the 'Open Class' can use either a maximum of 165kg Pb-acid batteries, 100 kg NiCd, 60 kg NiMH, or 30 kg Li-ion and Li-ion Alloy (polymer) batteries.

9.2.5 Motors and motor controllers

Using energy from the batteries, the motor provides mechanical torque to turn the drive wheel and thus propel the car. High torque electric motors that generally provide an equivalent of 750 to 1500 W are far more efficient at 80–95% than internal combustion engines (ICE) at 20–30%. By using additional power from the batteries, an eight to ten-fold increase is possible for short periods [100].

Many solar cars use three rather than four wheels as a means of reducing road friction and rolling resistance. The single wheel is located either to the front or rear of the car although this may be seen as a trade-off because four-wheeled cars tend to be more stable. In some models the wheels are linked to the motor by a drive belt although many teams now utilise in-wheel motors to minimise transmission losses and improve efficiency. Finally the motor and so vehicle speed is controlled by the motor-controller, which essentially modulates and conditions the electric current.

9.3 Solar car speeds

Developments in solar car technologies during the past twenty years have led to higher speed as is evident in the average speeds attained by winners of the World Solar Challenge (Table 4). This 3000 km race bisects Australia from Darwin in the north to Adelaide in the south and the first race in 1987 was won at an average speed of 66.9 km/h rising to 91.8 km/h in 2001. As team names suggest, three cars were built and raced by major motor-manufacturing companies, but impressive technological progress is equally evident among non-professional university teams even though their budgets tend to be smaller than those of commercial teams.

Table 4: World Solar Challenge winners – average speeds [101].

	Team	Car	Average Speed
1987	GM	Sunraycer	66.904km/h (41.8 mph)
1990	Biel School of Engineering	Spirit of Biel	65.184km/h (40.74 mph)
1993	Honda	Dream I	84.96km/h (53.1 mph)
1996	Honda	Dream II	89.76km/h (56.1 mph)
1999	Royal Melbourne Inst. Technology	Aurora	72.96km/h (45.6 mph)
2001	University of Delft	Nuna	91.81km/h (57.38 mph).

At 3600 km the 2001 American Solar Challenge along 'Route 66' from Chicago to Los Angeles was even longer than the WSC. However, because the cars are restricted to a speed limit of 80 km/h and because there are many more and steeper hills on that route, the overall average speed of the winning University of Michigan car, M-Pulse (Figure 67), was only 64 km/h.

Nevertheless over such distances, in addition to technical efficiency, factors such as team co-operation, race strategy, weather conditions and changes in terrain influence overall race results, while over shorter distances, speeds can be considerably higher. Having reached 107 km/h the Northern Territories University car 'Desert Rose' (Figure 68) was acclaimed as the fastest solar car in the world until the WSC 2001 time trials when 'M-Pulse' achieved 112.5 km/h.

Figure 67: M-Pulse 2001.   Figure 68: Desert Rose 1999.

9.4 Rolling resistance

Rolling resistance is influenced by: vehicle and driver weight, tyre type, tyre size and speed but can be minimised with use of appropriate tyres. The earliest solar cars used bicycle tyres but several manufacturers, (Michelin, Bridgestone and Dunlop), have subsequently developed high-pressure tubeless tyres for solar cars so that the resultant average rolling-resistance is approximately thirty times lower than that of a standard car tyre [102]. The design and set-up of a vehicle's suspension have a significant influence on rolling resistance. Since the suspension is critical to vehicle stability, an efficient set-up is often compromised by the overriding safety considerations.

9.5 Solar car aerodynamics

We have already seen that aerodynamic drag increases exponentially with speed, but as might be expected drag-speed characteristics shown in Figure 69 (which relate to the performance of a typical solar car) differ from those shown in Figure 15 (which relate to the performance of a standard ICE production car).

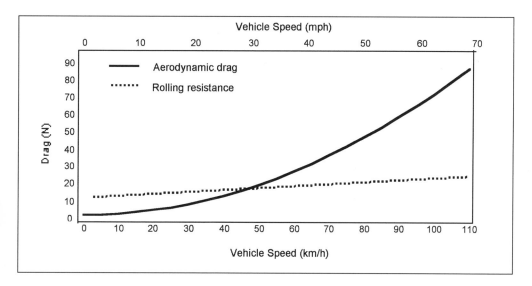

Figure 69: Aerodynamic drag and tyre rolling resistance as a function of vehicle speed for a solar car [103].

Specifications used for model:
$C_dA = 0.14$ m^2,
Gross Vehicle Weight = 3567 N (based on combined mass of car, driver, and battery pack of 364 kg)

Nevertheless it is evident that aerodynamic drag becomes the dominant factor in road resistance at speeds above 50 km/h and accounts for approximately 75% of total drag at racing speeds.

288 *Nature and Design*

Aerodynamic drag is comprised of four principle elements: pressure drag, viscous friction, induced drag and interference drag and is defined as

$$D_{aero.total} = D_{pres.sep} + (D_{skin} + D_{pres.BL}) + D_{ind} + D_{int} \qquad (7)$$

Pressure drag ($D_{pres,sep}$) arises when airflow separates from the surface of the car body forming a low pressure void or vacuum and causing a backward force on the car (see Figure 70). A significant amount of energy is also expended as vigorous vortices are generated behind the car body: generic shapes that cause this type of airflow are known as *bluff bodies*. Most mainstream cars are bluff bodies, and between 50 and 90% of aerodynamic drag is attributable to separation pressure drag which is dependent on the frontal area.

The second component of drag, viscous friction, is affected by the boundary layer (the thin layer of airflow near the body surface) as shown in Figure 70.

Some forms, known as *streamlined bodies*, do not exhibit airflow separation, one notable example being the aerofoil from which many solar car forms derive. As illustrated in Table 5, the aerodynamic drag of a bluff body such as a cylinder is between five and ten times greater than that of an aerofoil with an equivalent frontal area.

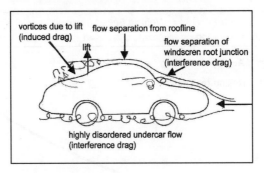

Figure 70: Airflow over a production car shows a large amount of separated flow starting near rear window.

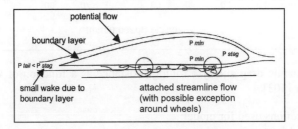

Figure 71: Separation-free airflow over a streamlined solar car body [104].

Viscous friction derives from the combination of skin friction (D_{skin}) and pressure drag due to the boundary layer ($D_{pres,BL}$) and is present in all vehicles. Viscous friction becomes the dominant type of drag for streamlined forms and therefore optimally designed solar cars (Figure 70) which do not generate separation.

Induced drag (D_{ind}) is caused by lift and may be upwards (as in an aeroplane) or downwards (as in Formula 1 type racing cars). As the body lifts, the pressure difference between the upper and lower body surfaces generates vortices. In certain instances these can be used to advantage and some solar cars, for example 'Sunswift' (Figure 63), effectively use a fin-like 'sail' on the driver's bubble to increase speed.

Table 5: Coefficient of drag measured over frontal area ($C_d A$) [105].

(rectangular block)	2
(cylinder)	1.2
(rounded block)	1.1
(airfoil)	0.15

Finally there is interference drag. This derives from a variety of factors, including the imperfect matching of body panels, driver and other ventilation inlets, surface imperfections in manufacture or surface roughness such as advertising stickers, holes where the wheels protrude from the body and at the corners between the driver canopy or fairings and the main body. Where two different forms meet, they create junction drag. If a car body is derived from two different forms, the overall aerodynamic drag will exceed the sum of the aerodynamic drag of the two separate forms. In many instances, the interference drag is actually pressure drag and can account for as much as 50% of total vehicle drag.

Because the drag of a bluff body is more or less proportional to its frontal area, C_dA is the most useful reference when considering the drag coefficient of standard vehicles as stated in section 6. It is also appropriate to use C_dA when comparing dissimilar forms such as standard ICE cars with solar cars as in Table 6.

Table 6: Comparison of the drag coefficient over frontal area (C_dA) for standard road and solar cars [106].

C_dA m² ICE cars		C_dA m² solar cars	
1928 'Sedan'	>2	GM Sunraycer	0.125
BMW X5 SUV	1.16	Spirit of Biel	0.13
Porsche Boxster	0.71	Honda Dream 1	0.1
Honda Insight	0.57	Honda Dream II	0.12
Smart Car	0.57	Aurora	0.1
GM Precept	0.52	Nuna	0.1

In contrast Table 6 compares the drag of similar streamlined bodies where viscous friction is the dominant type of drag. Here it is useful to quantify drag over the planview (sometimes described as planform) or wetted area. Such information indicates the success of the boundary-layer flow management. It is apparent that the coefficient of drag measured over the solar car planviews $(C_{d,plan})$ is greater than that of any of the aerofoil forms in Table 7.

Table 7: Comparison of coefficient of drag measured over planview area – solar cars [107].

	Planview area m^2	**C_d based on planview area $(C_{d,plan})$**
MIT Manta	8.8	0.0136–0.0148
MIT Manta GT	9.5	0.0095–0.0105
Honda Dream II	10.3	0.0111

Table 8: Coefficient of drag measured over planview $(C_{d,plan})$ and frontal areas (C_dA) aerofoils [108].

Aerofoil forms	$C_{d,plan}$	C_dA
	0.006	0.10
	0.007	0.06
	0.008	0.05

9.6 Streamlining in the animal world

Optimum streamlined bodies are generally torpedo-shaped and are evident within the man-made environment in armament and space technologies; aircraft are essentially streamlined body-forms with aerofoil-shaped wings for lift. Streamlined forms are also found throughout the natural world and two examples (the sailfish when hunting and the falcon when diving for prey) were cited in Section 2. Many other aquatic mammals have streamlined bodies and several examples are shown in Table 9. It is evident that the coefficient of drag measured over the wetted area $(C_{d,wet})$ for man is somewhat higher than the other examples because humans are primarily land animals while the low $C_{d,wet}$ of the other examples suggests that these animals are extremely well adapted to the environment in which they spend the majority of their lives.

The various images of solar cars in this chapter and data relating to coefficients of drag indicate that they are not pure aerofoil forms, but in order to assess whether they exhibit maximum aerodynamic efficiency, it is necessary to review the evolution of solar car form.

Table 9: Coefficient of drag measured over wetted area ($C_{d,wet}$) for aquatic forms [109].

Man (towed in water)		0.035
Penguin		0.0044
Seal		0.004
Sea-lion		0.0041
Dolphin		0.004

9.7 The evolution of solar car form

The earliest solar cars were very simple vehicles in comparison with contemporary models. Based on bicycle and light aircraft technologies the primary aim of the engineers and designers was to evaluate the potential of solar power for transport purposes. Consequently, emphasis was on the development of electrical and mechanical components and systems rather than aerodynamics and so the flat PV array panels and driver 'cockpit' were generally separate elements.

Figure 72: Early solar car.

Figure 73: Early solar car.

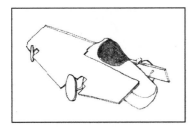

Figure 74: University of Mankato 'Northern Light' 1989.

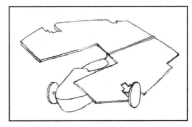

Figure 75: University of Mankato 'Northern Light' 1989.

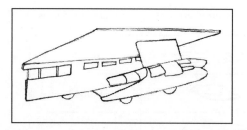
Figure 76: Western Washington University car, 1989.

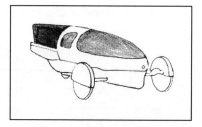
Figure 77: 'TNE-3'.

Cockpit design developed comparatively quickly and became more aerodynamically efficient (Figures 72 and 73) while various means of maximising energy generation were explored. These include hinged and tilting arrays (Figures 74–76) and location of PV panels on the sides of cars, as in the University of Michigan catamaran 'Sunrunner'.

In 1995 one team adopted the opposite approach by minimising aerodynamic drag rather than maximising power generation. TNE-3 (Figure 77) resembles an aerofoil or torpedo more than any other model and consequently has a C_dA of 0.07. Because the PV panels are carried behind the driver they cannot charge the batteries while in motion. Despite this factor, by definition the TNE-3 is a solar car and was placed seventh out of thirty-eight entries in the 1995 GM Sunrayce but subsequent developments in solar and other technologies, however, mean that this car is no longer as competitive.

In 1987 under the auspices of Paul MacCready, GM developed 'Sunraycer', one of, if not the first, solar cars to utilise the now familiar teardrop-shape initially investigated during the 1930s (Figure 24). This car demonstrated the importance of the trade-off between power and aerodynamics and that increased energy generation through tilted and angled arrays was not as beneficial as achieving lower coefficients of drag. Consequently the teardrop and aerofoil, and therefore more natural forms, became predominant while 'TNE-3' clearly demonstrates that using current PV technologies, it is not feasible to construct genuine solar cars with greatly reduced drag coefficients. This is because a large, more or less horizontal, surface area is required to accommodate the array and thus power source.

Figure 78: Junkyard 1999.

As stated in Section 9.2.1 the maximum overall size of solar cars and solar arrays is limited by the ISF regulations but some teams have built short cars in an attempt to improve performance by reducing drag and weight: Junkyard (Figure 78), for example, is 2.83 m long and 1.23 m wide.

Because the array is smaller, power generation is obviously reduced but the resulting compromise between lower weight and lower power is not necessarily beneficial. This is best illustrated by considering the solar viscosity ratio. The term viscosity ratio is developed from the conventional fluid property, viscosity, that is resistance to shear rate. The potential for higher speed is reflected by higher ratio and can be defined as follows:

$$Solar\ Viscosity\ Ratio = \frac{Solar\ Array\ Planview\ Area}{C_d A} \qquad (8)$$

Table 10: Solar Viscosity Ratio [110].

	$C_d A$ m^2	Solar Planview Area m^2	Solar-Viscosity Ratio
GM Sunraycer	0.165	8.0	48.55
Spirit of Biel	0.143	7.8	44.4
Honda Dream I	0.114	7.2	54.5
MIT Manta	0.125	7.5	63.2
Honda Dream II	0.15	9.84	60.1
Junkyard	0.114	3.23	28.3
MIT Manta GTX	0.09	6.5	72.76

Although the solar planview area for 'Manta GTX' (Figure 79) is smaller than that of the other examples, the $C_d A$ is also lower and the solar-viscosity ratio is therefore higher. Because Junkyard is so short, airflow over the body is affected resulting in a C_d of 0.2, and although the frontal area is only 0.57 m^2, the resultant $C_d A$ is 0.114 m^2 [111]. This, in conjunction with such a small solar viscosity ratio, indicate that although these benefits derive from lower weight, small size can create aerodynamic disadvantages and compromise overall efficiency and competitiveness.

Figure 79: MIT Manta GTX 1999.

9.8 Summary

In summary, it is evident that since the early 1990s the form of most solar cars has become increasingly sophisticated as design solutions have moved from being profile-based to truly three-dimensional and thus incorporating complex curves. This was facilitated by developments in computer-aided design and computational fluid dynamics software as well as the aforementioned advances in PV technology that enable laminated cells to follow complex forms. In conjunction with more efficient arrays, batteries, electrical components and systems, the development of solar car form and reduction of drag in particular have all contributed to improved performance and higher speeds.

Evidence shows that, although the primary intention of these vehicles is to improve performance by minimising energy consumption, this type of vehicle will inevitably be less aerodynamically efficient than pure aerofoil forms because it is necessary to incorporate a more or less horizontal array area within the overall body form. Nevertheless, when all above factors are taken into consideration and because all advantages and disadvantages are evaluated in relation to each other, solar powered cars are truly optimised design solutions.

10 Conclusion

This chapter has investigated the form, aerodynamics and energy efficiency of various means of natural and man-made locomotion. We have seen that the fastest animals in the air, in water and on land have evolved so that each is perfectly adapted to the environment in which it lives. In addition to patterns of behaviour, the development of particular physiological and anatomical attributes, levels of energy expenditure and overall body form all enable these creatures to move at high speed with maximum efficiency. We have briefly examined how human beings developed into bipedal hominids and the benefits, efficiencies and eventual limitations of human locomotion.

By comparing the travel speeds of humans with animal assisted transport, we then suggested why the latter became the preferred means of land transport over distance and how it thus contributed to the development of, and became integral to, many societies.

We then recorded the development of the bicycle as an example of human-powered transport and have seen that despite the fact that a bicycle and rider are not aerodynamically efficient, conversion of leg-muscle power to rotary pedals is highly energy-efficient and that the bicycle increased travel speeds beyond those of horses and carriages.

The greatest proportion of this chapter has explored the development of both internal combustion engine (ICE) and solar powered cars and highlights the significant differences between these two types of automobile. Invented in the 19th century, the internal combustion engine car became enormously popular and influential during the 20th century, although it has become a victim of its own success and in many congested cities for example, travel by bicycle and even on foot is quicker than by car. Use of the ICE car has also contributed to the depletion of natural resources and created localised, national and global pollution. Despite these factors most people with a car could not imagine changing their lifestyles to live without one [112] and many of those without a car aspire to ownership [113]. Consequently the car or an equivalent mode of personal transport is here to stay.

One of the main intentions of this chapter was to discover whether there are parallels between the evolution of mass-produced ICE cars, and in particular car form, and the natural world. By examining the relationship between fuel consumption and the development of vehicle aerodynamics we conclude that the ICE car is not efficient. Further research revealed that the form of ICE cars is subject to ergonomic constraints, fashion, style and taste and consequently mainstream cars continue to be *bluff bodies* rather than *streamlined forms*. While elephants may be described as *bluff bodies* and require a large energy intake (and like cars, can carry large loads), they only travel at speed over comparatively short distances and for short periods of time. However, like all natural life-forms, their energy and material needs are minimised in order to survive [114]. Mass-produced cars however are, (subject to refuelling), intended to travel at speed and carry loads for many thousands of kilometres, but they do not minimise energy and material needs and in this sense can be said to contradict this law of nature.

Although some small cars with lower fuel consumption were introduced in the late 1950s, there was no significant attempt to produce more fuel-efficient cars until the late 1970s. Since then incremental reductions in fuel consumption have occurred but in the main these derive from improved engine technologies and reductions in vehicle weight.

Since the mid-1990s, however, due in part to concern about increasing pollution, the eventual depletion of oil reserves and desire for fuel autonomy, government and other legislation has encouraged some motor manufacturers and independent organisations to develop considerably more fuel-efficient vehicles. In addition to use of hybrid and electric technologies, particular emphasis on lower aerodynamic drag is manifest and we conclude therefore that as the need for greater fuel economy increases, the design of mainstream cars will reflect the laws of nature more closely.

Investigation of the evolution of solar cars and associated technologies during the past two decades has revealed a very different approach. Admittedly these are single function (as opposed to multi-purpose) racing vehicles and the engineers and designers involved in their development have benefited from the knowledge and expertise gained from mainstream auto-manufacture. Nevertheless, from the outset these vehicles, have had a limited power supply and energy-efficiency is therefore a necessity. We have shown here that significant improvements in efficiency and performance derive from adoption of streamlined forms similar to those found in the natural world. We conclude therefore that solar powered cars are essentially optimised design solutions and that they adhere far more closely to the laws of nature than internal combustion engine vehicles.

References

[1] Darwin, C., *On the Origin of Species by Means of Natural Selection, or the Preservation of Favoured Races in the Struggle for Life*, Senate: London, 1994.
[2] Cook, R., *Animal Cognition & Learning*, Tufts University, Medford, MA, http://www.pigeon.psy.tufts.edu/psych26/birds.html
[3] Goodall, J., *The Chimpanzees of Gombe: Patterns of Behavior*, Belknap Press of Harvard University Press: Cambridge, MA, 1986.
[4] Clutton-Brock, J., *Horse Power*, British Museum (Natural History): London, pp. 68, 73, 12, 1992.
[5] Deuel, N.R., *Typical Speeds*, http://www.horse-science.com/horsespeeds.htm
[6] *Horse Power*, ibid.
[7] Gillmer, T.C., *A History of Working Watercraft of the Western World*, 2nd Edition, International Marine Camden Maine, 1994 and Landels, J.G., *Engineering in the Ancient World*, Constable, London, 1997.
[8] Seifert, U. & Walzer, P., *Automobile technology of the Future*, Society of Automotive Engineers, Warrendale, PA, 1992, http://www.webcom.com/~trw/London/59473610.html
[9] Zahn S., Rich, T. & Rich, C., *The Cyber Zoomobile*, 2000, http://www.primenet.com/~brendel/cheet.html
[10] McNeill A.R., *Exploring Biomechanics/Animals in Motion*, W.H. Freeman and Company: New York, 1992.

[11] Blood, D., *Hinterland Who's Who, Peregrine Falcon*, Minister of Public Works and Government Services Canada, http://www.cws-scf.ec.gc.ca/hww-fap/hww-fap.cfm?ID_species=32&lang=e
[12] Amadon, D. & Brown, L., *Eagles, Hawks and Falcons of the World*. McGraw-Hill Book Co.: New York, 1968.
[13] Maslin, P.E., *Ichthyology*, Biology Department, California State University, Chico, CA http://www.csuchico.edu/~pmaslin/ichthy/loco.html
[14] Johnson, G.D. & Gill A.C., in: *Encyclopedia of Fishes*. eds. J.R. Paxton & W.N. Eschmeyer, New South Wales University Press: Sydney, Academic Press: San Diego, 1995, p. 240. 1994.
[15] Chang, K., *On Two Feet 4 Million Years Ago*, ABC World of Sci/Tech.com, May 7, 2000, http://www.abcnews.go.com/sections/science/DailyNews/hominid980506.html
[16] Russell, J., *Homo Sapiens Neandertalensis: Forcing a Radical Reconsideration of Human Origins* Knowledge Integration Environment (KIE) Project, Graduate School of Education, University of California at Berkley, Integrated Science Environment Project. http://www.kie.berkeley.edu/ned/data/E01-980521-009/E01-980521-009.html, 1997.
[17] McMahon, T.A., *Muscles, Reflexes, and Locomotion*, Princeton University Press: Oxford, pp. 189–228, 1984.
[18] *Human Locomotion*, Department of Orthopaedic Surgery at the University of Aberdeen, Phase 1 Anatomy (locomotion), http://www.abdn.ac.uk/orthopaedics/phase1.htm
[19] *Human Locomotion*, ibid.
[20] Morrison, P. & Morrison, P., *Walk, Run and Skip*, Scientific American http://www.sciam.com/1999/0399issue/0399wonders.html, 1999.
[21] Elert, G. (ed), *The Physics Factbooook*, http://hypertextbook.com/facts/2000/KatarzynaJanuszkiewicz.shtml, 2000.
[22] McKeever, S., *Marathon 1:59:59*, http://www.scmtny.org/marathon15959.html
[23] McClelland, J., *Give Your Heart a Healthy Beat*, North Carolina State University, http://www.ces.ncsu.edu/depts/fcs/heart/hhb5-8.html
[24] *Muscles, Reflexes, and Locomotion*, ibid.
[25] Mureika, J.R., How Good Can We Get? Using mathematical models to predict the future of athletics. *Athletics*, Canada's National Track and Field/Running Magazine, Table 5: Predicted ultimate performances for men and women, April/May 1998.
[26] Bijker, E.W., *Of Bicycles, Bakelites and Bulbs*, M.I.T. Press: London, 1997, p. 20.
[27] Uhlig, R., (ed), *The History of Great Inventions*, The Daily Telegraph, London, 2000, p. 11.
[28] Of Bicycles, Bakelites and Bulbs, ibid., pp. 20–30, 54–73, 77–80.
[29] Malewicki, D.J., *152 mph Pedal Cycle*, http://www.canosoarus.com/08LSRbicycle/LSR%20Bike01.htm
[30] Stull, P., *Recumbent History*, http://www.bicycleman.com/History.html
[31] Poindexter, M., *We're not 'blocking' traffic, we are traffic!* http://www.msu.edu/user/corcora5/velo/bikefacts.html
[32] Rafoth, R., *Cycling Performance Tips*, http://www.cptips.com/energy.htm
[33] Lamancusa, J.L., *Cycling Bio-Mechanics*, http://www.me.psu.edu/lamancusa/ProdDiss/Bicycle/bike3.ppt
[34] Macari, Pallis, J., *The Ride of the Cyclist*, Cislunar Aerospace, Inc. 1997, http://wings.avkids.com/Curriculums/Sports/cyclist_summary.html
[35] Thompson, P. & Lawrence D., *Chris Boardman.com*, http://www.lunt.demon.co.uk/boardman.htm#hour and *152 mph Pedal Cycle*, ibid.

[36] Newell, M., *Recumbents*, Human-powered Vehicle Association, http://www.ihpva.org/FAQ/
[37] *Recumbent History*, ibid.
[38] Olds, *et al*, Modelling road-cycling performance, *Journal of Applied Physiology*, **78**, pp. 1596–1611, 1995.
[39] Capelli & di Prampero, Effects of altitude on top speeds during 1 hour unaccompanied cycling, *European Journal of Applied Physiology*, 1995, **71**, pp. 469–471.
[40] Rehling, U., *Energy and Development,* Chapter 3, What is Energy? http://www.uni-flensburg.de/sesam/Download/ Energy
[41] Exploratorium, *The Physics of Bicycles*, http://www.exploratorium.edu/cycling/humanpower1.html and Sharkey, B.J., *Physiology of Fitness*, Human Kinetics Publications, 1986.
[42] Perkin, H., *The Age of the Automobile,* Quartet Books; London, 1976.
[43] Sperling, D., *Future Drive: Electric Vehicles and Sustainable Transportation*, Island Press: Washington DC, p. 1, 1995 and Nieuwenhuis, P., Cope, P. & Armstrong, J., *The Green Car Guide*, Green Print: London, p. 1, 1992.
[44] Faiz, A., *et al.*, *Automotive Air Pollution: Issues and Options for Developing Countries* Infrastructure and Urban Development Department, The World Bank, August 1990
[45] *The Age of the Automobile*, ibid.
[46] Schiffer, M.B., *Taking Charge: The Electric Automobile in America*, Smithsonian Institution Press: Washington DC, 1994.
[47] Riva, J.P., Oil Distribution and Production Potential, *Oil & Gas Journal*, pp. 58–61 January 18, 1988 and MacKenzie, J.J., *Oil as a finite resource: When is global production likely to peak?* World Resources Institute 1996, updated 1999, http://www.lgc.org/wri/climate/jm_oil_001.html
[48] Ingersoll, J.G., *Natural Gas Vehicles*, The Fairmont Press Inc.: London, 1996.
[49] Intergovernmental Panel on Climate Change, *Climate Change - Second Assessment Report*, Cambridge University Press: Cambridge, 1996.
[50] The Department for the Environment Transport and Regions (DETR), *New Deal for Transport: Better for Everyone,* The Sationery Office; London, July 1998 and *Kyoto deal: Success or unsatisfactory compromise*, http://www.usnews.com/usnews/issue/971222/22trea.html, 1s/8/98 and *Kyoto: Economy or the Environment*, http://www.rose-hulman.edu, 12/5/97 and http://news.bbc.co.uk/hi/english/world/analysis/newsid_38000/38567.stm
[51] *Future Drive: Electric Vehicles and Sustainable Transportation*, ibid. pp. 27, 114.
[52] United States Council for Automotive Research, *About PNGV*, http://www.uscar.org/pngv/index.htm
[53] DETR (2001), *Graduated Vehicle Excise Duty Reforms*, http://www.dvla.gov.uk/newved.htm
[54] Directive 2000/53/EC of the European Parliament and of the Council of 18 September 2000 on end-of life vehicles.
[55] TransportAction PowerShift, *Clean Fuel Vehicles Market Report*, DETR: Stationery Office, London, Spring 2001.
[56] Sperling, D. & Shaheen, S. (eds.), *Energy Strategies for a Sustainable Transportation System,* American Council for an Energy-Efficient Economy, cited in *Future Drive: Electric Vehicles and Sustainable Transportation*, ibid.
[57] Nieuwenhuis, P. & Wells, P. *The Death of Motoring? Car Making and Automobility in the 21st Century*, John Wiley and Sons: Chichester, 1997.

[58] Tamai, G., *The Leading Edge, Aerodynamic Design of Ultra-streamlined Land Vehicles,* Bentley, Cambridge, MA, pp. 10–11.
[59] Hucho, W-H., *Aerodynamics of Road Vehicles, From Fluid Mechanics to Vehicle Engineering,* Butterworths: London, 1987, pp. 34–46.
[60] Seifert, U. & Walzer, P., *Automobile technology of the Future,* Society of Automotive Engineers, Warrendale, PA, 1992, pp. 27–30.
[61] Porter, L., *Fiat Punto 1994–1999,* The Porter Publishing Editors, 2001.
[62] Sanders M.S. & McCormick, E.J., *Human Factors in Engineering and Design,* 7th ed., McGraw-Hill: New York, 1993.
[63] *Aerodynamics of Road Vehicles, From Fluid Mechanics to Vehicle Engineering,* ibid. pp. 10–27.
[64] *Tracing History,* http://www.zagato-cars.com
[65] *The Age of the Automobile,* ibid. p. 109
[66] Marsh, P. & Collett, P., *Driving Passion, The Psychology of the Car,* Jonathan Cape: London, 1986.
[67] Bayley, S., *Sex, Drink and Fast Cars,* Faber: London, 1986.
[68] *Aerodynamics of Road Vehicles, From Fluid Mechanics to Vehicle Engineering,* ibid. pp. 256–258.
[69] *Driving Passion, The Psychology of the Car,* ibid.
[70] O'Donoghue, S. & Sims, M., http://www.carenthusiast.co.uk/shows/bmw_2.htm
[71] Ford Haven, *2001 Ford builds on initial success,* http://www.fordhaven.com/specs/2001_focus.html
[72] Autocar.com, *World Wide Wheels,* http://www.specialcar.com/8_97/boxster/boxster.html
[73] *AutoCar,* Alu-ise July 3, 1996, http://www.sandsmuseum.com/cars/elise/press/magazine/magazine1999/aluise.html
[74] Yeoman, http://www.yeomanstoyota.co.uk/yaris.htm
[75] *Alpha Newspapers,* Hyundai launch new Amica and Coupe, http://www.ulsternet-ni.co.uk/motoring/moto04/motopage.htm
[76] Helsinki University of Technology Department of Computer Science, *Mini List,* January 1998. http://www.niksula.cs.hut.fi/~mdobruck/siililand/mini/diy/1/drag.html
[77] Sparke, P., *Italian Design 1870 to the Present,* Thames and Hudson: London, 1988.
[78] Chaikin, D., *Small World,* http://www.popularmechanics.com/popmech/auto2/9804AUNCBM.html
[79] Cowan, R. & Hulten, S., (eds.), *Electric Vehicles,* Socio-economic prospects and technological challenges, Ashgate, Aldershot, pp. 44–47, 2000.
[80] USCAR, *PNGV,* http://www.uscar.org/pngv/index.htm
[81] Online Fiat Archive, *Fiat Ecobasic,* http://www.carsfromitaly.com/fiat/popups/ecobasic.html
[82] *Toyota Prius,* Marketing Brochure, Toyota (GB) PLC, Redhill, 2000.
[83] *Honda Insight,* Marketing Brochure, Honda Cars (UK), Slough, 2000.
[84] *Electric Vehicle Range,* Solectria Corporation, Wilmington, MA. http://www.solectria.com/force.html
[85] Peugeot Innovation, *The Peugeot 106 Electric,* http://www.peugeot.ru/eng/innovat/electro.asp, Product review, http://www.tin.it/electrictimes/a101.htm
[86] *Design and performance details,* Environment and Health Protection Administration of Stockholm, 1998, http://www.zeus-europe.org/seicento.html

[87] Ford Motor Co., *Think- a New Way of Moving*, http://www.thinkmobility.com/global/home.asp?TID=38798, 2000
[88] Southern California Edison Co., *Electro-drive History - 1990: GM Impact and Beyond* http://www.sce.com/electrodrive/004g4d_gm_impact.shtml *GM Impact Specifications*, http://www.innerauto.com/innerauto/text/fuel97.html
[89] *The Hypercar Concept*, Rocky Mountain Institute, http://www.rmi.org/sitepages/pid386.php, 2000.
[90] Hawken, P., Lovins, A. & Lovins, H.S., *Natural Capitalism, Creating the next industrial revolution*, Little, Brown & Co.: Boston, pp. 22–47, 1999.
[91] Society of Motor Manufacturers and Traders, *Motor Industry PR Campaign: Five Year Strategic Direction*, 21:07:98.
[92] Riley, R.Q., *Alternative Cars in the 21st Century, A New Personal Transportation Paradigm*, SAE, Warrendale PA, 1994.
[93] Sherman, J., *Charging Ahead*, Oxford University Press: Oxford, 1998.
[94] *Charging Ahead*, ibid. pp. 55, 68–75.
[95] King, R. & King, M., *Sunracing*, Human Resource Development Press, Amhurst MA, 1993, p. 10.
[96] *Sunracing*, ibid., pp. 5–11.
[97] Hexel Corporation, *Markets and Products: The Aerospace Industry*, http://www.hexelcomposites.com/markets/aerospace/civil.html
[98] Rand, D.A.J., Woods, R. & Dell, R.M., *Batteries for Electric Vehicles*, Research Press Studies, 1998.
[99] *Alternative Cars in the 21st Century: A New Personal Transportation Paradigm*, ibid. pp. 225–238.
[100] Duke, M.D. & Andrews, S.D., The Feasibility of a Solar Electric Passenger Vehicle (SEPV) for Personal Transportation in the UK, *Congress Proceedings*, World Solar Congress, Adelaide, Nov. 2001.
[101] *The Leading Edge, Aerodynamic Design of Ultra-streamlined Land Vehicles*, ibid. p. 169 and World Solar Challenge 2001, http://www.wsc.org.au/Results/2001/day4.solar
[102] *The Leading Edge, Aerodynamic Design of Ultra-streamlined Land Vehicles*, ibid., pp. 5–6.
[103] *The Leading Edge, Aerodynamic Design of Ultra-streamlined Land Vehicles*, ibid., Fig 1.1.1., p. 3.
[104] *Aerodynamics of Road Vehicles, From Fluid Mechanics to Vehicle Engineering*. ibid., p. 7.
[105] *The Leading Edge, Aerodynamic Design of Ultra-streamlined Land Vehicles*, ibid., (based on length – ref. White, 1986 and Hoerner) Table 3.1:1, p. 70.
[106] Based on Table 5.1:1, *The Leading Edge, Aerodynamic Design of Ultra-streamlined Land Vehicles*, ibid., pp. 15–17, 218.
[107] Based on Table 1.3.1, *The Leading Edge, Aerodynamic Design of Ultra-streamlined Land Vehicles*, ibid., p. 14.
[108] *The Leading Edge, Aerodynamic Design of Ultra-streamlined Land Vehicles*, ibid., based on Table 3.1.1. ref. Hoerner, p.70.
[109] *The Leading Edge, Aerodynamic Design of Ultra-streamlined Land Vehicles*, ibid., based on Table 3.1.1. ref. Vogel, p. 71.
[110] *The Leading Edge, Aerodynamic Design of Ultra-streamlined Land Vehicles*, ibid., based on Table 1.3.2, pp. 15–17.

[111] Cotter, J.E., Roche, D.M., Storey, J.W.V., Schinkel, A.E.T. & Humphries, C.P., *The Speed of Light 2*, Key Centre for Photovoltaic Engineering, University of New South Wales, 2001.
[112] 1995 Lex Report on Motoring, *What Drives the Motorist?* Lex Service PLC, with research by MORI, London, 1995.
[113] Clegg, M.W. & Dumoulin, H., The Barriers to Technological Change: the Case of Transportation, *International Journal of Vehicle Design,* **13(5/6)**, pp. 443–448, 1992.
[114] Hawken, P., Lovins. A. & Lovins, H.L., *Natural Capitalism, Creating the next industrial revolution*, Little Brown & Co.: Boston, 1999.

CHAPTER 14

Creativity and nature

C. Dowlen
Department of Architecture and Design, London South Bank University, London, UK

Abstract

Creativity is placed onto an overall perception map that describes the way in which design might be perceived within the natural and man-made worlds and the real and thought worlds. In fact, creativity can be seen to be relevant to each of these areas. This is juxtaposed with a description of the sometimes confusing and challenging nature of human creativity, and a brief overview of modern exploration of creativity is covered. Definitions and measures of human creativity are developed from those in literature.

The development of creativity is tackled through a specific example that serves to demonstrate the use of natural analogies and the way that these can help develop various types of thinking process.

1 Creativity

> Creativity is a magical thing that you are born with. It presents itself early and like a virus is unstoppable.
> It inhabits the kids who daydream and are good at drawing, but not much else. If I knew where it came from, I would bottle it up and make a fortune.
>
> Michael Dempsey [1]

2 The scope of creativity

In the introductory chapter on design definitions, Mark Atherton and myself developed an overall picture in order to place design and nature in context.

At the end of this chapter the picture looked like Figure 1.

The world view was split horizontally into real and thought worlds and again vertically into natural and man-made worlds. At the cusp of these four worlds is design: that process that takes ideas from the man-made (or human) thought world and changes material from the natural world into man-made artefacts – in the real man-made world.

The place of science is on the left of the picture, where observation of nature and experimental work develop hypotheses and these develop into scientific theories, served by mathematics.

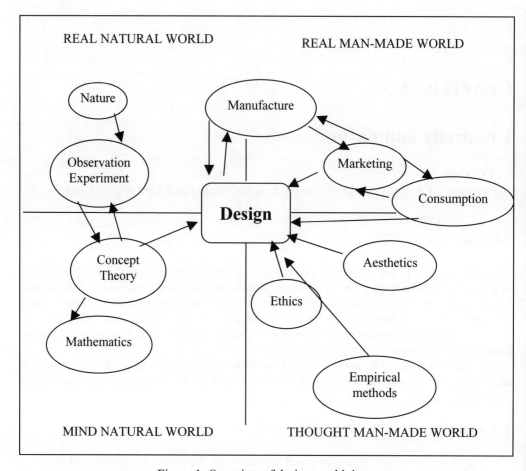

Figure 1: Overview of design worldview.

Manufacture links the real natural world with the real man-made world as natural materials are manufactured into artefacts, which are consumed, and then market research provides information that in turn feeds the design process. Empirical methods, ethics and aesthetics also feed the design process. The picture is not meant to be particularly precise, but is more of a useful way of perceiving a worldview.

What it doesn't show is the position of creativity in the worldview. Perhaps it is most obviously inherent in the human ideas that form in the thought man-made world of the lower right and are formed through the design process to become the ways in which the real natural world is changed to become the man-made natural world.

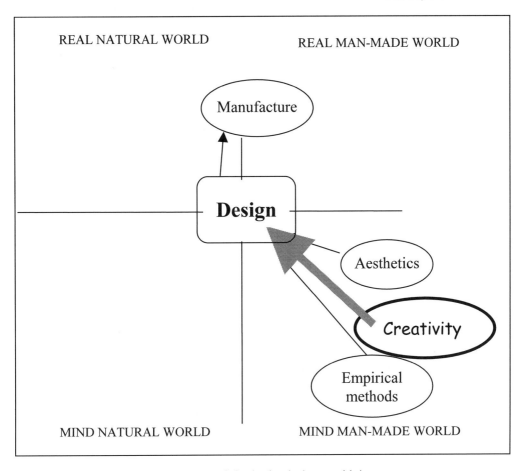

Figure 2: Creativity in the design worldview.

Figure 2 shows the general position of creativity in the generally accepted position in the overall plan.

Creativity is part of the human input to the design process, both forming the ideas in the mind and helping to develop them so that they become the manufactured items – whether these manufactured items are those conventionally thought of such as the artefacts that we use day by day, or something more ephemeral such as lines on paper, art objects, two-dimensional graphics, fashion items, computer programs and so on.

It is an in-built part of the process. Without it nothing changes from being part of the natural world to being part of the designed man-made world, however that transformation occurs.

But that is only part of the story. Creativity is not solely diametrically opposed to the real natural world, meeting it only in the cusp of design. It does indeed meet it at that point, but there are innumerable cross links that are not easily recorded on the diagram and in fact probably end up destroying the neat overview that the figure represents.

For creativity is not only evident in the human ideas in the design process. It is also evident in the way in which scientists form the theories and hypotheses by which they seek to understand the reality of the natural world, and as they develop processes that test and probe the reality of these hypotheses. These hypotheses, however real they are and however much they become part of the generally accepted scientific understanding, have been formed by human perception and from the spark of seeing that a particular space or gap in understanding existed that could be filled by a fresh way of seeing natural behaviour.

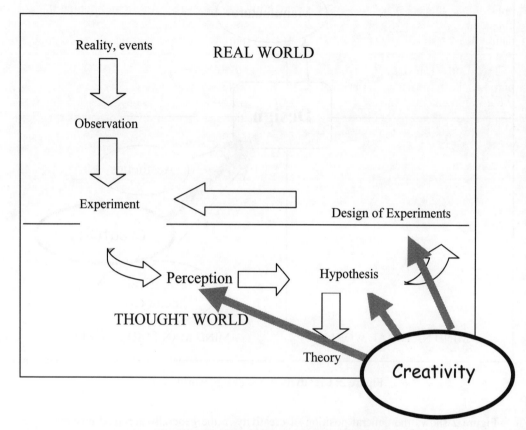

Figure 3: Input of creativity to the scientific process of thought.

Probably the best-known example of creativity in scientific discovery is that of Archimedes discovering his principle in the moment of climbing into his bath. This included a problem to solve and which was taking up his conscious energy, and an experiment that just happened to come along through what he was doing to link his thinking with the perception that the water rose and the hypothesis that he formed: initially concerning a means of measuring irregular volumes, and hence their density, but later generalised to the determination of the upthrust produced by displacement of fluid. Eureka!

Another well-known story is that of Louis Kerkulé's discovery of the ring structure of the benzene molecule. This is frequently cited as an example of the way in which creative thinking can take place and of the use of analogies and trance-like states in the process [2, 3]. This, likewise, did not see its fruition as a man-made product, but as a fresh understanding of the way in which carbon atoms link themselves together. In this case the experiment was not ready-made: the creative insight needed to be formed into a clear hypothesis that could be tested and further developed into a theory.

Creativity can also be perceived in nature itself, apart from mankind. This is much harder to put into a diagram, as it only really appears as a circle pointing at the top left corner.

If we argue that creativity is about creating things, then non-human nature is creative. Birds build nests, animals make dens, and so on. Although they are not exactly building things, chimpanzees' use of stones and sticks as tools could also be regarded as being creative in nature. How instinctive this behaviour is, and how purposeful, is another matter, but it is clear that creativity, at least of that type, is not absent in the natural world.

Frequently, Darwinian evolution is also deemed to be creative. It may not be deemed to be purposeful – for instance, birds didn't set out on purpose to learn how to develop wings to fly – but to the onlooker there can seem to be some sort of direction and developmental process going on as change is initiated and develops variations to the previous organism. Over time this change is seen, and new organisms come into being – or, one could argue, created.

This does not satisfy everyone, of course, and there is a reasonable minority opinion that direct creation has been responsible for the way we see the flora and fauna at the moment. In this instance, it is not that nature is developing itself in a creative manner, but that it has been developed by divine creativity.

In both categories, the effect of creativity is clearly present in nature.

Figure 4: Ardèche Gorge.

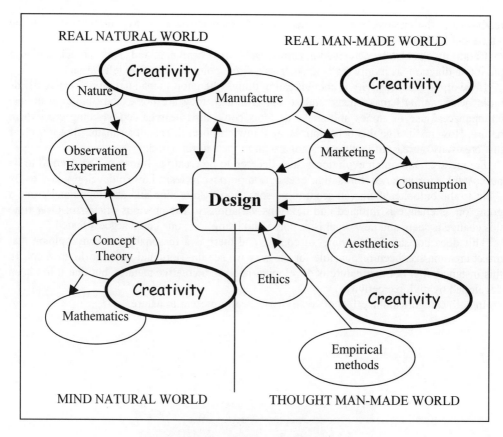

Figure 5: Creativity in the worldview.

Not only is creativity present in the animal and plant kingdoms, but it is also present in the most primeval forces within nature, if we take as our basis that creativity is about creating things. We could argue that creativity is not just about creating, but also about changing things. Take a picture of nature such as the Ardèche Gorge shown in Figure 4. When such forces as erosion, wind, rain, and the battering of the elements of the weather produce such elegant effects as this, there is no doubt that something has been created, and that something is deemed to have, in certain circumstances, a beauty that is not removed from the human generation of beauty in some human-developed product – and can only be regarded as having been created. And yet it is clearly the product of a very natural set of processes.

But our difficulty of inventing a diagram is compounded because these natural "creations" still remain as the real natural world and not as the real man-made world. What we created was an artificial vertical line separating the human aspects from nature. It is a useful distinction that we can make if we wish, but it tends to separate out ourselves from nature.

When we look at nature we still have to remember that we are part of it. Thus we have another aspect to the nature of creativity: that our human creativity is part of our human nature. Still harder to show on the diagram.

If we are naturally *Homer Faber*, as we suggested in the earlier chapter, then we are naturally creating things, and have always been doing that. We are then, presumably, naturally creative. It is part of our nature. It is argued that no matter how uncreative a person may appear, they are never devoid of creativity [4]. Some of us may be more creative than others, but part of our purposeful nature is to create, knowingly, by developing ideas that grow into whatever they grow into – products, theories and the like.

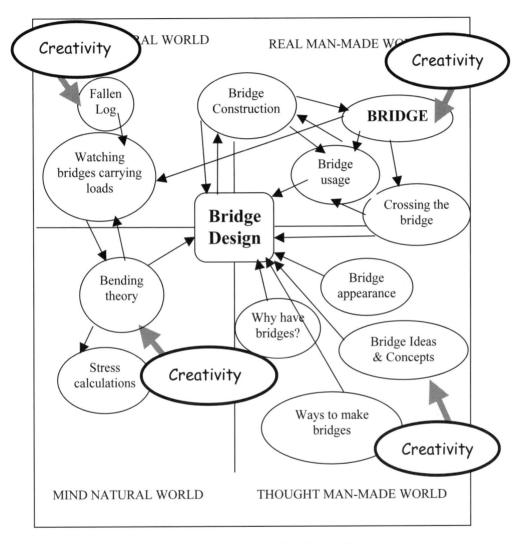

Figure 6: Bridge: an example of creativity.

So in a sense creativity is all encompassing. It fits into every corner of our original picture and is evident in the real natural world in the way it changes, the real man-made world in the products and artefacts that we make, the real mind world in the theories and hypotheses that we propose and, perhaps supremely, in the human mind world in the way that our ideas are formed. We cannot escape from creativity and nature, even if we wanted to. They are inextricably linked.

If we now take as an example our bridge design that we showed as Figure 10 at the end of the chapter on what design is, we can put creativity in each quadrant, as in Figure 6.

In the lower right quadrant, creativity is what we humans use to come up with bridge ideas and concepts. In the lower left, creativity was used in order for someone to develop bending theory: in the upper right, creativity is evident in the final bridge product, and in the upper left, nature's creativity is demonstrated in the developmental process that led to the existence of the tree that formed the log before it fell, the processes of erosion that placed the stream where it was, and in the new creation of the bridge-like structure when the log fell across the stream.

3 The nature of creativity

This is all very well and ordered. But the nature of creativity is to be anything but ordered and understood. That's probably why so many people have written books about it and about how to develop it. In spite of creativity being a natural part of human life, it is one that is not evident as much as it might be. "We need to invest in creativity, develop it in ourselves, nurture it in others, identify its value everywhere (except in accountancy) and use it to nourish the new things we create" [5].

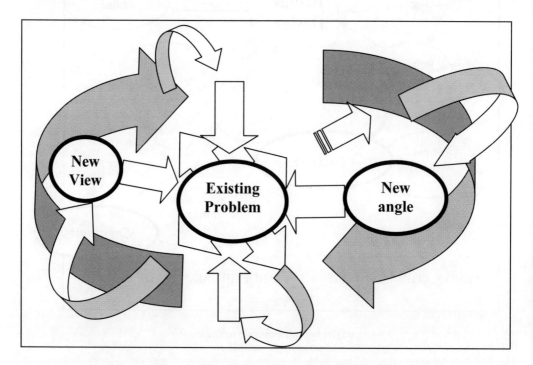

Figure 7: Creative thinking for an existing problem.

Creativity throws wobblies and upsets our perceived and understood order so that it, by nature, creates something that is new and different. And the way that it does it is crucial to our understanding of creativity.

It is the juxtaposition and combination of the seemingly unexpected that seems to account for creativity and develop it, and thus if we can find ways of doing things that initially seem to be contrary to our current understanding, then that may result in the creative urge to develop a new understanding.

Someone stops to ask the way. The response is "Yes, I can tell you. But I wouldn't start from here, if I were you."

Creativity can be a bit like that. In order to get there, we need to be somewhere else. When we know that we have some sort of creation to do, whether an essay for the university course we are taking, a new bridge, a television advertisement or whatever, we need to be in a position of deliberately starting from somewhere else: heading off into the distance, going round in ever-increasing circles from there and then heading back in to sort out whatever it was we were trying to do in the first place.

In fact, in common parlance, we can enlarge on this particular diagram, and end up with Figure 8 – where creative thinking is known by its requirement to think "outside the box" – although in this diagram it seems as if it is the box that has changed size rather than the thinking that has got bigger.

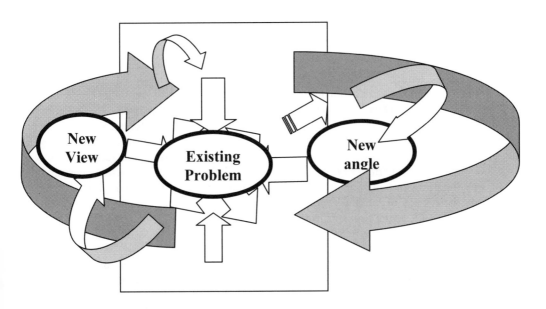

Figure 8: Creative thinking outside the box.

With this there is another danger – well, it might be, or could be an opportunity, depending on your point of view. This is the danger of going off at a tangent, and when outside the box that has a particular danger in that you remain outside of the box and never get back into it. The problem remains unsolved and you have moved on to greater glory and the next item that you have come across, which may not be a problem at all. See Figure 9.

Creativity has a habit of doing that, particularly when combined with human forgetfulness. Not only can it challenge our ways of thinking, but it can also move us into totally new ways of doing things that are not only remote from where we were before we started, but which we can prove are ways that almost have their own being. The challenge, of course, is to find time to catch up with all the bits and pieces that we have left and to enable us not only to enjoy the various places that we are led to, but to manage to return to the box and the original problem.

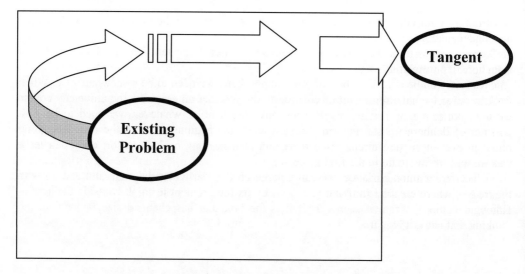

Figure 9: Creativity at a tangent.

So that is the nature of creativity. In being so apparently disorganised and divergent it spawns numerous books and documents and tends to enlarge the problem rather than confront and attack it directly. Once one starts to investigate problems creatively, the number of solutions and avenues that need to be investigated expands dramatically, even if all that you were trying to do was to understand what was going on. Thus, for instance, while using the stepping up / down / sideways methodology developed in the Clarification section of Petty's excellent book *How to be Better at... Creativity* [6] is likely to result in a better understanding of the way the problem you were attacking is structured, it is also likely that you will end up with many more ways of looking at it, and hence very many more possible solutions than if you took a direct hit at the problem from the most obvious angle. It does work significantly well, however, with problems where the direct hit lands in the middle of a large protective piece of armour plating and no solution.

This can and has resulted in a number of books on creativity, particularly those that can be used to develop creativity, appearing to have a somewhat zany character, being disorganised, sometimes to the point of distraction. But I shall continue to recommend books such as *A Whack on the Side of the Head* [7] and *The Art of Looking Sideways* [8] on the grounds that you won't always be wanting to focus on the problem in hand, and some things are just more interesting when they aren't done conventionally.

4 Creativity history

People must have always been creative, especially as it would seem to be part of being a natural human. And evidence would clearly back that up. In Mesopotamia they had the hanging gardens of Babylon; the Parthenon is quite old, to say the least; there were libraries in Alexandria; the Romans had plumbing – and so on. Although society was largely agricultural in nature, people still behaved in much the same way. They still lived somewhere, ate food, had children and had enough similarity to us to see that we are very much the same.

But we don't know a lot about how they looked at the topic of creativity. We do know that they developed ways of doing things such as building cathedrals that were incredible works of construction with very little in terms of paperwork and plans, and that trades developed largely through craft techniques and apprenticeship methods [9]. Rather than things developing in leaps and bounds, this process meant that they developed and changed slowly and evolved bit by bit as small developments were incorporated into the way that the craftsman "did things".

It is unlikely that things are significantly different now. "New forms do not come from nothing, not for us as humans at any rate; they come from prior forms, through mutations, whether unsought or invited. In a fundamental sense, there are no theories of creation; there are only accounts of the development of new forms from earlier forms." [10]. We always have the past around us. We take what we like and what works well for us and develop it to become slightly different and slightly novel and we believe we are being creative. But it is always from the base of where we are in society and from the products and artefacts that we have around us that we develop.

Starting around 1950, there was a considerable amount of research into the mechanisms of creativity carried out in the United States, and this generated a multiplicity of definitions and approaches. The research commenced by averring that the United States came about through creativity – not that the creativity meant here was divine creativity, and not that the term "United States" meant the physical land mass either. Rather, the whole approach towards the building of nationhood was based on creative principles and on a creative approach. This approach in the early years of the country generated what would appear to us looking from the outside as a way of living and working that engendered self-thinking, self-confidence and the development of ways of living and solutions to problems that perhaps hadn't even managed to exist before. It was the aim of this 1950s group of researchers to develop ways that this spirit of free-thinking could be developed and encouraged, find its way into the warp and woof of the American educational system, and hence disseminate itself throughout society. They spent some considerable time and effort trying to understand what creativity is and might be, how we might develop creative thinking, how creativity might be measured, and still trying to keep the essential *Je ne sais quoi* and mystique of the subject without demystifying it. Their stories make fascinating reading [10, 11].

Things have developed since then. Whether creativity has disseminated itself throughout American society is not for me to judge, but certainly the subject has become popular in Britain, with people such as Edward do Bono (for instance with his book *Serious Creativity* [12]) and Tony Buzan [13]. Popularity and dissemination are probably two different things. Certainly creativity should be enjoyed. It should be popular. Job advertisements say that companies want creative people. But I suppose I remain to be convinced on that last comment, to some extent.

5 Definitions of creativity

What is creativity?

One of the feelings within this American group of researchers was that they didn't want the definition and research into creativity to move the way that the research into human ability and intelligence had moved – into a tightly defined field, concerned largely with psychometric measures to determine a well-defined entity known as Intelligence Quota – or IQ for short. What they wanted was for the open nature of this entity roughly called creativity to be understood and fostered, and hence the existence and acceptance of a multiplicity of definitions for the term [11].

So, with that background, dare we try to achieve a definition? Again, there have been so many definitions produced that virtually anything might be considered as a definition. They range from being rather basic dictionary definitions through those that are completely esoteric to those couched in allegories and even in pictures or fairy stories. Petty in his otherwise very useful book *How to be Better at ... Creativity* [6] doesn't even bother with a working definition; he just talks about how creativity can be used and what the potential benefits are of being better at ... creativity.

Much of the work amongst this American group of researchers has been published in a volume edited by Sternberg [14]. In this text there are sections on the role of the environment and on the individual, together with a section on the role of the environment-individual interaction in creativity. These sections to some extent define the extent and breadth of the subsequent definitions of creativity, stretching between those involved with the determination of a creative environment, sensing whether individuals are creative, groups are creative, products or processes are creative, or whether a combination of all of these is legitimate and required, not to mention the interactions of all of them with all of the others. That gives us thirty different combinations, and we are just talking about areas for definitions, rather than definitions themselves here.

5.1 Research definition

Torrance [15], within this collection of papers, although he feels that creativity defies precise definition and that it doesn't bother him particularly, chooses to use the concept of newness, as opposed to conformity, as the criterion of creativity, and then goes on to develop a number of working definitions, starting with a research definition: *"The process of sensing difficulties, problems, gaps in information, missing elements, something askew; making guesses and formulating hypotheses about those deficiencies; evaluating and testing these guesses and hypotheses; possibly revisiting and retesting them, and finally communicating the results"*. A process definition, in this instance. It bears a remarkable similarity to the vast array of creative problem-solving process recommendations such as Edward de Bono's TO / LO / PO / SO / GO [16].

5.2 Artistic definition

He then goes on to look at (and the term look at is significant) what he terms an artistic definition but may better be called a host of artistic definitions. One that instead of defining creativity as *"creativity is..."* instead redefines the term as *"creativity is like..."* and then uses a whole host of drawn images with the theme "Creativity is..." in a manner similar to the "Love is..." cartoons. He then moves on to an analogical definition, from Prince [17], *"Creativity: an arbitrary harmony, an expected astonishment, a habitual revelation, a familiar surprise, a generous selfishness, an unexpected certainty, a formable stubbornness, a vital triviality, a disciplined freedom, an intoxicating steadiness, a repeated initiation, a difficult delight, a predictable gamble, an ephemeral solidity, a unifying difference, a demanding satisfier, a miraculous expectation, an accustomed amazement."*

5.3 Survival definition

His next approach is to tackle what he calls a survival definition, *"When someone has no learned or practiced solution to a problem, some degree of creativity is required."* This ties up with the need for self-reliance and discovery, but somehow seems to remain somewhat enigmatic, although Torrance states that he thinks it is probably one of the most satisfactory definitions. One can argue that this definition also has a process base.

5.4 Definitions from a list collated by Taylor

In the same volume, the appendix to Taylor's chapter [11] lists definitions of creativity from Repucci of Dow Chemicals [18]. This states that there had been some fifty to sixty definitions up to that date and that these could be classified into about six classes – except that class number six had the heading *Varia* simply because they didn't fall into the other five groupings. There are in all probability many more definitions and probably more classes by now.

The first of these classes is termed gestalt or perception, where the emphasis is on the combination of ideas to produce an obvious sort of whole, perceived at once.

Second is that where the definition focuses on the end result or on an innovation produced as a result of that creativity. Repucci quotes a definition of Stein's, *"Creativity is that process which results in a novel work that is accepted as tenable or useful or satisfying by a group at some point in time"* as exemplifying this approach.

The third class of definitions is regarded as aesthetic, and here the emphasis is on the use of self-expression, particularly of a unique manner and that which is special and particular to the individual who is being creative.

Fourth, we have a psychoanalytically based class of definitions, largely through the use of Freudian-type expressions and coming from the interaction of the id, ego and superego. This may mean, for instance, that the ego must be subjugated in order for the subconscious or unconscious material to be allowed to emerge as the creative idea.

Class five deals with definitions that can be categorised as solution-based definitions, where the emphasis is on the thinking process rather than the outcome of that process. Thus creativity becomes a divergent thinking phase within the context of a complete problem-solving process.

The sixth group is that which defies proper classification, being classified by Repucci under the heading of *Varia*. This means that they just do not fit anywhere else in his classification scheme, although from the list of definitions in the brief paper some of them would appear to be similar to ones in the first category, that of the gestalt definition.

5.5 A range of definitions

Outside of this collection of papers, Dewulf and Baillie [19] mention a range of definitions, coming down in favour of a working definition of creativity as *shared imagination*. For me this definition seems to be strangely divorced from the overall scope of the term and feels like an attempt to obtain a lowest common denominator as a definition. Similarly to Torrance and the American researchers, Dewulf and Baillie question whether creativity is a quality of products, processes or people, whether it is personal or social, universal or special and ask how dependent it is on circumstances.

5.6 My own preference

I prefer to side with a dictionary definition such as *The ability to make or otherwise bring into existence something new, whether a new solution to a problem, a new method or device, or a new artistic object or form* [20]. This has something to do with the product – that creativity is evident in that something has been created – although I am happy with the notion that the "something" can be abstract, such as an idea. A similar definition, but focusing on the imaginative aspects might be one such as *The ability to contemplate something that hasn't existed before* [21]. But a single definition, whichever one is used, hardly does justice to the scope of creativity and although a working definition is perhaps desirable, the vast coverage of the various collated definitions that Repucci uses – or even Torrance's assertion – remind us of the breadth of the topic and also of the way in which it clearly crosses boundaries to become an all-encompassing and embracing theme.

Even with this working definition there seem to be differing perceptions. In the first instance, it has something to do with creating, with bringing something into existence. On the other hand, there is something about the term creativity that implies or may imply some judgement of the quality of that creation. If we ask whether someone's work is creative, we are not usually questioning whether they have brought something into being, but whether that something is perceived as original, and how original it is perceived to be.

For instance, Bohm [22] contrasts creativity with mediocrity and what he calls *mechanicalness*. I don't think he means that people acting with a mechanical behaviour do not create things, whether abstract or concrete, but that *real* creativity is something that needs to rise above the normal state of things. It is similar to Edison's well-known comment that invention is one percent inspiration and ninety-nine percent perspiration. Again, he didn't mean that things weren't created through perspiration, but that inspiration (in whatever way he defined it) lifted the quality of the development to a different level.

For my own part, I prefer to think of creativity *per se* as something that is involved in creating anything, however banal or mechanical that thing might be. I would then prefer to investigate the level of that creativity in some way; to qualify and describe it as being of a higher or lower level. Creativity simply implies the creation of something. So looking up design data in code books in order to develop a boring new reinforced concrete structure that does its job but is severely constrained by cost is still creative; maybe not as creative as it could have been if there had been a more enlightened search for solutions at an early design stage, or maybe not as creative as it might have been if the client had been more interested in the product than in the income gained from its use. The mere existence of creativity does not imply a high level of novelty, and the existence of novelty does not automatically assume a high level of creativity.

6 Measures of creativity

It would be useful also for us to be able to measure creativity in some way, both in terms of its quantity and its quality and character. In a similar way to the breadth of the definitions of creativity, we have a problem in deciding what we actually need to measure. We also have a problem, in that as soon as we introduce our measuring system we tend to devalue creativity into "that which can be measured by our creativity measuring system", in much the same way as intelligence became defined as that which can be measured by IQ tests (although it can be argued that they measure something called IQ rather than intelligence). Sometimes we are trying to measure the creativity of a person, sometimes the creativity evident in a product, sometimes the efficacy or efficiency of a creative process, sometimes some form of novelty; and sometimes the degree to which an environment aids or hinders the development of creativity evidenced otherwise. It is difficult to measure the psychological dimensions in the formation of ideas that are not shared or in any way made public, but this is still perceived as a dimension of creativity.

Another aspect of measuring creativity, and in particular measuring creativity as evidenced in people, is whether we can measure actual creative output, or whether (and in what way) we can measure creative potential.

And then there are the obvious questions of whether there should in fact be a single dimension of creativity or whether creativity, being defined in a multi-faceted manner, actually has a significant number of dimensions and whether some form of factor analysis is needed n order to generate a respectable reduction of the multi-faceted nature into two or three combined measures.

Torrance [15] has a multiplicity of measurement systems in place, to match his multiplicity of definitions, perhaps. He starts off talking about different levels of creativity, quoting Taylor. These are:

1. Expressive creativity
2. Productive creativity
3. Inventive creativity
4. Innovative creativity
5. Emergenative creativity

In spite of these being stated as different levels of creativity, it would seem that they do not necessarily measure the same thing. I cannot see how expressive creativity can be measured as the same thing as innovative creativity, and I do not know whether it equates as more or less creativity on a scale of 1 to 5. I think these are rather different types of creativity.

Torrance's measures of creativity tend to measure incipient creativity rather than actual creativity, and seem to be measuring creative potential in people. He develops a number of aspects to creativity that he can test, measuring as many as eighteen different factors, some of which seem to need expert analysis to perform a quantified measure. He carried out his tests over a long period of time, and was thus also able to measure actual performance of people in jobs versus their career aspirations when children, and these were also taken into account. In the end, he suggests that it is a labour of love that defines creativity and is the best at predicting future creative performance, and produces a list of items that he describes as a manifesto on how to grow up creatively gifted.

That doesn't really answer the question of how we can measure creativity. It just gives us some way of assessing whether an individual may have creative potential. Eysenck suggests that measurement of creative traits or potential only has a moderate correlation to creative achievement [23], and suggests that in order for creative output to be significant the creative traits need to be aligned to a persuasive and hard-skinned personality in order that inevitable resistance to new ideas and thinking can be almost ignored. George Bernard Shaw perhaps sums it up when he states,

> *Reasonable men adapt themselves to the world.*
> *Unreasonable men adapt the world to themselves.*
> *That's why all progress depends on unreasonable men.* [6]

Alternatively, some measures could be used that relate to the breadth of an individual's insights, the extent to which the insights are recognised and developed by colleagues and acquaintances, and the way in which an individual's creative insights have opened up areas for further work. These, again, only measure an individual's creativity and have been used to reward individuals for the quality of their work rather than the quantity of their output. These measures are somewhat limited in their scope and are specific in terms of their aims. The difficulty with measuring creativity is that it is like comparing chalk with cheese – the outputs are so variable that it is difficult to arrange a reasonable yardstick.

On a more pragmatic level, many people working in the field have suggested that the quality of an individual's creativity is linked to the quantity of creative work that they produce. If they wish to produce good ideas, then this ability is linked to the ability to produce a lot of ideas. Thus there are links between the quantity and quality measures: but it is doubtful that these links have been adequately explored in spite of the significant amount of research that has been developed in the area. Another difficulty with measurement of creativity is that it almost always requires some output. It is very difficult to measure the quantity or quality of what are simply ideas in a person's brain without there being some recourse to an output, either written, drawn, spoken, discussed or whatever else may be appropriate.

When assessing the level of creative output, such as that from the conceptual design phase in a design project, I tend to favour something of the approach of Petty [6], where ideas are favoured on the grounds of where they are likely to lead and whether they are where the person who did the design wants to go (or is able to go) or not. This latter criterion does not measure the level of creativity – in fact, the first one only measures it in an informal way. Petty is fairly clear that progress is due largely to ideas that develop into products where the end result cannot be judged properly when the idea is first mooted, and that the ideas that seem the most impractical and absurd are often those that contain the most potential. This may describe them as being the most creative. In the context of a student design project, those ideas that make the best choice are usually those that are the most extreme and far-fetched. They are the ones that will hold the students' attention and keep them enthralled with trying to obtain a workable solution to something that, on the face of it, may seem to be presented initially as an impossibility. In this instance, there is no need for a proper creativity measure to be taken. All that is needed is to identify the one or perhaps two ideas out of the (hopefully) vast array of ideas presented that hold the most promise. Although Pugh [24] would suggest that a process of careful evaluation should take place in a controlled fashion, he cites a significant number of criteria other than creativity or innovation, and it would be quite difficult to identify significant creative criteria unless doing so for a specific project.

Looking at the broader picture, I often suggest that students use a very quick selection process to select their ideas, based on a criteria net known as NAF – Novelty, Attractiveness and Function. The concepts or ideas should go through the successive filters that only dictate whether or not they have novelty value, are attractive and then, lastly, whether they can obviously be made to work or not. Novelty, however, becomes the prime criterion, as progress and developments in most fields depends primarily on this [25]. Lose novelty, however it is determined, and you don't have progress.

A quick system such as this can be vital for those involved with a multiplicity of concepts from which to judge, particularly if they are using the process within a creative concept-development design phase. Within this sort of phase there is a strong likelihood that the mood of the designer will be such that rather than seeking out those few examples that should be worked on, they are in an active mode for creating more concepts. With this multiple approach, a quick and easy approach is absolutely essential, as a methodology such as Pugh's comparative selection process [24] that uses a significant number of criteria is going to take far too long.

7 Creativity in practice

7.1 A creativity development exercise

This topic of the measurement of creativity links with the relatively frequently sensed need to improve one's creative skills. In some senses, what is important is not necessarily that creativity can be measured, but if it is both important (perhaps even to a country) and enjoyable to be creative, then what we should be doing is finding ways to develop, nurture and use it.

It is into this scenario that I come to report on an exercise that I trust will be able to show significant links between nature and creativity, and which was also setting out to develop creative thinking – but of course whether that was there in the first place and only needed developing is a matter for significant debate.

Within the teaching of Engineering and Design, which is what I spend most of my time doing, creativity needs to be introduced at an early stage of the course so that students have a chance of making the creative ways of thinking and behaving part of their normal, usable, enjoyable "design toolkit".

Thus we need to develop a way of not just introducing the theory of creativity, but of providing a practical hands-on experience where these tools are not just promoted but are actually used.

I have used a specific design project over a number of years. It is one that relates well to the topic of creativity and nature in that it utilises significant interaction between the two in order to develop students' creative thinking and output.

First, perhaps, I need to outline some of my own background and this may help to explain what I have been trying to do.

I'm not someone who studies nature. That would be a scientist, I suppose: someone who tries in earnest to understand the laws behind what goes on in the natural world, who builds and develops theories about how it works and why things are and which came first – the chicken or the egg.

Neither am I one of the creative professionals: one who can throw together a three-hour film in twenty minutes or do a sculpture to rival Henry Moore (in a very long time) or produce an advertisement that's going to sell a million and a half pens or deodorant canisters (full, of course) in the two weeks before Christmas. Nor am I an artist, using nature as inspiration for the next artistic production or trying to capture something of my emotional response to what I perceive in nature.

I'm an engineer. Someone who deals not with the natural world, but with the man-made world. In my reckoning, engineers deal with understanding how that goes together and comes apart and what makes that tick. In some senses the natural world is just *there*, making all the rules I work to and trying its utmost to throw its rule book at me so that I can't make a perpetual motion machine or a higher monument than anyone else has, or so that I can't make a car run round corners on rails because of the behaviour of the interface between the rubber and the road.

But I'm also a designer: someone who likes to take what is there in nature and change it to something I want it to be. In my forlorn hope I would like to think that what I create may be just as inspiring as what it was created from, but actually I am more interested in getting it to work and in sorting out how to make it work better. Perhaps some of my creations are inspired, but then, perhaps they aren't.

I'm also an educator. I teach students at university. My students are those who are learning to become designers and engineers. They do this by developing not just head knowledge, but person knowledge in developing whole-person skills, particularly in making and perceiving the way that they might make them better [26]. Some of these skills come under the heading of getting them to develop their creativity. Perhaps it's too late at university. We need to try to get them a bit earlier, when they aren't afraid of being children any more. So I need to get them to be happy about regressing into childhood and happy to do things that appear to be just a little bit wacky. After all, when they leave me they won't get the chance, in many cases.

The exercise I shall be describing is one that has been given to first-year students shortly after they arrive at the university so that they have a toolkit of ways of tapping into creative thought. It's also meant to be a toolkit that they know how to use, where the tools have got a little bit of use on them and will not sit there unused, but become loved, cherished and effective well-used items, like the favourite screwdriver that fits snugly into the hand or the tweezers with every small bend known from constant use.

This exercise has been expounded twice in the past, both times as part of conferences. In one, the whole tenor of the conference was on the development of product design education [27], whilst in the other, only a small part was devoted to this topic [28]. This proportion also applies, as it happens, to the proportion of each paper devoted to the description of the particular exercise. It is to be hoped that some of the thinking behind the development of these ways of being creative will have whetted the students' appetites and they will be able to ask specific questions and develop their own ways of continuing to be creative beyond the specific exercise.

The first tenet of the exercise derives from an old Chinese proverb, which is credited to Confucius and is variously reported:

Tell me and I will forget:
Show me and I will remember:
Involve me and I will understand:
Take one step back and I will act. [29]

Although the reference comments in particular about the lack of application of the last line of the proverb, the important one from the point of view of the development of this particular exercise was the third line.

Creative methods are not things that can be shouted from the front of the classroom and are immediately understood. There is some speaking from the front, of course, but they are apt to be forgotten if that is the case (see the proverb!). The idea was that students were expected to arrive in the class and be told that they had to perform tasks using a series of particular methods. The project ran over a period of a number of weeks, with the students having the appropriate classes on a one morning a week basis.

7.2 The assignment

This took its cue from nature, and was that the students were asked to design an animal. It was introduced in the context of the allegorical story of Noah's Ark, with some animal that hadn't quite made it because it had been too busy with its own concerns to hear the calls to safety. Students were told that the outcome of the assignment, eventually, was to be a model of the animal that they were designing. They were also told that they would have to experience some particular methods and that there would be an element of surprise in what they would be given.

They would need to turn up on the morning of the class and then find out what they were supposed to be getting on with.

What they were not told was that part-way through the assignment the outcome was going to change, and that instead of designing an animal they were going to be designing an animal toy. The argument for doing this was that the better and more coherent the animal, the better the toy would be. Correlations between the two have, unfortunately, not been possible to establish, as the students were not formally assessed before the change was made.

7.3 The methods

A mixture of group and individual methods were used. Group ones tended to be used in the class, with some individual ones used as "homework", students bringing the results of their homework searches and methods into the class for group discussion. In the final analysis and assessment, they were to be producing an individual product.

Initial exposure to a problem of this sort, of course, tends to produce a shock horror response from students, who can't grasp where to start. From the start they need to know that they need not worry at this point about what the eventual outcome will be; they just have to realise that they will be asked to carry out a selection of methodologies that should, if they carry them out correctly, result in some sort of reasonable outcome to the project. But the "if" seems to be quite important here.

Group methods included the obvious one of brainstorming [30] developed as early as 1939 by Alex Osborn, but also included games such as visual animal consequences, where rather than telling a story, each person develops part of an animal and then the next person carries the "story" on so that they complete the adjacent part, that links up with the first. The completed animal is given a name – before the "consequence" is seen by unwrapping the drawing done in stages. This provides the sort of freewheeling atmosphere of enjoyment that is needed for creative exercises and is done before the first brainstorming session. In the middle of the brainstorming session a randomly generated word is thrown into each group of students, and they are asked to generate twenty different animals based on that one word. At the end of the session, they are asked to identify the most way-out animal idea from those they have "designed".

For homework, students are asked to find as many pictures of animals as possible and to decide what it is they like and dislike about each, why they've chosen that particular one, and what their emotional response is to it. They bring the pictures of animals to the next session.

At the next session, the task is to separate out the animal ideas, both real and developed, using affinity diagrams, with each animal picture stuck on a board and then moved around so that it fits in relationship to the other ones on the group board. In theory (but not usually in practice) this activity is supposed to be carried out without any talking: pictures are just moved around without reasons being given and the process continues until they don't move any more when anyone's back is turned.

A group that finishes this task early is allowed to go round all the other groups and see what they've done, stealing the ideas (not the pictures!) and seeing if they can gather ideas that appear to be better than their own ones. A bit of industrial espionage, and linked to the idea of "trawling" put forward by Petty [6].

Again, the most way-out animal is chosen for special consideration.

For homework, students are introduced to the idea of a Morphological Chart [31, 32] and are asked to complete one before the next class.

In the next lesson they are expected to do some creative evaluation processes, in groups. NAF is the first one to be used, and the request is that they find the six most novel animals, the three most attractive ones and the most functional one, giving them ten animals in total as a group. These ten animal examples now form a breeding stock, and they are now expected to use a rather basic genetic process to develop others: they are asked to breed them – whether or not any DNA could possibly do anything like this. The expectation, stated clearly, is that they would manage to achieve five generations of animals: this is not just five animals, but five generations, and they are expected to produce not just one bred animal from each liaison, but a number of siblings.

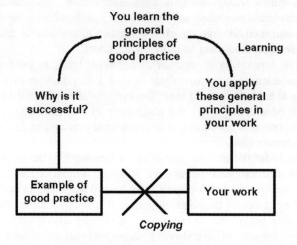

Figure 10: Trawling (after [6]).

Following on from this, the methods become more specific and less group in character. From this point on they are working individually, on an animal that has been given to them by the group (in theory, although in practice they tend to have chosen the one they would like to work on).

The next task is to sort out how the animal behaves. The students start off by writing an imaginary "day in the life" of the animal so that they can work out what its habits are, how it moves, how it interacts, feeds, defecates, loves and cares for its progeny (perhaps). On from this, students try to develop out any weaknesses they have found in the animal, and then start drawing and developing systems diagrams to explain what happens to food, fluids, materials and energy transfer processes, although these last are fairly complex things to think through. They also develop structural diagrams showing how loads are carried and how joints (perhaps) move and stay together, including where things such as muscles attach and how they stay warm.

It is at this point that the animal suddenly changes to an animal toy, and they need to come back down to earth a little, think how best to turn it into a toy, what the toy has to do, what age-group it relates to, and how the parts of the toy are put together for mass production to achieve an effect similar to those described for the imaginary animal. They write a design specification for the toy – itself not a particularly creative task, but one that they need to be involved in.

After this they need to sort out the practicalities of making the model and describing how the toy parts are arranged using some sort of layout drawing work: making a model is not optional: they need to sort out what materials they will need, where they are going to get them from and manage to achieve some sort of result overall.

7.4 The results

So, what sort of things do they come up with? Does the process work? And how do they manage to get bits of the toolkit under their skins, as it were?

Students generally respond well to the topic and produce some high-quality work, with few disappointing results. Those results that are disappointing tend to be those from students who do not make the commitment to consistent attendance that is demanded by the assignment, and as a result may miss out on significant experiences along the way.

The project also brings the students together at the start of their studies and forces them to work together, building up something of a team spirit among them.

From some groups there has been a lot of work with a science fiction theme, but on other occasions this theme is less apparent. There have been some animals with links to fairy tales, such as dragons, and some that have been developments of animals that exist or have existed, such as dinosaurs. At one point there was a group that ended up designing "party" animals, equipped with large beer guts, heavy boots to make lots of noise, mouths adapted for beer-swigging, hangover-avoiding constitutions and sets of limbs adapted for easy opening of cans and the production of cocktails, as seen in Figure 11.

Figure 11: Two party animals.

In other cases it has been the story that has been compelling – descriptions of how some very different animals spend their days have been enthralling reading, sometimes translated into comic-strip cartoons to make them easy for those who prefer to look at the pictures.

In a number of cases students have continued to remember with affection their first creative project, and in one case there was even a student who wanted to continue development of his animal-based vacuum cleaner several years later, when he was about to finish his degree course.

The other encouraging thing is that students seemed to develop their creative methods reasonably well, and have found that being forced to utilise the methods early in the course has given them a useful and valuable experience that stands the test of time.

It may well be that selection of another project brief would not have the same effect: choice of an animal seems to draw out many links that are very useful in developing general creative behaviour in a positive manner.

7.5 Natural analogies

That particular creativity development project may not have worked without the forced analogies with the natural world. Such analogies are an extremely valuable way of developing an understanding of creative techniques and of applying creativity in other spheres. Looking at how an analogous situation is "designed" in the plant and animal world is a very powerful mechanism towards developing a new understanding of a situation or a product.

These analogies can be taken in many different ways. Firstly and most obvious is that of the direct analogy. In many cases these have strong links with nature. For instance, Marc Brunel, Isambard Kingdom Brunel's father, developed a tunnelling system based on his observation of how woodworm bore holes – and this tunnelling system is still used, in the form of the tunnelling shield [6], and it is well known that an investigation of the hooking principle of the burr and of goose grass developed into the hook and loop fastener, now commonplace.

Less obvious is the personal analogy. In this instance we invent ourselves as being the answer to the problem we have been given, describing how we feel and what our personal difficulties would be about being there. The relationship with nature is here not quite so obvious. We need to remember that we are part of nature and therefore what we feel is probably the way in which a material, initially from the natural world, would behave in the circumstances. For instance, I have been told of someone who was designing a helicopter, and when thinking about the rotor blades, felt that if he were a rotor blade he would probably feel very cold. As a result the rotor blades on his helicopter wore nylon fur covers. I understand that it stopped them from icing up. In another instance, designers of racing engines commonly imagine themselves as part of the air/fuel charge making its way around the pipes in the induction system and being pushed through the combustion chamber. This mimics the way a natural charge behaves as it moves around the passages and gives almost an emotional understanding of the difficulties of making the engine produce more power.

Another specific type of analogy that is useful for developing creativity appertains to fantasy. In this instance the relationship is not necessarily with the natural world, but more with the opposite: some sort of unnatural and fantastic arrangement is developed that fits the bill. This is then related to possibilities in the real world and to the way that real materials and systems behave to produce a solution to what is normally determined as a somewhat intractable problem.

7.6 Analogical thinking for creative products

This creativity development assignment is also useful for the way in which it gets students to develop a number of useful ways of design thinking, using the analogies that are needed to design the animal. Two of these stand out significantly.

The first is that of systems thinking. Students are significantly challenged to develop the detailed processes that need to take place within the animal's body in order for food assimilation and digestion, energy transfer through some sort of breathing, the way the nerves work and things such as fluid and blood circulation systems that go on in most animals. These are mimicked by many other products that the students may well have to design, where material, in solid, liquid or gaseous form, is moved around from one place to another within the system boundary, and where the system needs to have energy to perform its function and supplies control and signals of different sorts in order to make things work properly. Such things as system feedback are essential elements both to animal behaviour and to the way in which products such as robotic spray-painting facilities need to function – and, as designers of these have found, providing the right sort of feedback is never easy. Somehow, for students to tackle some of these issues in a first-year project with little need for an accurate outcome prepares them well for the need to do it for real, perhaps when they are doing a final-year project or when they are working in industry.

The second type of thinking that is very important is structural thinking. This is not the narrow type of structural analysis that is only concerned with stress levels and distribution, although this comes into it, but it is also concerned about what is connected to what, whether the structure is of a plated exo-skeleton like an insect or an internal one like many other animals – or whether the animal is jelly-like and largely supported in an aqueous environment with little need in the way of significant load-bearing structure. In the future, the vast majority of products that these students will be designing will need to have some sort of load-bearing arrangement, and to tackle the topic early on in their courses prepares them for the need to consider load-carrying capacity and arrangements in other work that they will tackle in the future [33].

These two types of thinking use the analogies that are apparent within the natural world in order to start to teach a certain methodology to students that they will be able to apply beyond natural systems and structures. But natural structures got there first, and in considering these in both concept and detail students are enabled to develop their own understanding of the way in which their own designs of systems and structures function.

8 Creativity and nature

So: how do creativity and nature relate to each other? The answer is that they do so in many and diverse ways; from being essential components within the make-up of the nature of *Homo Faber* to being something that perhaps carries on even as impersonal forces change the way the universe presents itself to us. Creativity is something that is entwined within human nature but which most of us could do well to develop more of.

As we saw at the start of this chapter, creativity, without forcing it too much, can be fitted into every corner of our perceptions. It is evident in the natural, non-human world: and it is evident in the artefacts and products that we humans make. It is a necessary ingredient in the formation of scientific hypotheses and theories, and is certainly part of the way that we develop ideas in our human thought world – that we can go on to flesh out as they become encapsulated into the artefacts in the real human world.

It is also something that has intrigued people over the years, but which only recently seems to have been studied to any great extent.

Not only is creativity evident in nature, particularly human nature, but nature is also a very useful ingredient to the development of creativity in terms of the use of analogies and the development of methods and products that utilise and mimic natural forms and systems. These all combine together to make creativity and its development, cultivation and nurture one of the most fruitful and enjoyable things in life. Without creativity at least somewhere in it, where would we be? The answer is probably all the same, and extremely boring. Or maybe not part of nature at all.

References

[1] Dempsey, M., Creative Accounts. In Creative Survey. *Design Week*, November 2002.
[2] Langley, P.J. & Jones, R., A Computational Model of Scientific Insight. in *The Nature of Creativity*, Ed. R.J. Sternberg, Cambridge University Press: Cambridge, pp. 177-201, 1988.
[3] Boden, M., *Dimensions of Creativity*, MIT Press: London, 1994.
[4] Kneller, G.F., *The Art and Science of Creativity*, Holt, Rinehart & Winston: New York, 1965.
[5] Grinyer, C., Creative Accounts. In Creative Survey. *Design Week*, November 2002.
[6] Petty, G., *How to be Better at ... Creativity*, Kogan Page: London, 1997.
[7] Von Oech, R., *A Whack on the Side of the Head*, Thorsons: London, 1990.
[8] Fletcher, A., *The Art of Looking Sideways*, Phaidon: London, 2001.
[9] Jones, J.C., *Design Methods: Seeds of Human Futures*, Van Nostrand Reinhold, 1992.
[10] Barron, F., Putting Creativity to Work. in *The Nature of Creativity*, Ed. R.J. Sternberg, Cambridge University Press: Cambridge, 1988.
[11] Taylor, C.W., Various Approaches to and Definitions of Creativity. In *The Nature of Creativity*, Ed. R.J. Sternberg, Cambridge University Press: Cambridge, 1988.
[12] de Bono, E., *Serious Creativity*, Harper Collins: London, 1992.
[13] Buzan, T., *Use Your Head*, BBC Books: London, 1974.
[14] Sternberg, R.J., *The Nature of Creativity: Contemporary Psychological Perspectives*, Cambridge University Press: Cambridge, 1988.
[15] Torrance, E.P., Creativity as Manifest in its Testing. In *The Nature of Creativity*, Ed. R.J. Sternberg, Cambridge University Press: Cambridge, 1988.
[16] de Bono, E., *Teach your Child to Think*, Viking: London, 1992.
[17] Prince, G.M., *The Practice of Creativity*, Harper & Row: New York, 1970.
[18] Repucci, L.C., *Definitions and Criteria of Creativity*, Dow Chemical Company 1960.
[19] Dewulf, S. & Baillie, C., *How to Foster Creativity*, Department for Industry and Education: London, 1999.
[20] In *Encyclopaedia Britannica*, Encyclopaedia Britannica 2000.
[21] Harth, E., *The Creative Loop*, Penguin: Harmondsworth, 1995.
[22] Bohm, D., *On Creativity*, Routledge 1998.
[23] Eysenck, H.J., The Measurement of Creativity. in *Dimensions of Creativity*, Ed. M. Boden, MIT Press: London, 1994.
[24] Pugh, S., *Total Design*, Addison Wesley 1991.
[25] Martindale, C., *The Clockwork Muse*, Basic Books 1990.
[26] Dowlen, C. & Cook, D., Design Realisation: a Vital Skill for Product Design Students. In *E&PDE*. University of Sussex: Professional Engineering Publications. 2000.
[27] Dowlen, C., Teaching Creative Thinking to First Year Students: a Report on a Specific Experiment. In *Product Development in Engineering Education*. University of Limerick, Ireland. 1994.
[28] Dowlen, C., Hurst, K. & Gilchrist, I., Creativity in Engineering Design; Can it be Taught? In *ICED'95*. Prague: Heurista. 1995.

[29] Vostracy, Z., Vacek, J., Skalicky, J. & Sulak, M., Higher Education for Designers and an Example of Creativity in Electrical Engineering Innovation. In *Engineering Design and Creativity*. Pilsen, Czech Republic: Heurista. 1995.
[30] Osborn, A,F., *Applied Imagination*, Scribner: New York, 1953.
[31] Cross, N., *Engineering Design Methods: Strategies for Product Design*, John Wiley & Sons Ltd: Chichester, 2000.
[32] Hubka, V., *Principles of Engineering Design*, Butterworth: London, 1982.
[33] Dowlen, C., Structural design: Three Alternative Paradigms. In *ICED'95*. Prague, Czech Republic: Heurista. 1995.

New Journal from WIT Press

Design & Nature
A Transdisciplinary Journal Relating Nature, Science and the Humanities

Editors: **C.A. BREBBIA**, Wessex Institute of Technology, UK and **S.C. BURGESS**, University of Bristol, UK

This journal will act as a forum for researchers studying nature and its significance to various aspects of modern scientific thought and design. Today, the huge increase in biological knowledge and developments in design engineering systems, together with the virtual revolution in computer power and simulation modelling, have all made possible more comprehensive studies of nature. Scientists and engineers now have at their disposal a vast array of relationships for materials, mechanisms and control.

Providing a channel of communication between all those working in this exciting new discipline, the **Design & Nature** journal will be published quarterly both online (ISSN: 1744-3679) and in paper format (ISSN: 1744-3687).

For further information please contact: *harnold@witpress.com*.

Design and Nature II
Comparing Design in Nature with Science and Engineering

Editors: **M.W. COLLINS**, London South Bank University, UK and **C.A. BREBBIA**, Wessex Institute of Technology, UK

Throughout history, many leading thinkers in disciplines such as mathematics and engineering have been inspired by the parallels between nature and human design. Today the huge increase in biological knowledge, developments in design engineering systems together with the virtual revolution in computer power and simulation modelling, have all made possible more comprehensive studies of nature.

Containing the proceedings of the Second International Conference on Design and Nature, this book brings together contributions from researchers working around the world on a variety of studies involving nature and its significance for modern scientific thought and design.

Over 55 papers are featured and these span the following broad range of topics: Architectural Design and Structures; Architecture and Sustainability; Acoustics; Biology; Biomimetics; Design Philosophy and Methods; Human Biology and Medicine; Materials; Nature and Architectural Design; Space.

Series: Design & Nature, Vol 6
ISBN: 1-85312-721-3   2004   apx 300pp
apx £99.00/US$158.00/€148.50

WIT eLibrary

Home of the Transactions of the Wessex Institute, the WIT electronic-library provides the international scientific community with immediate and permanent access to individual papers presented at WIT conferences. Visitors to the WIT eLibrary can freely browse and search abstracts of all papers in the collection before progressing to download their full text.

Visit the WIT eLibrary at
http://www.witpress.com

Find us at
http://www.witpress.com

US Mirror Site
http://www.compmech.com

Save 10% when you order from our encrypted ordering service on the web using your credit card.

Compliant Structures in Nature and Engineering

Editor: **C.H.M. JENKINS**, South Dakota School of Mines and Technology, Rapid City, SD, USA

Biological organisms are exceptionally efficient structures, having evolved over millions of years to achieve high load carrying capacity per unit weight. Compliance (flexibility) plays a significant role in the efficacy of natural structures. Examples range from cell walls and microorganisms, to bat wings, to human muscles.

Long ago, engineers adopted compliant structures for solutions where structural efficiency was critical while in modern times, compliance has seen increasing use in civil structures such as temporary storage facilities and large-span roofs. Recently, there has been considerable interest in large membrane/inflatable structures for aerospace applications.

This unique volume provides a concise summary of the important features of these interesting structures. Written by experts in the field, it covers topics on compliant structures in nature, compliant materials, mechanics of compliant structures, applications in engineering, and design for compliance. The text will contribute greatly to our continued learning from nature in order to design highly efficient engineering structures.

Series: Design & Nature, Vol 5
ISBN: 1-85312-941-0   2004   apx 400pp
apx £120.00/US$192.00/€180.00

Optimisation Mechanics in Nature

Editors: **M.W. COLLINS**, London South Bank University, UK, **D.G. HUNT** and **M.A. ATHERTON**, London South Bank University, UK

A detailed exploration of the behaviour and use of two major naturally occurring solids, wood and bone, and their relevance for engineering applications. The chapters are written by international specialists in the field.

Partial Contents: Wood as an Engineeering Material; Nature and Shipbuilding; The Structural Efficiency of Trees; Application of the Homeostasis Principle to Expand Gaudi's Funicular Technique; Bones - The Need for Intrinsic Material and Architectural Design; Restoration of Biological and Mechanical Function in Orthopaedics - A Role for Biomimesis in Tissue Engineering.

Series: Design & Nature, Vol 4
ISBN: 1-85312-946-1   2004   apx 170pp
apx £59.00/US$94.00/€88.50

Design and Information in Biology

From Molecules to Systems

Editors: **M.W. COLLINS**, London South Bank University, UK, **J.A. BRYANT**, University of Exeter, UK and **M.A. ATHERTON**, London South Bank University, UK

This book complements and extends the scope of Nature and Design (see third column).
Partial Contents: Genomes, Genes and Proteins; The Human Genome Project; Green Grass, Red Blood, Blueprint; Flight; The Company as a Living System; Insect Observations and Hexapod Design.
Series: Design & Nature, Vol 2
ISBN: 1-85312-853-8   2005   apx 300pp
apx £99.00/US$158.00/€148.50

We are now able to supply you with details of new WIT Press titles via
E-Mail. To subscribe to this free service, or for information on any of our titles, please contact the Marketing Department, WIT Press, Ashurst Lodge, Ashurst, Southampton, SO40 7AA, UK
Tel: +44 (0) 238 029 3223
Fax: +44 (0) 238 029 2853
E-mail: marketing@witpress.com